CONSERVATION CHECKLIST OF THE TREES OF UGANDA

James Kalema and Henk Beentje

CONSERVATION CHECKLIST OF THE TREES OF UGANDA

James Kalema and Henk Beentje

Mapping and computing of EOO and AOO by Steve Bachman

Kew Publishing

Royal Botanic Gardens, Kew

First published in 2012 by
Royal Botanic Gardens, Kew
Richmond, Surrey, TW9 3AB, UK

www.kew.org

ISBN 978 1 84246 377 2

Distributed on behalf of the Royal Botanic Gardens, Kew in North America by the University of Chicago Press, 1427 East 60th Street, Chicago, IL 60637, USA.

British Library Cataloguing in Publication Data
A catalogue record for this book is available from the British Library.

James Kalema:Herbarium, Department of Botany, Makerere University
Henk Beentje: HLAA, Royal Botanic Gardens, Kew
Steve Bachman: GIS section, Royal Botanic Gardens, Kew

Cover photograph: James Kalema
Title page photograph: James Kalema

Project editor: Ruth Linklater
Typesetting and page layout: Margaret Newman
Publishing, Design & Photography Department
Royal Botanic Gardens, Kew

Printed in the UK by Berforts Group Ltd

For information or to purchase all Kew titles please visit
www.kewbooks.com or email publishing@kew.org

Kew's mission is to inspire and deliver science-based plant conservation worldwide, enhancing the quality of life.

Kew receives half of its running costs from Government through the Department for Environment, Food and Rural Affairs (Defra). All other funding needed to support Kew's vital work comes from members, foundations, donors and commercial activities including book sales.

Contents

Boundary of Bwindi Impenetrable National Park and cultivated community land. (photo: James Kalema, Sep. 2010).

INTRODUCTION

Uganda is a country blessed with a high diversity of plant species, owing to moist and warm conditions in many parts, and altitude varying from about 600 to 5000 m — although most of the country lies between 1000 and 1500 m above sea level (Kalema 2008). These species are not distributed evenly over the country, however. Some areas are richer than others (Friis 2008, Kalema 2008). The country has quite varied habitats (Kalema & Bukenya 2005) supporting an array of plants: from open water systems, wetlands, tropical high moist forest, dry forest, woodland, thicket, bushland, steppe-like vegetation, grassland to semi-arid habitats (Langdale-Brown *et al.* 1964). The distribution of plants and levels of species richness and endemism vary across these various habitats (White 1983, Mittermeier *et al.* 1998, Groombridge & Jenkins 2002, Moore *et al.* 2002). Moist tropical forests are the richest and most highly diversified (WCMC 1992). Of the 22 vegetation types of Langdale-Brown *et al.* (1964), those richest in tree species are: B (High Altitude Forest), C (Medium Altitude Moist Evergreen Forest), D (Medium Altitude Moist Semi-Deciduous Forest) and N (Dry *Combretum* savanna) (Kalema *et al.* unpub., see Table 2 of this study). The most widely spread tree species across the vegetation types are *Acacia gerrardii*, *A. hockii*, *A. polyacantha*, *Albizia zygia*, *Balanites aegyptiaca*, *Bridelia scleroneura*, *Combretum collinum*, *C. molle* and *Lannea humilis* (Kalema *et al.* unpub.).

There are a number of threats to plants generally, and trees in particular, in Uganda. In some areas these threats are more serious than in others, through development of agriculture, town enlargement, industrialisation, or other causes. Plant resources in Uganda now face very heavy use and extraction for various uses (Peters 1994, NEMA 2007), partly because of the rapidly increasing population, lack of employment and limited alternative means of livelihood. As an example to illustrate this heavy direct dependence on plant resources, over 96% of Uganda's population depends on biomass for fuel (Syngellakis & Arudo 2006). By far the most important cause of vegetation change is subsistence farming (Forest Department 1995, 2003; Kalema & Ssegawa 2007) which has led to reduction in extent, degradation and sometimes total loss of some habitats. Areas at very high elevations, particularly vegetation type A (Moorland and Heath), have had the least interference. Densely populated areas e.g. Kigezi region, Lake Victoria crescent, Bugisu-Sebei region and Arua, have registered the highest level of degradation and loss of vegetation in the country e.g. the *Prunus* Forests in Kigezi have been significantly reduced (through conversion to farmland); yet the *Juniperus-Podocarpus* forests of the less densely populated Karamoja have barely been affected (Kalema *et al.* unpub.). There is an appreciable number of areas outside the protected area network that are important for tree and other biodiversity conservation in Uganda (Kalema 2006, 2008; Friis 2008).

The threats to plants and conservation issues need to be addressed, as the plant richness of the country is an important resource. It supports community livelihoods in general (Tabuti *et al.* 2003, Ssegawa & Kasenene 2007, Lye *et al.* 2008), and with a vast potential for possible sources of future crops to ensure food security, sustainable tourism, pharmaceutical development, creation of jobs and ultimately alleviation of the looming poverty in the country. About 35% of Uganda's population still lives below the poverty line (Uganda Government 2002). The trend of plant species loss must be reversed, especially as the country adopts a decentralisation policy and liberalises the economic environment. Over-exploitation reduces the resource base and alters species composition. Some species may become more abundant at the expense of others. This scenario can be disastrous to biodiversity and the economy if the changes result in the wanton spread of invasive species. The use of plant resources therefore needs to be regulated for both ecosystem stability and human posterity; we would like our children to inherit an Earth with a sustainable natural environment, rather than a degraded one.

Availability of accurate and up-to-date information is essential and requisite to proper conservation planning. In this vein, we have tried to make a start by making a list of all the species of tree occurring in Uganda, and by indicating which of these have restricted distribution areas. Such restricted species may be threatened by either habitat destruction or more specific threats, and we have highlighted and illustrated such species. Product development, bio-trade and bio-prospecting all need to be guided by information on the status, trends and patterns of the very resource targeted. We hope that this checklist will be useful for both the future development of Uganda, and for the conservation of one of the country's important resources — its trees.

MATERIALS AND METHODS

We have adopted the definition of a tree to be "a perennial woody plant with many secondary branches forming a distinct elevated crown supported clear of the ground on a single distinct main stem or trunk" (Harris & Harris 1994), or "a perennial woody plant with secondary thickening, with a clear main trunk" (Beentje 2010). Building on these working definitions, we have further considered the description of plants in the *Flora of Tropical East Africa* (various authors, 1952–2011) and categorised the species according to the Flora's description. Using this working definition, we listed all the trees of Uganda by going through various existing publications: the *Flora of Tropical East Africa* (both published and unpublished parts); Hamilton's (1981) *Field guide to the trees of Ugandan forests*; and Eggeling & Dale's (1952) *Indigenous trees of the Uganda Protectorate*, were key resources to this effort. This initial list was updated through more recent taxonomic literature such as monographs and revisions, the International Plant Names Index website (http://www/ipni.org/ipni/), the African Plant Database (http://www.ville-ge.ch/musinfo/bd/cjb/africa/index/php), annotations on herbarium covers, and through consulting expert colleagues.

For all the tree species, we checked the distribution area, from both literature and herbarium specimens. For those species with more restricted distribution (that is, restricted to Uganda, or occurring in Uganda and adjacent parts of neighbouring countries, but not widespread outside Uganda), we georeferenced the specimens from relevant herbaria; in our case the Makerere University Herbarium of Uganda (MHU), the East African Herbarium in Kenya (EA), the Herbarium of the Royal Botanic Gardens, Kew (K), the herbarium of the British Museum of Natural History (BM), and the national herbaria of Belgium and France in Brussels (BR) and Paris (P). From the georeferenced files, and our estimates of population sizes and threats, we made global conservation assessments following version 3.1 of IUCN (2001) (see p. 221). For every species with really restricted distribution, and those with a conservation category of Vulnerable or worse, we prepared species information accounts which include maps, illustrations, data on how we arrived at the conservation assessment, and notes on local uses and local names.

Regarding the maps, these are centred on Uganda; some of the dots (denoting points where specimens have been collected) fall outside the mapping area; e.g. if specimens have been collected in central DRC. The alternative, mapping every specimen, would have meant loss of information on the Ugandan scale, so we have opted for the current solution; occurrence outside Uganda is, of course, given in the text of the species.

For all those species with wide distribution and a Red List Category that we estimated as 'Least Concern', we have disregarded varieties, where the main species had a conservation assessment of Least Concern; in the case of subspecies, we have treated these when the subspecies had a high-category assessment, even if the full species had not.

Altitudes and habitats given in the species information accounts are those for the whole of the distribution area, not just for Uganda or tropical East Africa. Altitudes are rounded off to the nearest 50 m, the minimum downwards, the maximum upwards. In the main list (for the 'Least Concern' species), however, both habitat and altitude given are for Uganda only.

In the main checklist we give details on which districts of Uganda the species is distributed (these are the old districts, as used in the *Flora of Tropical East Africa*, one of our main reference tools); the bioclimate and/or habitat type, the elevation in 100 m increments — in this case altitude range in East Africa, not over the whole distribution area.

The species are arranged in families according to Angiosperm Phylogeny Group (2009).

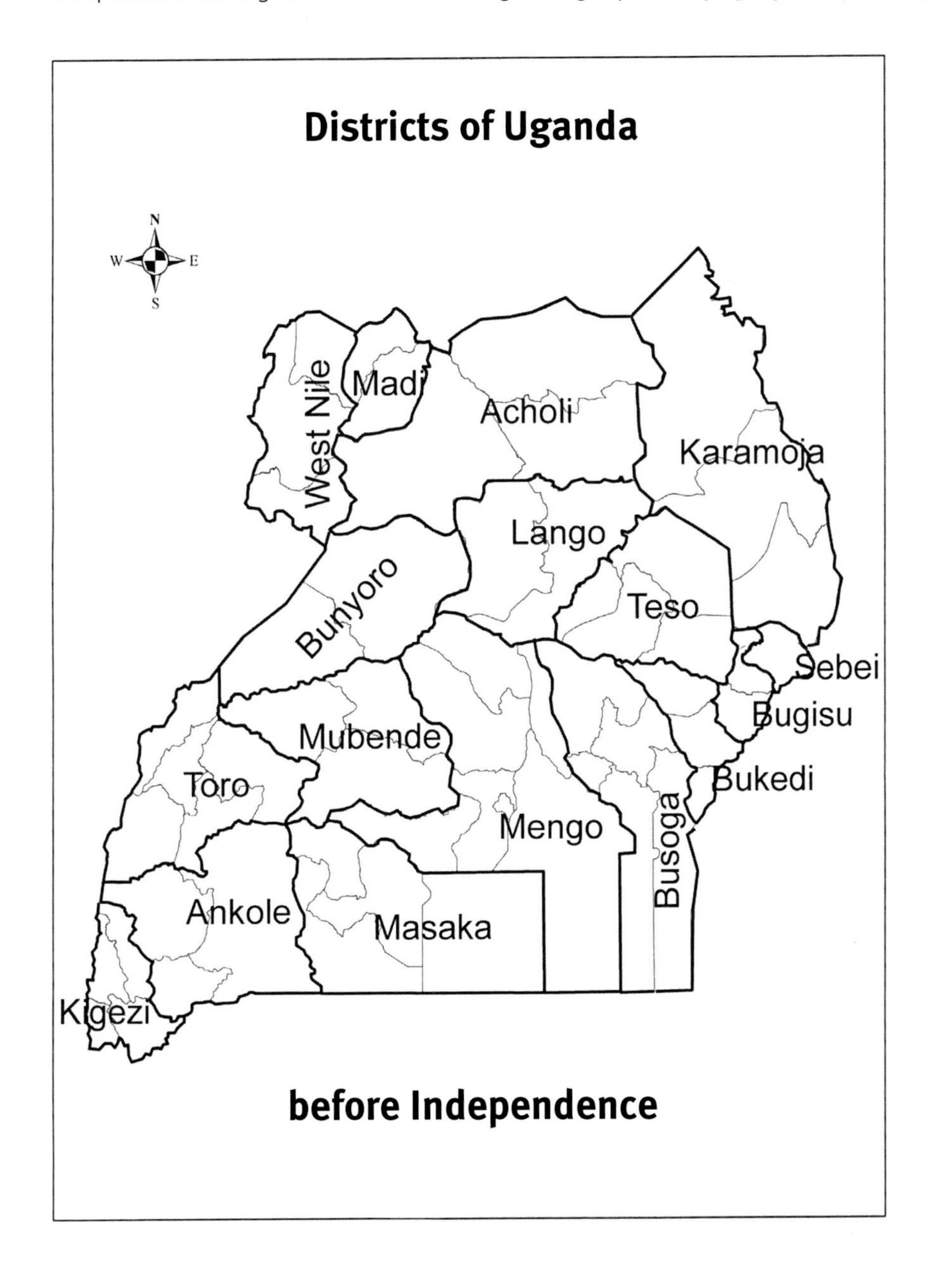

Constraints

i) We had inadequate information for some neighbouring countries on a few taxa, especially about their distribution and levels of threat. We have consulted expert colleagues who are nationals, or who have recently been in these countries.
ii) Some taxa are poorly represented in herbarium collections even though they are known to occur more commonly in their natural habitats. In such cases of inadequate herbarium data, we have sought to get additional information from literature or colleagues.
iii) It is not unusual to have only a few recent collections of trees. We broadened our knowledge on such species by widely consulting people who are known to us as having visited the known areas of occurrence of these species. We accordingly based our assessment on both available herbarium records and observational records/citings as far as we could.
iv) Population size data was grossly lacking on most trees. This made the use of IUCN Red List criteria A and C quite limited.
v) Absence of monitoring programmes on population size leads to lack of trends and patterns for such species in most cases. We have extrapolated from the extent of coverage and quality of habitats in which the species occur, as a proxy for assessment of trends in abundance of species.

Results

A total of 829 tree species in more than 300 genera and 92 families have been recorded in Uganda. The most speciose families are Fabaceae/Leguminosae, Rubiaceae, Euphorbiaceae, Moraceae and Malvaceae in that order (see Table 1). Eleven families have 20 or more species each while 54 families have less than five species each (Table 1).

The genera with the largest numbers of tree species are *Ficus* (41 species), *Acacia* (21), *Albizia* (15), *Grewia* (14) and *Euphorbia*, *Ochna* and *Rytigynia* (each with 9 species)
Of the 829 tree species, four are Critically Endangered, four Endangered and four Vulnerable. This number of species with conservation ratings of Vulnerable and worse is relatively low. This is just over one percent of the total number of tree species in the country. Why is this? Are there hardly any conservation problems in Uganda, is there no habitat destruction or fragmentation? Yes, of course there are problems, and of course there is erosion of pristine habitat, like in most other countries. So why are there so few species under threat?

There are, in fact, many species under threat in Uganda through habitat loss or habitat erosion, through over-harvesting, and through other reasons. But many species of tree occurring in Uganda have a wide distribution, and are therefore less likely to be flagged up as being in danger at global level. If a species has a small distribution area, *any* threat will significantly affect such a species. On the other hand, if a species has a large distribution area, it will take quite a while before such a threat has an impact; or, more worryingly, before such a threat becomes clear. A species may be slowly disappearing over its entire range, but if such a range encompasses twelve or twenty different countries, such a slow threat may only become clear when it is too late. What can we do about this? We can communicate better and more with our close and far neighbours, of course; and in a way, this book is a step in that direction.

Table 1: The distribution of tree species by their families in Uganda

Family	Number of species	Percentage of total
Fabaceae/Leguminosae	96	11.58
Rubiaceae	91	10.98
Moraceae	48	5.79
Euphorbiaceae	46	5.55
Malvaceae	33	3.98
Meliaceae	28	3.38
Phyllanthaceae	23	2.77
Sapindaceae	23	2.77
Annonaceae	21	2.53
Anacardiaceae	20	2.41
Sapotaceae	20	2.41
Rutaceae	18	2.17
Apocynaceae	16	1.93
Ochnaceae	15	1.81
Salicaceae	15	1.81
Asteraceae/Compositae	14	1.69
Capparaceae	13	1.57
Celastraceae	13	1.57
Combretaceae	13	1.57
Loganiaceae	12	1.45
Rhamnaceae	11	1.33
Lamiaceae/Labiatae	10	1.21
Ulmaceae	10	1.21
Boraginaceae	9	1.09
Burseraceae	9	1.09
Clusiaceae	9	1.09
Solanaceae	9	1.09
Araliaceae	8	0.97
Ebenaceae	8	0.97
Oleaceae	7	0.84
Proteaceae	7	0.84
Thymeleaceae	7	0.84
Violaceae	7	0.84
Achariaceae	6	0.72
Ericaceae	6	0.72
Myrtaceae	6	0.72
Olacaceae	5	0.6
Primulaceae	5	0.6
Arecaceae/Palmae	4	0.48
Balanitaceae	4	0.48
Bignoniaceae	4	0.48
Cecropiaceae	4	0.48
Connaraceae	4	0.48
Cyatheaceae	4	0.48
Dracaenaceae	4	0.48
Melastomataceae	4	0.48
Putranjivaceae	4	0.48

Family	Number of species	Percentage of total
Rhizophoraceae	4	0.48
Zamiaceae	4	0.48
Hypericaceae	3	0.36
Lauraceae	3	0.36
Poaceae/Gramineae	3	0.36
Podocarpaceae	3	0.36
Simaroubaceae	3	0.36
Acanthaceae	2	0.24
Apiaceae/Umbelliferae	2	0.24
Cardiopteridaceae	2	0.24
Chrysobalanaceae	2	0.24
Dichapetalaceae	2	0.24
Irvingiaceae	2	0.24
Lythraceae	2	0.24
Myricaceae	2	0.24
Myristicaceae	2	0.24
Passifloraceae	2	0.24
Pentaphylacaceae	2	0.24
Pittosporaceae	2	0.24
Polygalaceae	2	0.24
Rosaceae	2	0.24
Salvadoraceae	2	0.24
Alangiaceae	1	0.12
Aloaceae	1	0.12
Aquifoliaceae	1	0.12
Canellaceae	1	0.12
Cornaceae	1	0.12
Cupressaceae	1	0.12
Erythroxylaceae	1	0.12
Hamamelidaceae	1	0.12
Icacinaceae	1	0.12
Lecythidaceae	1	0.12
Linaceae	1	0.12
Melianthaceae	1	0.12
Monimiaceae	1	0.12
Moringaceae	1	0.12
Oliniaceae	1	0.12
Pandaceae	1	0.12
Pandanaceae	1	0.12
Peraceae	1	0.12
Santalaceae	1	0.12
Scrophulariaceae	1	0.12
Stilbaceae	1	0.12
Urticaceae	1	0.12
Verbenaceae	1	0.12
Σ	829	100.00

Table 2: Habitat preferences for Ugandan trees

Habitat	Number of tree species
Moist forest	455
Riverine/lakeshore forest	232
Wooded grassland	157
Woodland	119
Bushland (dry and evergreen types)	98
Swamp forest	61
Forest edges	50
Thicket	49
Drier forest types	44
Secondary bush/forest	21
Afromontane/alpine vegetation	20
Rocky outcrops	14
Naturalised	13
Bushed grassland	7
Ruderal	4

(Note: percentages are not indicated, as many tree species occur in more than one type of vegetation)

Many of Uganda's trees have large distribution areas; Uganda is at a phytogeographical crossroads. In other words, many habitat types or biogeographical zones occur in Uganda, and widespread phytochoria (geographic areas with relatively uniform composition of plant species) such as the Sudanian Regional Centre of Endemism, the Guineo-Congolian Regional Centre of Endemism, the Somali-Masai Regional Centre of Endemism, the Afromontane archipelago-like Regional Centre of Endemism are found in Uganda; as are the Guineo-Congolia/Sudania Regional transition zone and the Lake Victoria Regional transition zone. This latter transition zone occupies much of Uganda, and is a reason why we have so (relatively) few endemic tree species, as well as why many of our species are so widespread. Or rather, because so many of our species are so widespread, much of our vegetation has been classified as a transition zone!

Out of the seven tree taxa endemic to Uganda, four occur in protected areas, leaving three (43%) outside of any protected area. But even out of the four in protected areas, only *Balsamocitrus dawei* and *Ficus katendei* occur in National Parks (Kibale and Bwindi Impenetrable respectively); the rest occur only in Forest Reserves and Wildlife Reserves. Some of the Forest Reserves are now planted with exotic plantations, particularly of pines, eucalyptus and cypresses while others are heavily degraded through encroachment.

While our level of endemics, species confined to Uganda and not occurring anywhere else, is relatively low, our species richness, the sheer number of plant species occurring in our country, is high. These species have varying affinities; some are West African, some Central African, some East African, and a few have affinities with the Horn of Africa.

Major threats and recommended conservation actions

The tree species with Near Threatened, Vulnerable, Endangered and Critically Endangered assessments number eighteen. All of these have narrow distribution ranges with the exception of *Dalbergia melanoxylon*, *Prunus africana*, *Vitellaria paradoxa* and *Lychnodiscus cerospermus*.

CR: Four species*: Diospyros katendei, Encephalartos equatorialis, Encephalartos whitelockii and Ficus katendei.*

EN: Four species: *Gomphia mildbraedii, Encephalartos macrostrobilus, Ochna leucophloeos* subsp. *ugandensis* and *Uvariodendron magnificum.*

VU: Four species: *Cyathea mildbraedii, Desplatsia mildbraedii*, *Dicranolepis incisa* and *Vepris eggelingii*.

NT: Six species, of which two are restricted-range, thus: *Afrocarpus dawei*, *Cnestis mildbraedii* and *Pandanus chiliocarpus* while the rest are not: *Dalbergia melanoxylon*, *Lychnodiscus cerospermus*, *Prunus africana* and *Vitellaria paradoxa*.

DD: Five species: *Grewia* sp. A of FTEA, *Mussaenda microdonta* subsp. *odorata, Ochna* sp. 40 of FTEA, *Rytigynia* sp. B of FTEA and *Vernonia calvoana*.

LC: 804 species.

NE: Five species: *Cassia spectabilis*, *Broussonetia papyrifera* and three *Solanum* species.

The percentage of tree taxa that are classified as threatened (Vulnerable, Endangered and Critically Endangered) is 1.45%. This percentage is higher than that of 0.84% strictly Ugandan endemic trees (7 taxa) where Pitman & Jorgensen (2002) estimate this should be similar to the percentage of threatened taxa; however, this is probably a result of several narrowly endemic taxa occurring in transborder areas such as the Virunga, the Ruwenzoris and Mt Elgon. If such narrow transborder endemics are taken into account, the percentage is about the same for threatened, and for endemic taxa.

A single species (*Diospyros katendei*) is only known from the type.

From these data, it can be seen that the major risk factor for trees across Uganda, and our species under threat in particular, is habitat degradation: for the vast majority of all plant species in Uganda, habitat degradation is also the main threat (Kalema *et al.* 2010). Such degradation usually takes the form of general degradation through forest destruction, encroachment for cultivation land, and burning.

Occasionally, there is a threat actually or potentially affecting a whole community of tree and other species, such as large-scale sugarcane cultivation around Mabira Forest Reserve, or a specific one such as a hydro-electric power scheme e.g. the Mpanga small-scale electricity dam, affecting the endemic and restricted-range *Encephalartos whitelockii* in Mpanga River Gorge.

Only three species are specifically targeted by over-use across their entire range: *Cyathea mildbraedii, Cordyla richardii* and *Tricalysia bagshawei*. In Uganda, however, there are other trees specifically targeted for exploitation especially for the supply of timber e.g. *Milicia excelsa* (Mvule), *Cordia millenii, Piptadeniastrum africanum, Newtonia buchananii, Afrocarpus dawei*, Meliaceae such as *Entandrophragma* species, *Lovoa* species and *Khaya* species; for medicine e.g. *Warbugia ugandensis*, *Prunus africana*, *Zanthoxylum chalybeum* and *Spathodea campanulata*; or for fuelwood e.g. *Vitellaria paradoxa* and *Combretum* species.

Some species such as *Warburghia ugandensis* and *Prunus africana* are steadily declining in abundance in the country owing to over-exploitation of these trees for medicinal purposes, both within Uganda and beyond (Kalema *et al.* 2010).

Table 3: Tree species with restricted range of distribution: endemics or near-endemics

Species	**Former assessment** (in brackets: preliminary)	**Our assessment**					
		DD	**LC**	**NT**	**VU**	**EN**	**CR**
Grewia sp. A of FTEA	None	√					
Ochna sp. 40 of FTEA	None	√					
Rytigynia sp. B of FTEA	None	√					
Vernonia calvoana	None	√					
Aeglopsis eggelingii	None		√				
Allanblackia kimbiliensis	None		√				
Balsamocitrus dawei	None		√				
Balthasaria schliebenii	None		√				
Baphia wollastonii	None		√				
Brazzeia longipedicellata	None		√				
Chrysophyllum muerense	None		√				
Dasylepis eggelingii	None		√				
Dendrosenecio adnivalis	(LC)		√				
subsp. *adnivalis* var. *adnivalis*	(LC)		√				
subsp. *adnivalis* var. *petiolatus*	(LC)		√				
subsp. *friesiorum*	(LC)		√				
Dendrosenecio elgonensis	(LC)		√				
subsp. *barbatipes*	(LC)		√				
subsp. *elgonensis*	(LC)		√				
Dendrosenecio erici-rosenii	(LC)		√				
subsp. *alticola*	(LC)		√				
subsp. *erici-rosenii*	(LC)		√				
Drypetes ugandensis	None		√				
Duvigneaudia leonardii-crispi	None		√				
Encephalartos septentrionalis	DD		√				
Euphorbia bwambensis	VU		√				
Euphorbia magnicapsula	None		√				
Grewia rugosifolia	None		√				
Grewia ugandensis	None		√				
Hypericum bequaertii	None		√				
Ixora seretii	None		√				
Leptonychia mildbraedii	(NT)		√				

Table 3: continued

Species	**Former assessment** (in brackets: preliminary)	**Our assessment**					
		DD	**LC**	**NT**	**VU**	**EN**	**CR**
Lijndenia bequaertii	None		√				
Ochna leucophloeos	None		√				
Pavetta urundensis	None		√				
Psychotria bagshawei	None		√				
Pseudagrostistachys ugandensis	None		√				
Rinorea beniensis	None		√				
Rinorea tschingandaensis	None		√				
Rytigynia acuminatissima	None		√				
Rytigynia kigeziensis	None		√				
Rytigynia ruwenzoriensis	None		√				
Sesbania dummeri	None		√				
Zanthoxylum mildbraedii	None		√				
Afrocarpus dawei	NT		√				
Cnestis mildbraedii	None			√			
Cordyla richardii	VU		√				
Pandanus chiliocarpus	None			√			
Tricalysia bagshawei	None						
subsp. *bagshawei*	None		√				
Cyathea mildbraedii	None				√		
Desplatsia mildbraedii	None				√		
Dicranolepis incisa	None				√		
Vepris eggelingii	None				√		
Encephalartos macrostrobilus	VU					√	
Gomphia mildbraedii	None					√	
Ochna leucophloeos	None						
subsp. *ugandensis*	None					√	
Uvariodendron magnificum	None					√	
Diospyros katendei	CR						√ (PEx)
Encephalartos equatorialis	CR					√	
Encephalartos whitelockii	LC – VU						√
Ficus katendei	None						√

Vitellaria paradoxa, which was a dominant species in parts of northern Uganda and Teso wooded grasslands has now declined drastically as a result of burning for charcoal (Okullo *et al.* 2004). Only scattered or lone individual trees may be seen in gardens in Kumi, Soroti and Katakwi Districts of Teso (James Kalema, Mary Namaganda & John Wasswa Mulumba, personal observation in 2010). Whereas the species is reported to still be in healthier populations further north in the Districts of Gulu, Kitgum and Pader, this situation is bound to change rapidly with pacification of that part of Uganda and the return of people to their homes, thus depleting the trees as they continue to open land for cultivation (Kalema *et al.* 2010).

One of the most notable introductions of aliens has been of the Paper Mulberry, *Broussonetia papyrifera*. Large populations of this exotic tree are visible in Mabira and Budongo Forest Reserves. It grows and spreads pretty fast, occupying large expanses in a very short time, and forms near-pure stands; thus it has a huge potential to reduce species diversity. Tracts of forest that are impoverished through tree felling and the subsequent creation of large gaps are most susceptible to invasion by this species; this is the reason *Broussonetia* is abundant in Mabira presently. Small, scattered populations may also be seen on private land as small groves as people spare it for shade.

Nearly all Uganda's threatened (Vulnerable to Critically Endangered) species are from moderate altitude (750–1500 m), with only *Cyathea mildbraedii* from higher altitude; this species, occurring between 2700 and 3300 m, is one of the few taxa threatened by over-harvesting throughout its range of global distribution.The Near-Threatened taxa of trees range from 600 to 1800 m, except *Prunus africana* which has a wider range (900–3000 m). The areas in the altitude range of 600–1800 m are most prone to degradation and habitat loss, arising from conversion to cultivation land.

Site conservation plans focusing on reversal of degradation and loss of habitats need to be worked out, or strengthened. For purposes of conserving trees, emphasis ought to be placed on areas 600 to 1800 m above sea level as this is the range most susceptible to degradation, especially through cultivation. Landscape approaches with programmes targeting sites that are linked through features such as dispersal corridors would go a long way towards minimising habitat fragmentation and the resultant isolation leading to rapid loss of species and diversity.

Some of the species that have been reduced to very low population size need appropriate specific Species Action Plans to rescue them and restore viable populations. These may include encouraging local communities to plant the species they tend to extract intensively and widely, e.g. *Cyathea mildbraedii*. Considering the congruence of agricultural activity and occurrence of threatened trees, on-farm planting of trees needs to be promoted through support of agroforestry programmes under the Government's Plan for Modernization Agriculture (PMA). This should then be backed up by encouraging and supporting efforts and initiatives for *ex situ* conservation through planting the threatened species in botanic gardens and arboreta. Furthermore, we strongly recommend making arrangements for having genebank resources for preservation of the germplasm of taxa in danger, and these should be duplicated elsewhere as a back-up.

Finally, sixteen of our tree species are already listed on CITES, the Convention on International Trade in Endangered Species. All our *Encephalartos* species are listed on Appendix I, the highest category. On Appendix II are listed *Aloe volkensii*, *Prunus africana*, all four *Cyathea* species and six succulent tree *Euphorbia* species: *E. candelabrum, E. dawei, E. gratii, E. magnicapsula, E. teke* and *E. tirucalli*.

Format of species list

Species name plus author	*Rawsonia lucida* Harv. & Sond.
Main literature	Literature: Sleumer, FTEA 1975: 4, fig. 1. UFT: 128.
Synonyms	Synonym: *Rawsonia ugandensis* Dawe & Sprague of ITU: 150.
Habit, size, habitat plus elevation range	Shrub or tree to 20 m high. Rainforest, dry forest, riverine forest; 50–1900 m.
Distribution	Bunyoro, Toro, Busoga, Ankole. Widespread in Africa.
Conservation assessment	Here assessed preliminarily as Least Concern (LC) because of its wide distribution and habitat/altitude range.

Note: *Synonym name* [in the sense of ITU: page number] — means that this is not the proper use of that name, merely the ITU interpretation of that name.

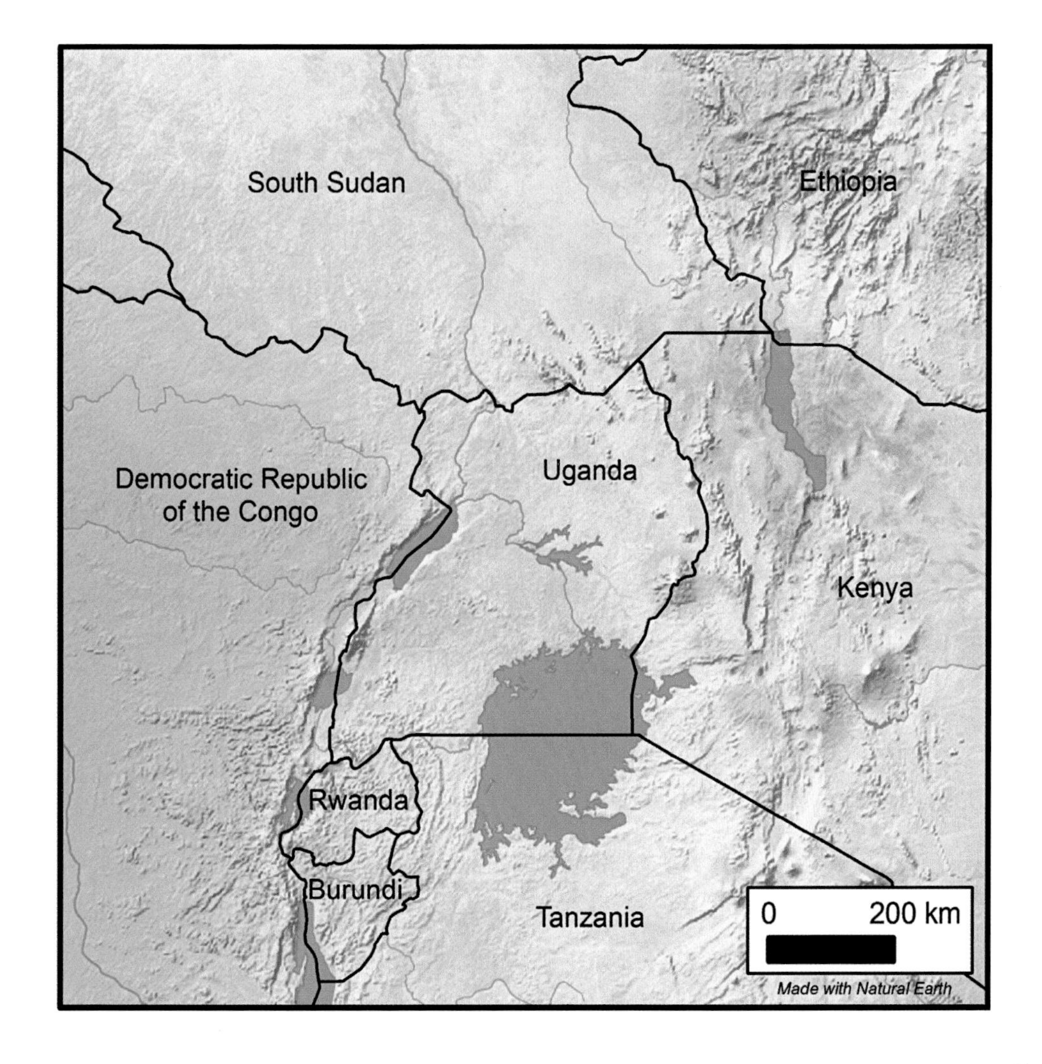

Acacia woodland in Queen Elizabeth National Park (photo: James Kalema).

ABBREVIATIONS USED IN SPECIES LIST

AOO Area of Occupancy in the sense of IUCN (2001) (see p. 221).
EOO Extent of Occurrence in the sense of IUCN (2001) (see p. 221).
FTEA: Flora of Tropical East Africa.
ITU: Eggeling, W. J. & Dale, I. R. (1952). The indigenous trees of the Uganda Protectorate, 2nd edition.
L&S: Lebrun, J.-P. & Stork, A. L. (1991–2010). Enumeration des plantes a fleurs d'Afrique tropicale, vol. 1–5. (vol. 1, 1991; vol. 2, 1992; vol. 3, 1995; vol. 4, 1997; vol. 5, 2010).
UFT: Hamilton, A. (1981). A field guide to Uganda forest trees.
WCMC World Conservation Monitoring Centre.

Mount Muhavura, Kigezi, June 2008 (photo: James Kalema).

Species list

ACANTHACEAE

Justicia maxima S. Moore

Literature: Darbyshire *et al.*, FTEA 2010: 511.

Shrub or tree 3–6 m high. Rainforest; 1150–1200 m.

Masaka, Bunyoro, Mengo. Only three specimens known from Uganda but widespread in Africa.

Here assessed preliminarily as Least Concern (LC) because of its wide distribution.

Mimulopsis arborescens C. B. Clarke

Literature: Vollesen, FTEA 2008: 220. ITU: 2.

Stout, sub-woody shrub, sometimes called a tree, to 8 m high. Forest; 2100–2400 m.

Kigezi, Toro, Bugisu; also in DRC, Rwanda, Burundi, Kenya and S Tanzania.

Here assessed preliminarily as Least Concern (LC) because of its wide distribution.

ACHARIACEAE

Caloncoba crepiniana (De Wild. & T. Durand) Gilg

Literature: Sleumer, FTEA 1975: 24, fig. 8.

Synonym: *Caloncoba schweinfurthii* Gilg; ITU: 144, fig. 31. UFT: 126.

Shrub or tree to 25 m high. Rainforest and forest edges, riverine forest, wooded grassland; 850–1500 m.

West Nile, Bunyoro, Toro, Mubende, Madi, Lango, Acholi. Also in CAR, South Sudan, DRC.

Here assessed preliminarily as Least Concern (LC) because of its wide distribution and habitat range.

Dasylepis eggelingii J. B. Gillett

See restricted list, page 160.

Dasylepis racemosa Oliv.

Literature: Sleumer, FTEA 1975: 7. UFT: 128.

Synonym: *Dasylepis leptophylla* in the sense of ITU: 146.

Tree to 20 m high. Rainforest; 1250–2450 m.

Kigezi. Widespread in Africa.

Here assessed preliminarily as Least Concern (LC) because of its wide distribution and altitude range.

Lindackeria bukobensis Gilg

Literature: Sleumer, FTEA 1975: 25, fig. 9. UFT: 126.

Synonym: *Lindackeria bequaertii* De Wild.; UFT: 126.

Synonym: *Lindackeria mildbraedii* Gilg; ITU: 149.

Shrub or tree to 12 m high. Forest and forest edges; 900–1550 m.

Bunyoro, Kigezi, Masaka, Ankole, Toro. Widespread in Africa.

Here assessed preliminarily as Least Concern (LC) because of its wide distribution and habitat range.

Lindackeria schweinfurthii Gilg

LITERATURE: Sleumer, FTEA 1975: 27. UFT: 126.

Shrub or tree to 7 m high. Rainforest and riverine forest; 600–1200 m.

Bunyoro, Masaka, Mengo. Widespread in Africa.

Here assessed preliminarily as Least Concern (LC) because of its wide distribution.

Rawsonia lucida Harv. & Sond.

LITERATURE: Sleumer, FTEA 1975: 4, fig. 1. UFT: 128.

SYNONYM: *Rawsonia ugandensis* Dawe & Sprague; ITU: 150.

Shrub or tree to 20 m high. Rainforest, dry forest, riverine forest; 50–1900 m.

Bunyoro, Toro, Busoga, Ankole. Widespread in Africa.

Here assessed preliminarily as Least Concern (LC) because of its wide distribution and habitat/altitude range.

ALANGIACEAE

Alangium chinense (Lour.) Harms

LITERATURE: Verdcourt, FTEA 1958: 3, fig. 1. ITU: 4. UFT: 114.

Deciduous tree 9–24 m high. Moist forest, a pioneer in clearings and margins.

Kigezi, Mbale, Mengo, Ankole, Toro, Bunyoro, Karamoja. Widespread in Africa and S Asia.

Here assessed preliminarily as Least Concern (LC) because of its wide distribution and common habitat.

ALOACEAE

Aloe volkensii Engl.

LITERATURE: Carter, FTEA 1994: 54.

SYNONYM: *Aloe* sp. of ITU: 163, photo 27.

Shrub or tree 3–9 m high. Bushed grassland, usually near river; 1150–1950 m.

Ankole. Fairly widespread in East Africa, also in Kenya, Tanzania, Rwanda.

Here assessed preliminarily as Least Concern (LC) because of its wide distribution.

ANACARDIACEAE

Antrocaryon micraster A. Chev. & Guill.

LITERATURE: Kokwaro, FTEA 1986: 49, fig. 10. ITU: 5. UFT: 210.

Deciduous tree to 45 m high. Moist forest; 1050–1500 m.

Bunyoro, Kigezi, Mengo, Madi, Bunyoro. Sierra Leone to Uganda with a number of sub-populations in protected areas.

Here assessed preliminarily as Least Concern (LC) because of its wide distribution.

[*Heeria pulcherrima* (now *Ozoroa pulcherrima*) of ITU is really a shrub.]

Lannea barteri (Oliv.) Engl.

Literature: Kokwaro, FTEA 1986: 21.

Synonym: *Lannea kerstingii* Engl. & K. Krause; ITU: 8, fig. 2.

Tree 5–15 m high. Wooded grassland, forest edges; 900–1200 m.

Acholi, Teso, Mbale, Bunyoro, West Nile, Madi, Lango. From Guinee in W Africa to South Sudan and SW Ethiopia.

Assessed as Least Concern (LC) by Sacande, Sanou & Beentje in *Guide de terrain des arbres du Burkina Faso* (in press).

Lannea fruticosa (A. Rich.) Engl.

Literature: Kokwaro, FTEA 1986: 23. ITU: 7.

Shrub or tree 3–10 m high. Wooded grassland, rocky hillsides, sometimes on black cotton soils; 900–1800 m.

Acholi, Karamoja, West Nile, Madi, Teso, Bugisu. From Niger in W Africa to Ethiopia and Yemen.

Here assessed preliminarily as Least Concern (LC) because of its wide distribution and common habitat.

Lannea fulva (Engl.) Engl.

Literature: Kokwaro, FTEA 1986: 15. ITU: 8.

Shrub or tree 3–10 m high. Wooded grassland, often on rocky hills; also in thicket and dry forest edges; 900–1600 m.

Acholi, Karamoja, Teso, Ankole. Also in Kenya, Tanzania, Rwanda, Burundi, Congo.

Here assessed preliminarily as Least Concern (LC) because of its wide distribution and common habitat.

Lannea humilis (Oliv.) Engl.

Literature: Kokwaro, FTEA 1986: 18. ITU: 8.

Deciduous shrub or tree to 6 m high. Woodland, bushland, wooded grassland, where it may be locally common; 750–1700 m.

Acholi, Teso, Karamoja, Ankole. Widespread in Africa.

Here assessed preliminarily as Least Concern (LC) because of its wide distribution and common habitat.

Lannea schimperi (A. Rich.) Engl.

Literature: Kokwaro, FTEA 1986: 19, fig. 2. ITU: 10.

Tree 2–9 m high. Woodland, wooded grassland; 750–1900 m.

West Nile, Karamoja, Bunyoro, Mengo, Ankole, Mubende, Madi, Acholi, Lango, Teso, Bugisu, Busoga, Toro. Widespread in Africa.

Here assessed preliminarily as Least Concern (LC) because of its wide distribution and common habitat.

Lannea schweinfurthii (Engl.) Engl.

Literature: Kokwaro, FTEA 1986: 23, fig. 3

Synonym: *Lannea stuhlmannii* Engl.; ITU: 10.

Two varieties in Uganda: var. *stuhlmannii* (Engl.) Kokwaro and var. *schweinfurthii*

Shrub or tree 3–15 m high. Wooded grassland; 600–1550 m.

West Nile, Teso, Mengo, Ankole, Bunyoro, Madi, Acholi, Lango, Karamoja, Bugisu, Busoga. Widespread in eastern and southern Africa.

Here assessed preliminarily as Least Concern (LC) because of its wide distribution and common habitat.

Lannea triphylla (A. Rich.) Engl.

Literature: Kokwaro, FTEA 1986: 13. ITU: 11.

Deciduous shrub or small tree 2–5 m high. Wooded grassland, deciduous bushland, often on rocky hills; 300–1400 m.

Karamoja. Also in Kenya, Tanzania, Ethiopia, Somalia, Arabia.

Here assessed preliminarily as Least Concern (LC) because of its wide distribution and common habitat.

Lannea welwitschii (Hiern) Engl.

Literature: Kokwaro, FTEA 1986: 26. ITU: 11. UFT: 209.

Tree to 30 m high. Moist forest; ± 1100 m.

Bunyoro, Mengo, Toro. Widespread in Africa.

Here assessed preliminarily as Least Concern (LC) because of its wide distribution.

Ozoroa insignis Delile subsp. *reticulata* (Baker f.) J.B. Gillett

Literature: Kokwaro, FTEA 1986: 5, fig. 1.

Synonym: *Heeria reticulata* (Bak. f.) Engl.; ITU: 5, fig. 1.

Shrub or tree 3–15 m high. Wooded grassland, often in rocky sites, and in dry forest margins; 1–2200 m.

Karamoja, Teso, Mengo, Mubende, West Nile, Madi, Acholi, Lango, Bugisu, Ankole. Widespread in eastern and southern Africa.

Here assessed preliminarily as Least Concern (LC) because of its wide distribution and range of habitat.

Pistacia aethiopica Kokwaro

Literature: Kokwaro, FTEA 1986: 40, fig. 7.

Synonym: *Pistacia lentiscus* L.; ITU: 11.

Tree 5–15 m high. Dry evergreen forest or associated bushland, often associated with *Juniperus*; 1500–2500 m.

Karamoja. Also in Kenya, N Tanzania, Ethiopia, N Somalia.

Here assessed preliminarily as Least Concern (LC) because of its wide distribution and common habitat.

Pseudospondias microcarpa (A. Rich.) Engl.

Literature: Kokwaro, FTEA 1986: 53, fig. 12. ITU: 12. UFT: 209.

Tree 10–40 m high. Lake shores, riverine, swamp forest, moist forest edges; 900–1700 m.

Ankole, Busoga, Mengo, Masaka, Kigezi, Toro, Mubende, Bunyoro, West Nile, Madi, Acholi, Bugisu. Widespread in Africa.

Here assessed preliminarily as Least Concern (LC) because of its wide distribution and range of habitats.

Rhus anchietae Hiern
LITERATURE: Kokwaro, FTEA 1986: 35.

Shrub or tree 1–8 m high. Riverine, swamp edges; 900–1600 m.

Mengo. Also in Tanzania, Congo, Malawi, Zambia, Angola.

Here assessed preliminarily as Least Concern (LC) because of its wide distribution.

Rhus longipes Engl.
LITERATURE: Kokwaro, FTEA 1986: 36. ITU: 13.

Shrub or tree 2–9 m high. Evergreen bushland, forest margins, riverine, termite hills, thicket; 1–2100 m.

Bugisu, Kigezi, Karamoja. Widespread in eastern and southern Africa.

Here assessed preliminarily as Least Concern (LC) because of its wide distribution and range of habitats.

Rhus natalensis Krauss
LITERATURE: Kokwaro, FTEA 1986: 28, fig. 4. ITU: 13.

Shrub or tree 2–8 m high. Deciduous or evergreen bushland, woodland, riverine, forest edges; 1–3000 m.

Karamoja, Ankole, Mengo, Masaka, Mubende, Bunyoro, West Nile, Madi, Acholi, Lango, Teso, Bugisu, Busoga, Toro. Widespread in Africa.

Here assessed preliminarily as Least Concern (LC) because of its wide distribution and range of habitats.

Rhus ruspolii Engl.
LITERATURE: Kokwaro, FTEA 1986: 29.
SYNONYM: *Rhus bequaertii* Robyns & Lawalrée; ITU: 12.

Shrub or tree 2–7 m high. Upland evergreen bushland and forest margins, riverine; 1200–2450 m.

Acholi, Karamoja, Bugisu, Kigezi, Toro. Kenya, Ethiopia, E DRC.

Here assessed preliminarily as Least Concern (LC) because of its wide distribution and range of habitats.

Rhus pyroides Burch.
LITERATURE: Lebrun & Stork 1997: 299.
SYNONYM: *Rhus vulgaris* Meikle; Kokwaro, FTEA 1986: 31.

Shrub or tree to 10 m. Evergreen bushland, forest margins, riverine and along lakes, wooded grassland in thicket; 800–2700 m.

Karamoja, Ankole, Mengo, Toro, Kigezi, Mubende, Bunyoro, Madi, Acholi, Lango, Teso, Bugisu, Busoga, Masaka. Widespread in Africa.

Here assessed preliminarily as Least Concern (LC) because of its wide distribution and range of habitats.

Sclerocarya birrea (A. Rich.) Hochst. subsp. *birrea*
LITERATURE: Kokwaro, FTEA 1986: 42, fig. 8. ITU: 14, fig. 3.

Deciduous tree to 18 m, spreading. Woodland, wooded grassland, often on rocky hills; 500–1600 m.

Acholi, Karamoja, Teso, West Nile, Madi, Lango, Bugisu, Bunyoro. Senegal to Ethiopia.

Assessed as Least Concern (LC) by Sacande, Sanou & Beentje in *Guide de terrain des arbres du Burkina Faso* (in press).

Sorindeia juglandifolia (A. Rich.) Oliv.

LITERATURE: Breteler 2003: 105.

SYNONYM: *Sorindeia submontana* Van der Veken; Kokwaro, FTEA 1986: 48.

Shrub or tree 5–10 m. Moist forest, semi-swamp forest; ± 1200 m.

Masaka. Widespread in Africa.

Here assessed preliminarily as Least Concern (LC) because of its wide distribution.

Trichoscypha lucens Oliv.

LITERATURE: Breteler 2004: 113.

SYNONYM: *Trichoscypha sp.* in the sense of ITU: 14.

SYNONYM: *Trichoscypha ulugurensis* Mildbr. subsp. *submontana* (Van der Veken) Kokwaro; Kokwaro, FTEA 1986: 51.

Tree 6–25 m high with short bole. Moist forest; 1500–1800 m.

Ankole. Widespread in Africa.

Here assessed preliminarily as Least Concern (LC) because of its wide distribution.

ANNONACEAE

Annona senegalensis Pers. subsp. *senegalensis*

LITERATURE: Verdcourt, FTEA 1971: 113.

SYNONYM: *Annona chrysophylla* Boj.; ITU: 16, fig. 4.

Shrub or tree 1–10 m high. Scattered tree grassland, thicket, woodland, especially where burning is frequent; 500–1800 m.

Bunyoro, Teso, Mengo, Masaka, Toro, West Nile, Madi, Acholi, Lango, Karamoja, Bugisu. Widespread in Africa and Madagascar.

Here assessed preliminarily as Least Concern (LC) because of its wide distribution and range of habitats.

Artabotrys monteiroae Oliv.

LITERATURE: Verdcourt, FTEA 1971: 62, fig. 63.

Shrub, small tree or liana 1.5–6 m high. Forest edges, thickets in moist grassland, woodland; 900–1800 m.

Bunyoro, Busoga, Mengo, Ankole, Masaka. Widespread in eastern, central and southern Africa and Madagascar.

Here assessed preliminarily as Least Concern (LC) because of its wide distribution and range of habitats.

Cleistopholis patens (Benth.) Engl. & Diels

LITERATURE: Verdcourt, FTEA 1971: 31, fig. 5. ITU: 18. UFT: 146.

Tree 7–20 m high, the trunk to 90 cm DBH. Swampy riverine forest, evergreen forest.

West Nile, Bunyoro, Ankole, Toro, Acholi, Madi. West Africa to Angola, and CAR.

Here assessed preliminarily as Least Concern (LC) because of its wide distribution.

Greenwayodendron suaveolens (Engl. & Diels) Verdc. subsp. *suaveolens*
Literature: Verdcourt, FTEA 1971: 67, fig. 16. UFT: 146.
Synonym: *Polyalthia suaveolens* Engl. & Diels; ITU: 20.
Deciduous tree 10–30 m, to 50 cm DBH. Moist forest; 1100 m.
Bunyoro, Mengo, Masaka. Widespread in Africa.
Here assessed preliminarily as Least Concern (LC) because of its wide distribution.

Hexalobus monopetalus (A. Rich.) Engl. & Diels
Literature: Verdcourt, FTEA 1971: 46, fig. 10. ITU: 18.
Shrub or tree 2–9 m high. Wooded grassland, open woodland, forest remnants; 1050–1300 m.
West Nile, Acholi. Widespread in Africa.
Here assessed preliminarily as Least Concern (LC) because of its wide distribution.

Isolona congolana (De Wild. & T. Durand) Engl. & Diels
Literature: Verdcourt, FTEA 1971: 124. ITU: 19. UFT: 148.
Tree 10–30 m high. Riverine forest, degraded primary forest; altitude 470–1400 m over its whole area.
Toro (Bwamba); also in DRC, from 3° N to 7° S and 31° to 30°22′ E.
Here assessed preliminarily as Least Concern (LC) because of its wide distribution (EOO 654,356 km^2, AOO 128,146 km^2) but rare in Uganda, with two records; forest obligate.

Monanthotaxis buchananii (Engl.) Verdc.
Literature: Verdcourt, FTEA 1971: 97.
Shrub, liana or tree 1–8 m high. Riverine forest, evergreen forest, bushland, thicket, grassland, often in rocky sites; 100–1300 m.
Acholi, Busoga, Karamoja, West Nile. Widespread in eastern and central Africa.
Here assessed preliminarily as Least Concern (LC) because of its wide distribution and range of habitats.

Monanthotaxis ferruginea (Oliv.) Verdc.
Literature: Verdcourt, FTEA 1971: 106.
Shrub, liana or tree 3–8 m high. Evergreen forest where it borders on grassland, often on termite mounds; 1050–1200 m.
Busoga, Mbale, Mengo. Widespread in eastern and central Africa.
Here assessed preliminarily as Least Concern (LC) because of its wide distribution.

Monanthotaxis parvifolia (Oliv.) Verdc. subsp. *kenyensis* Verdc.
Literature: Verdcourt, FTEA 1971: 102.
Shrub or tree 1–6 m high. Riverine bushland, *Newtonia* forest, dry forest, usually in rocky places near water; 1200–1500 m.
Karamoja. Widespread in eastern and central Africa.
Here assessed preliminarily as Least Concern (LC) because of its wide distribution; subspecies endemic to Uganda and central Kenya, but also assessed as of Least Concern (LC).

Monodora angolensis Welw.

LITERATURE: Verdcourt, FTEA 1971: 119, fig. 28. UFT: 148.

SYNONYM: *Monodora sp. nov.* of ITU: 20.

Shrub or tree 3–20 m high. Evergreen forest, riverine forest; 950–1350 m.

Bunyoro, Mengo, Ankole. Widespread in eastern and central Africa.

Here assessed preliminarily as Least Concern (LC) because of its wide distribution.

Monodora myristica (Gaertn.) Dunal

LITERATURE: Verdcourt, FTEA 1971: 118. ITU: 19. UFT: 148.

Shrub or tree 3–30 m high, trunk to 1 m DBH. Evergreen forest, often near water; 1100–1800 m.

Ankole, Masaka, Mengo, Kigezi, Toro, Bunyoro, Mbale, Busoga. Widespread in West and central Africa.

Here assessed preliminarily as Least Concern (LC) because of its wide distribution.

Uvaria angolensis Oliv.

LITERATURE: Verdcourt, FTEA 1971: 15. UFT: 149.

SYNONYM: *Uvaria bukobensis* Engl.; ITU: 21.

Shrub, small tree or liana 1–12 m high. Riverine and secondary forest and forest edges, thicket; 1100–1350 m.

West Nile, Kigezi, Mengo, Masaka, Ankole. Widespread in Africa.

Here assessed preliminarily as Least Concern (LC) because of its wide distribution and range of habitats.

Uvaria scheffleri Diels

LITERATURE: Verdcourt, FTEA 1971: 19. UFT: 149.

Shrub, small tree or liana to 3 m high. Evergreen thicket, scrub or forest; 900–1800 m.

Karamoja. Widespread in Kenya and N Tanzania.

Here assessed preliminarily as Least Concern (LC) because of its wide distribution (EOO 841,753 km^2, AOO 362,593 km^2) in a range of habitats.

Uvaria schweinfurthii Engl. & Diels

LITERATURE: Verdcourt, FTEA 1971: 21. ITU: 21. UFT: 149.

SYNONYM: *Uvaria ugandensis* (Bagshawe & Baker f.) Exell; ITU: 21.

Climbing shrub or tree to 6 m high. Probably riverine forest, wooded grassland; ± 720 m.

Bunyoro; rare in Uganda, with four records from Murchison Falls. Also in Cameroon, DRC, CAR, South Sudan.

EOO = 841,753 km^2, AOO = 362,593 km^2; though rare in Uganda, it does not have major threats and occurs in a well protected area; here assessed preliminarily as Least Concern (LC) .

Uvaria welwitschii (Hiern) Engl. & Diels

LITERATURE: Verdcourt, FTEA 1971: 24. ITU: 22. UFT: 149.

Liana, shrub or tree 2–6 m high. Forest on rocky hills; 1150–1500 m.

Ankole, Busoga, Mengo, Masaka, Toro. Also in Kenya, Tanzania, E DRC, Zambia, Angola.

Here assessed preliminarily as Least Concern (LC) because of its wide distribution.

Uvariodendron magnificum Verdc.

See restricted list, page 161.

Uvariopsis congensis Robyns & Ghesq.

LITERATURE: Verdcourt, FTEA 1971: 71, fig. 17. UFT: 146.

SYNONYM *Uvariopsis sp. nov.* of ITU: 22.

Shrub or tree 4–7 m high. Secondary evergreen forest; 1050–1650 m.

Bunyoro, Toro, Mengo, Ankole, Busoga. Widespread in Kenya, Congo, South Sudan, Zambia, Angola.

Here assessed preliminarily as Least Concern (LC) because of its wide distribution and common habitat.

Xylopia aethiopica (Dunal) A. Rich.

LITERATURE: Verdcourt, FTEA 1971: 76, fig. 19. UFT: 148.

SYNONYM: *Xylopia eminii* Engl.; ITU: 22.

Tree or shrub 5–30 m high. Moist forest, swamp forest, woodland; 800–1200 m.

Masaka, Mengo. Widespread in Africa.

Here assessed preliminarily as Least Concern (LC) because of its wide distribution and range of habitats.

Xylopia parviflora (A. Rich.) Benth.

LITERATURE: Verdcourt, FTEA 1971: 79. ITU: 23. UFT: 148.

Tree 6–30 m high. Riverine forest, evergreen forest edge, woodland, thicket; 50–1200 m.

West Nile, West Madi, Busoga, Toro. Widespread in Africa.

Here assessed preliminarily as Least Concern (LC) because of its wide distribution and range of habitats.

Xylopia rubescens Oliv.

LITERATURE: Verdcourt, FTEA 1971: 76. ITU: 23.

Tree 4–30 m high, with stilt roots. Riverine forest; 1150–1350 m.

West Nile, Madi. Widespread in Africa.

Here assessed preliminarily as Least Concern (LC) because of its wide distribution.

Xylopia staudtii Engl. & Diels

LITERATURE: Verdcourt, FTEA 1971: 75. UFT: 148.

Tree to 45 m high with short buttresses and sometimes stilt roots. Moist forest; 1500–1800 m.

Kigezi (Bwindi). Widespread in Africa.

Here assessed preliminarily as Least Concern (LC) because of its wide distribution.

APIACEAE/UMBELLIFERAE

Heteromorpha arborescens Cham. & Schltdl.

SYNONYM: *Heteromorpha trifoliata* (Wendl.) Eckl. & Zeyh.

LITERATURE: Townsend, FTEA 1989: 38, fig. 12.

Herb, shrub or tree 1–8 m high. Forest edges, grassland, woodland; 450–2750 m.

West Nile, Kigezi, Mbale. Widespread in Africa.

Here assessed preliminarily as Least Concern (LC) because of its wide distribution and habitat/altitude range.

Steganotaenia araliacea Hochst.

LITERATURE: Townsend, FTEA 1989: 115, fig. 39. ITU: 440.

Tree 2–12 m high. Rocky slopes in dry woodland, dry bushland, grassland; 150–2200 m.

Karamoja, Busoga, Mengo. All Districts. Widespread in Africa.

Here assessed preliminarily as Least Concern (LC) because of its wide distribution and habitat/altitude range.

APOCYNACEAE

Acokanthera schimperi (A. DC.) Schweinf.

LITERATURE: Omino, FTEA 2002: 15, fig. 5/5–6.

SYNONYM: *A. friesiorum* Markgr.; ITU: 24.

Shrub or tree 1–10 m high. Dry forest, forest margins, wooded grassland, rocky bushland; 250–2200 m.

Karamoja, Mbale. Widespread in East Africa.

Here assessed preliminarily as Least Concern (LC) because of its fairly wide distribution in a range of common habitats.

Adenium obesum (Forssk.) Roem. & Schult.

LITERATURE: Omino, FTEA 2002: 67, fig. 23.

Shrub or tree 0.4–6 m high. Dry bushland, semi-desert scrub, dry woodland, wooded grassland; 500–1200 m.

Acholi, Karamoja. Widespread in Africa.

Here assessed preliminarily as Least Concern (LC) because of its wide distribution and range of habitats.

Alstonia boonei De Wild.

LITERATURE: Omino, FTEA 2002: 60, fig. 20. ITU: 24, t. 5. UFT: 163.

Tree 25–40 m high. Rainforest, groundwater or riverine forest; 800–1200 m.

Bunyoro, Mengo, Masaka, Toro, Ankole, Mbale, Busoga. Widespread in Africa.

Here assessed preliminarily as Least Concern (LC) because of its wide distribution.

Funtumia africana (Benth.) Stapf

LITERATURE: Omino, FTEA 2002: 86, fig. 29. ITU: 28. UFT: 164.

SYNONYM: *Kickxia latifolia* Stapf; ITU: 28.

Tree 8–30 m high. Moist, riverine or swamp forest; 500–1600 m.

Toro, Mbale, Masaka, Mengo. Widespread in Africa.

Here assessed preliminarily as Least Concern (LC) because of its wide distribution.

Funtumia elastica (Benth.) Stapf

LITERATURE: Omino, FTEA 2002: 88. ITU: 27. UFT: 164.

Tree to 35 m high. Moist forest; 1050–1200 m.

Toro, Bunyoro, Mengo. Widespread in Africa.

Here assessed preliminarily as Least Concern (LC) because of its wide distribution.

Picralima nitida (Stapf) T. Durand & H. Durand

LITERATURE: Omino, FTEA 2002: 36, fig. 11. ITU: 28. UFT: 166.

Shrub or tree 4–35 m high. Rainforest; 800–1200 m.

Toro, Bunyoro, Mengo. Widespread in Africa.

Here assessed preliminarily as Least Concern (LC) because of its wide distribution.

Pleiocarpa pycnantha (K. Schum.) Stapf

LITERATURE: Omino, FTEA 2002: 39, fig. 13. ITU: 29. UFT: 164.

Tree 1.5–30 m high. Evergreen forest, riverine or swamp forest; 500–2300 m.

Toro, Ankole, Mengo, Masaka. Widespread in Africa.

Here assessed preliminarily as Least Concern (LC) because of its wide distribution.

Rauvolfia caffra Sond.

LITERATURE: Omino, FTEA 2002: 61, fig. 21. ITU: 30.

SYNONYM: *Rauvolfia oxyphylla* Stapf; ITU: 30. UFT: 163.

Shrub or tree 2–40 m high. Riverine and swamp forest; 450–1950 m.

Toro, Mbale, Mengo, Acholi, Teso, Karamoja. Widespread in Africa.

Here assessed preliminarily as Least Concern (LC) because of its wide distribution.

Rauvolfia mannii Stapf

LITERATURE: Omino, FTEA 2002: 63.

Shrub or tree 0.3–8 m high. Moist forest; 300–2250 m.

Kigezi. Widespread in Africa.

Here assessed preliminarily as Least Concern (LC) because of its wide distribution.

Rauvolfia vomitoria Afzel.

LITERATURE: Omino, FTEA 2002: 65. ITU: 30. UFT: 163.

Shrub or tree 0.5–20(–40) m high. Moist forest and forest margins, thicket; 900–1200 m.

Ankole, Busoga, Mengo, Masaka, Toro, Bunyoro. Widespread in Africa.

Here assessed preliminarily as Least Concern (LC) because of its wide distribution and range of habitats.

Tabernaemontana odoratissima (Stapf) Leeuwenb.

LITERATURE: Omino, FTEA 2002: 46. UFT: 166.

SYNONYM: *Gabunia odoratissima* Stapf; ITU: 28.

Tree 5–15 m high. Moist forest, in woodland on rock; 450–1850 m.

Ankole, Toro, Mengo, Kigezi. Also in Tanzania and DRC, and quite widespread there.

Here assessed preliminarily as Least Concern (LC) because of its wide distribution and habitat range.

Tabernaemontana pachysiphon Stapf

Literature: Omino, FTEA 2002: 47, fig. 15.

Synonym: *Conopharyngia holstii* (K. Schum.) Stapf; ITU: 26.
Synonym: *Tabernaemontana holstii* K. Schum.; UFT: 164.

Shrub or tree 2–15 m high. Moist forest and forest margins, riverine and gallery forest; 500–2000 m.

Kigezi, Mengo, Masaka, Toro, Bunyoro, Mbale. Widespread in Africa.

Here assessed preliminarily as Least Concern (LC) because of its wide distribution and range of habitats.

Tabernaemontana stapfiana Britten

Literature: Omino, FTEA 2002: 49.

Synonym: *Conopharyngia johnstonii* Stapf; ITU: 27.
Synonym: *Tabernaemontana johnstonii* (Stapf) Pichon; UFT: 166.

Tree 5–35 m high. Forest; 1400–2300 m.

Mbale, Toro, Kigezi. Widespread in Africa.

Here assessed preliminarily as Least Concern (LC) because of its wide distribution.

Tabernaemontana ventricosa A. DC.

Literature: Omino, FTEA 2002: 49.

Synonym: *Conopharyngia usambarensis* (Engl.) Stapf; ITU: 27.
Synonym: *Tabernaemontana usambarensis* Engl.; UFT: 166.

Shrub or tree 3–15 m high. Moist forest and forest margins, riverine and groundwater forest, evergreen thicket; 500–1650 m.

Toro, Kigezi, Masaka, Mengo. Widespread in Africa.

Here assessed preliminarily as Least Concern (LC) because of its wide distribution and range of habitats.

Voacanga africana Stapf

Literature: Omino, FTEA 2002: 42. ITU: 30.

Shrub or tree to 10(–25) m high. Forest and riverine forest; 500–1200 m.

Madi. Widespread in Africa.

Here assessed preliminarily as Least Concern (LC) because of its wide distribution.

Voacanga thouarsii Roem. & Schult.

Literature: Omino, FTEA 2002: 44, fig. 14. UFT: 166.

Synonym: *Voacanga obtusa* K. Schum.; ITU: 31.

Tree 2–15 m high. Riverine forest or bush, swamps; 500–1600 m.

Mengo, Masaka, Ankole, Toro, Mubende, Bunyoro, West Nile, Mbale, Acholi. Widespread in Africa.

Here assessed preliminarily as Least Concern (LC) because of its wide distribution.

AQUIFOLIACEAE

Ilex mitis (L.) Radlk.

LITERATURE: Verdcourt, FTEA 1968: 1, fig. 1. ITU: 31. UFT: 135.

Shrub or tree 4–40 m high. Rainforest, dry forest, thicket; 900–3150 m.

Karamoja, Toro, Kigezi, Mbale, Masaka. Widespread in Africa.

Here assessed preliminarily as Least Concern (LC) because of its wide distribution and range of habitats.

ARALIACEAE

Cussonia arborea A. Rich.

LITERATURE: Tennant, FTEA 1968: 4, fig. 1. ITU: 32, fig. 6.

Tree to 13 m high. Woodland, grouped tree grassland; 300–2500 m.

Acholi, Mbale, Mengo, Toro, Bunyoro, West Nile, Madi, Lango, Teso, Karamoja, Busoga. Widespread in Africa.

Here assessed preliminarily as Least Concern (LC) because of its wide distribution.

Cussonia holstii Engl.

LITERATURE: Tennant, FTEA 1968: 8. ITU: 34. UFT: 198.

Tree to 20 m high. Dry forest, grouped tree grassland, semi-evergreen bushland; 1100–2500 m.

Karamoja, Kigezi, Ankole, Mbale. Widespread in Africa.

Here assessed preliminarily as Least Concern (LC) because of its wide distribution and range of habitats.

Cussonia spicata Thunb.

LITERATURE: Tennant, FTEA 1968: 3. ITU: 34. UFT: 198.

Tree to 17 m high. Rainforest, dry forest, grouped tree grassland; 1450–2250 m.

Karamoja, Mbale. Widespread in Africa.

Here assessed preliminarily as Least Concern (LC) because of its wide distribution and range of habitats.

Polyscias fulva (Hiern) Harms

LITERATURE: Tennant, FTEA 1968: 12, fig. 4. ITU: 34, photo 3. UFT: 209.

Tree to 30 m high. Rainforest, riverine forest, grassland; 1150–2200 m.

Kigezi, Mbale, Masaka, Mengo, Ankole, Toro, Mubende, West Nile, Acholi, Busoga. Widespread in Africa.

Here assessed preliminarily as Least Concern (LC) because of its wide distribution and range of habitats.

Schefflera abyssinica (A. Rich.) Harms.

LITERATURE: Tennant, FTEA 1968: 20. ITU: 35. UFT: 199.

Epiphyte or tree to 30 m high. Rainforest; 1800–2800 m.

Karamoja, Mbale. Widespread in Africa.

Here assessed preliminarily as Least Concern (LC) because of its wide distribution.

Schefflera barteri Harms

Literature: Tennant, FTEA 1968: 17, fig. 5. UFT: 199.

Synonym: *Schefflera goetzenii* Harms; ITU: 35.
Synonym: *Schefflera urostachya* Harms; ITU: 36.

Shrub, liana or tree to 30 m high. Rainforest; 900–2000 m.

Kigezi, Mengo, Toro, Mengo, Masaka, Busoga. Widespread in Africa.

Here assessed preliminarily as Least Concern (LC) because of its wide distribution.

Schefflera myriantha (Baker) Drake

Literature: Lebrun & Stork 1992: 234.

Synonym: *Schefflera polysciadia* Harms; Tennant, FTEA 1968: 21. ITU: 36.

Liana or tree to 16 m high. Rainforest, bamboo thickets; 1500–2800 m.

Kigezi, Toro. Widespread in Africa.

Here assessed preliminarily as Least Concern (LC) because of its wide distribution and protection.

Schefflera volkensii (Engl.) Harms

Literature: Tennant, FTEA 1968: 21. ITU: 36. UFT: 199.

Shrub, liana or tree to 30 m high. Rainforest, dry forest; 1600–3250 m.

Mbale, Karamoja. Also in Kenya, Ethiopia, N Tanzania.

Here assessed preliminarily as Least Concern (LC) because of its wide distribution.

ARECACEAE/PALMAE

Borassus aethiopum Mart.

Literature: Dransfield, FTEA 1986: 19, fig. 2. ITU: 291, photo 46.

Tree to 30 m high. Along streams, on hillslopes and in forest; 500–3000 m.

Mengo, Kigezi, Toro, Mubende, Bunyoro, West Nile, Madi, Acholi, Lango, Teso, Karamoja, Mbale, Busoga. Widespread in Africa.

Assessed as Least Concern (LC) by Sacande, Sanou & Beentje in *Guide de terrain des arbres du Burkina Faso* (in press).

Elaeis guineensis Jacq.

Literature: Dransfield, FTEA 1986: 50, fig. 11. ITU: 292.

Tree to 30 m high. Gallery forest; 500–1500 m.

Toro. Widespread in Africa.

Here assessed preliminarily as Least Concern (LC) because of its wide distribution and altitude range.

Phoenix reclinata Jacq.

Literature: Dransfield, FTEA 1986: 15, fig. 1. ITU: 293, photo 47. UFT: 75.

Clustering tree to 10 m high. Along streams, or where water-table is high; 500–1200 m.

In most districts of Uganda. Widespread in Africa.

Assessed as Least Concern (LC) by Sacande, Sanou & Beentje in *Guide de terrain des arbres du Burkina Faso* (in press).

Borassus wooded grassland in Murchison Falls National Park (photo: James Kalema).

Raphia farinifera (Gaertn.) Hyl.

Literature: Dransfield, FTEA 1986: 38, fig. 7. UFT: 76.

Synonym: *Raphia monbuttorum* in the sense of ITU: 293, photo 48.

Clustering tree to 25 m high. Gallery and swamp forest; 500–2500 m.

Mengo, Masaka. Widespread in Africa.

Here assessed preliminarily as Least Concern (LC) because of its wide distribution and habitat/altitude range.

ASTERACEAE/COMPOSITAE

Brachylaena huillensis O. Hoffm.

Literature: Jeffrey & Beentje, FTEA 2000: 4, fig. 1.

Synonym: *Brachylaena hutchinsii* Hutch.; ITU: 95.

Shrub or tree to 40 m high. Dry forest, bushland; 1–2000 m.

Busoga, Mengo. Widespread in eastern and southern Africa. In Uganda now very rare, known only from very old collections.

Here assessed preliminarily as Least Concern (LC) because of its wide distribution and habitat range.

Conyza vernonioides (A. Rich.) Wild

Literature: Beentje, FTEA 2002: 505.

Synonym: *Nidorella arborea* R. E. Fr.; UFT: 135.

Synonym: *Nidorella vernonioides* A. Rich.; ITU: 95.

Shrub or tree 1.5–9 m high. Bamboo, *Hagenia* woodland, giant heath, forest edges; 2300–3600 m.

Kigezi, Mbale. Also in Kenya, Ethiopia, N Tanzania, DRC.

Here assessed preliminarily as Least Concern (LC) because of its wide distribution and habitat range.

Dendrosenecio adnivalis (Stapf) E. B. Knox (two subspecies)

See restricted list, page 162.

Dendrosenecio elgonensis (T. C. E. Fr.) E. B. Knox (two subspecies)

See restricted list, page 164.

Dendrosenecio erici-rosenii (R. E. Fr. & T. C. E. Fr.) E. B. Knox (two subspecies)

See restricted list, page 166.

Solanecio mannii (Hook. f.) C. Jeffrey

LITERATURE: Beentje *et al.*, FTEA 2005: 678

SYNONYM: *Crassocephalum mannii* (Hook. f.) Milne-Redh.; ITU: 95. UFT: 136.

Shrub or tree 1–10 m high. Forest, forest edges, bushland, grassland; 700–2700 m.

Karamoja, Kigezi, Mengo. Widespread in Africa.

Here assessed preliminarily as Least Concern (LC) because of its wide distribution and habitat range.

Tarchonanthus camphoratus L.

LITERATURE: Jeffrey & Beentje, FTEA 2000: 8, fig. 2. ITU: 98.

Shrub or tree to 9 m high. Dry forest, bushland, woodland, wooded grassland; 1200–2750 m.

Karamoja, Mbale. Widespread in Africa.

Here assessed preliminarily as Least Concern (LC) because of its wide distribution and habitat range.

Vernonia amygdalina Delile

LITERATURE: Jeffrey & Beentje, FTEA 2000: 178, fig. 40. ITU: 99. UFT: 136.

Shrub or tree 0.5–10 m high. Riverine forest, forest edges, woodland, wooded grassland, secondary bushland; 300–1950 m.

West Nile, Toro, Mubende, Mengo, Ankole, Kigezi, Bunyoro, Madi, Acholi, Lango, Teso, Karamoja. Mbale, Busoga. Widespread in Africa.

Here assessed preliminarily as Least Concern (LC) because of its wide distribution and habitat range.

Vernonia auriculifera Hiern

LITERATURE: Jeffrey & Beentje, FTEA 2000: 196. ITU: 99.

Herb, shrub or tree 1–7.5 m high. Forest edges, riverine forest, secondary bush, woodland, thicket; 750–3000 m.

Toro, Mbale, Mengo, Mubende, Ankole, Kigezi, Busoga. Widespread in Africa.

Here assessed preliminarily as Least Concern (LC) because of its wide distribution and habitat range.

Vernonia calvoana (Hook. f.) Hook. f.
LITERATURE: Jeffrey & Beentje, FTEA 2000: 236.

Woody herb, shrub or tree 1.5–12 m high.

subsp. *adolfi-friederici* (Muschl.) C. Jeffrey — a Virunga endemic, also in DRC and Rwanda

SYNONYM: *Vernonia sp. aff. adolfi-friderici* Muschl. of ITU: 99. UFT: 136.

Montane woodland or bushland, 2300–3600 m.

Because of the common habitat and wide altitude range, Least Concern (LC)

subsp. *ruwenzoriensis* C. Jeffrey — a Ruwenzori endemic, also in DRC

No data, once 'near river' and presumably a moist forest taxon; 2250–3200 m. We have seen nine Ugandan collections, and from this relatively large number over a range of altitudes, surmise this is fairly widespread; however, due to the lack of recent data (all specimens are over 60 years old) we assess this taxon as data Deficient (DD).

Vernonia conferta Benth.
LITERATURE: Jeffrey & Beentje, FTEA 2000: 198. ITU: 99. UFT: 135.

Tree 6–14 m high. Evergreen forest, swamp forest; 1100–1850 m.

Kigezi, Ankole, Masaka, Toro. Widespread in Africa.

Here assessed preliminarily as Least Concern (LC) because of its wide distribution.

Vernonia hochstetteri Walp.
LITERATURE: Jeffrey & Beentje, FTEA 2000: 212, fig. 43.

Herb, shrub or tree 1.5–4.5 m high. Forest, forest edges, secondary bushland; 1200–2400 m.

Karamoja, Ankole, Mengo, Kigezi, Toro, Teso. Widespread in Africa.

Here assessed preliminarily as Least Concern (LC) because of its wide distribution and habitat range.

Vernonia hymenolepis A. Rich.
LITERATURE: Jeffrey & Beentje, FTEA 2000: 241.

Woody herb, shrub or tree 1–4 m high. Forest and forest margin, riverine, secondary vegetation, old cultivations; 1200–2950 m.

Karamoja, Mbale. Fairly widespread in Africa.

Here assessed preliminarily as Least Concern (LC) because of its wide distribution and habitat range.

Vernonia myriantha Hook. f.
LITERATURE: Jeffrey & Beentje, FTEA 2000: 195.

Shrub or tree 1.5–6 m high. Forest edges, riverine forest, grassland, cultivated areas; 850–2400 m.

Karamoja, Kigezi, Mbale. Widespread in Africa.

Here assessed preliminarily as Least Concern (LC) because of its wide distribution and habitat range.

BALANITACEAE

Balanites aegyptiaca (L.) Delile

Literature: Sands, FTEA 2003: 6, fig. 1. ITU: 405, photo 54, fig. 84.

Shrub or tree to 12 m high. Wooded grassland, dry bushland; 350–1800 m.

Bunyoro, Acholi, Teso. Widespread in Africa.

Assessed as Least Concern (LC) by Sacande, Sanou & Beentje in *Guide de terrain des arbres du Burkina Faso* (in press).

Balanites pedicellaris Mildbr. & Schltr.

Literature: Sands, FTEA 2003: 8. ITU: 407.

Shrub or tree to 8 m high. Dry bushland, thicket, woodland, wooded grassland; 500–1450 m.

Karamoja. Widespread in Africa.

Here assessed preliminarily as Least Concern (LC) because of its wide distribution and range of habitats.

Balanites rotundifolia (Tiegh.) Blatt.

Literature: Sands, FTEA 2003: 10, fig. 2.

Synonym: *Balanites orbicularis* Sprague; ITU: 407.

Shrub or tree to 6 m high. Dry bushland, wooded grassland; 500–800 m.

Karamoja. Widespread in NE Africa.

Here assessed preliminarily as Least Concern (LC) because of its wide distribution and habitat range.

Balanites wilsoniana Dawe & Sprague

Literature: Sands, FTEA 2003: 3. ITU: 407. UFT: 194.

Tree 30–50 m high. Moist forest; 1100–1200 m.

Bunyoro, Toro, Mengo. Widespread in Africa.

Here assessed preliminarily as Least Concern (LC) because of its wide distribution.

BIGNONIACEAE

Kigelia africana (Lam.) Benth.

Literature: ITU: 38, frontispiece, photo 4. UFT: 203. Bidgood *et al.*, FTEA 2006: 43.

Synonym: *Kigelia aethiopica* Decne.

Includes two subspecies: subsp. *africana* and subsp. *moosa* (Sprague) Bidgood & Verdc. which includes the synonyms *K. lanceolata* Sprague, *K. moosa* Sprague.

Tree to 18(–24) m high. Stream-banks, swamp forest, wooded grassland; 0–2250 m.

Mengo, Masaka, Ankole, Toro, Mubende, Bunyoro, West Nile, Mbale, Madi, Acholi, Lango, Teso, Karamoja, Busoga. Widespread in Africa.

Assessed as Least Concern (LC) by Sacande, Sanou & Beentje in *Guide de terrain des arbres du Burkina Faso* (in press).

Markhamia lutea (Benth.) K. Schum.

Literature: Bidgood *et al.*, FTEA 2006: 34.

Synonym: *Markhamia platycalyx* (Baker) Sprague; ITU: 41, fig. 8. UFT: 202.

Tree to 24 m high. Forest edges, riverine forest, wooded grassland, also planted; 1500–1900 m.

Mengo, Masaka, Ankole, Kigezi, Toro, Mubende, Bunyoro, West Nile, Madi, Acholi, Busoga, Mbale. Widespread in Africa.

Here assessed preliminarily as Least Concern (LC) because of its wide distribution and range of habitats.

Spathodea campanulata P. Beauv. subsp. *nilotica* (Seem.) Bidgood

Literature: ITU: 42, plate 1. UFT: 203. Bidgood *et al.*, FTEA 2006: 31.

Tree to 30 m high. Riverine forest, secondary scrub, widely planted; 750–1500 m.

Mengo, Masaka, Ankole, Kigezi, Toro, Bunyoro, Mubende, Madi, Acholi, Lango, Mbale, Busoga. Widespread in Africa.

Here assessed preliminarily as Least Concern (LC) because of its wide distribution and range of habitats.

Stereospermum kunthianum Cham.

Literature: ITU: 42, plate 2, fig. 9. FTEA: Bidgood *et al.*, FTEA 2006: 37.

Shrub or tree 2–20 m high. Wooded grassland; 20–2050 m.

Mengo, Toro, Bunyoro, West Nile, Madi, Acholi, Lango, Teso, Karamoja, Mbale, Busoga. Widespread in Africa.

Here assessed preliminarily as Least Concern (LC) because of its wide distribution.

BORAGINACEAE

Cordia africana Lam.

Literature: Verdcourt, FTEA 1991: 31. UFT: 116.

Synonym: *Cordia abyssinica* in the sense of ITU: 46, pl. 3, fig. 11.

Shrub or tree to 10(–30) m high. Forest edges, secondary forest, gallery forest, scattered tree grassland; 450–2100 m.

West Nile, Kigezi, Mengo, Masaka, Ankole, Toro, Bunyoro, Madi, Acholi, Lango, Karamoja, Mbale, Busoga. Widespread in Africa and Arabia.

Here assessed preliminarily as Least Concern (LC) because of its wide distribution and habitat range, although it is used for timber and its roots and bark for medicine in Tanzania (Mbuya *et al.* 1994).

Cordia crenata Delile

Literature: Verdcourt, FTEA 1991: 19.

Synonym: *Cordia sp. 1* of ITU: 49.
Synonym: *Cordia sp. 2* of ITU: 49, partly.

Shrub or tree 1.5–9 m high. Dry bushland, riverine woodland; 50–1500 m.

West Nile, Acholi, Karamoja, Madi. Widespread in NE Africa.

Here assessed preliminarily as Least Concern (LC) because of its wide distribution and habitat range.

Cordia millenii Baker

Literature: Verdcourt, FTEA 1991: 13, fig. 3. ITU: 48. UFT: 116.

Shrub or tree 4–32 m high. Rainforest; 900–1650 m.

Bunyoro, Masaka, Mengo, Toro, Madi. Widespread in Africa.

Here assessed preliminarily as Least Concern (LC) because of its wide distribution.

Cordia monoica Roxb.

LITERATURE: Verdcourt, FTEA 1991: 15.

SYNONYM: *Cordia ovalis* DC. & A. DC.; ITU: 17.

Shrub or tree 1.5–8 m high. Moist forest, woodland, dry bushland, thicket, often riverine; 500–1850 m.

Karamoja, Bunyoro, Teso, Masaka, West Nile, Acholi, Lango, Mbale, Busoga, Ankole, Toro. Widespread in Africa and Asia.

Here assessed preliminarily as Least Concern (LC) because of its wide distribution and habitat range.

Cordia myxa L.

LITERATURE: Verdcourt, FTEA 1991: 9. ITU: 24.

Tree 6–12 m high. Bushland (naturalised); 500–1050 m.

Acholi. Widespread in Africa and Asia.

Here assessed preliminarily as Least Concern (LC) because of its wide distribution.

Cordia quercifolia Klotzsch

LITERATURE: Verdcourt, FTEA 1991: 22.

SYNONYM: *Cordia gharaf* in the sense of ITU: 48.
SYNONYM: *Cordia sp. 3* of ITU: 49.
SYNONYM: *Cordia sp. 4* of ITU: 50.

Shrub or tree 0.3–5 m high. Scattered tree grassland, dry bushland, thicket, lava desert; 50–1500 m.

Karamoja, Acholi, Mbale. Widespread in Africa and Asia.

Here assessed preliminarily as Least Concern (LC) because of its wide distribution and habitat range.

Cordia sinensis Lam.

LITERATURE: Verdcourt, FTEA 1991: 21.

Shrub or tree 3–10 m high. Riverine, also dry bushland; 500–1800 m.

Bunyoro, Teso. Widespread in Africa and Asia.

Assessed as Least Concern (LC) by Sacande, Sanou & Beentje in *Guide de terrain des arbres du Burkina Faso* (in press).

Ehretia cymosa Thonn.

LITERATURE: Verdcourt, FTEA 1991: 37, fig. 9. UFT: 116.

SYNONYM: *Ehretia silvatica* Guerke; ITU: 50.

Shrub or tree 2–9(–20) m high. Moist or dry evergreen forest and derived vegetation; 950–2250 m.

Bunyoro, Toro, Kigezi, Teso, Masaka, Mengo, Ankole, Mubende, Mbale. Widespread in Africa.

Here assessed preliminarily as Least Concern (LC) because of its wide distribution and habitat range.

Ehretia obtusifolia A. DC.

LITERATURE: Verdcourt, FTEA 1991: 35.

SYNONYM: *Ehretia sp.* of ITU: 50.

Shrub or tree 1.8–6 m high. Woodland, riverine woodland, wooded grassland, thicket, thornbush, often in rocky places; 700–1500 m.

Karamoja. Widespread in Africa and Asia.

Here assessed preliminarily as Least Concern (LC) because of its wide distribution and habitat range.

BURSERACEAE

Boswellia neglecta S. Moore.

LITERATURE: Gillett, FTEA 1991: 7.

SYNONYM: *Boswellia elegans* Engl.; ITU: 51.

Shrub or tree to 8 m high. Dry bushland; 200–1350 m.

Karamoja. Widespread in East Africa.

Here assessed preliminarily as Least Concern (LC) because of its wide distribution and common habitat.

Boswellia papyrifera (Delile) Hochst.

LITERATURE: Gillett, FTEA 1991: 5. ITU: 51.

Tree to 10 m high. Wooded grassland; 1200–1500 m.

West Nile, Acholi, Karamoja, Madi. Widespread in Africa.

Here assessed preliminarily as Least Concern (LC) because of its wide distribution and habitat range.

Canarium schweinfurthii Engl.

LITERATURE: Gillett, FTEA 1991: 2, fig. 1. ITU: 51, photo 6. UFT: 210.

Tree to 40 m high. Secondary forest, often by lakes or rivers; 1000–1600 m.

Ankole, Kigezi, Masaka, Mengo, Toro, Bunyoro, West Nile, Busoga, Mbale. Widespread in Africa.

Assessed as Least Concern (LC) by Sacande, Sanou & Beentje in *Guide de terrain des arbres du Burkina Faso* (in press).

Commiphora africana (A. Rich.) Engl.

LITERATURE Gillett, FTEA 1991: 45. ITU: 53.

SYNONYM: *Commiphora pilosa* Engl.; ITU: 54.

Shrub or tree to 10 m high. Bushed grassland, dry bushland; 500–2100 m.

Karamoja, Ankole, Busoga, West Nile. Widespread in Africa.

Assessed as Least Concern (LC) by Sacande, Sanou & Beentje in *Guide de terrain des arbres du Burkina Faso* (in press).

Commiphora edulis (Klotzsch) Engl. subsp. *boiviniana* (Engl.) J. B. Gillett

LITERATURE: Gillett, FTEA 1991: 64.

SYNONYM: *Commiphora boiviniana* Engl.; ITU: 53.

Shrub or tree 2–8(–10) m high. Dry bushland; 500–1400 m.

Karamoja. Widespread in E and NE Africa.

Here assessed preliminarily as Least Concern (LC) because of its wide distribution and common habitat.

Commiphora habessinica (O. Berg) Engl.

Literature: Gillett, FTEA 1991: 24, fig. 5.1–7.

Synonym: *Commiphora abyssinica* in the sense of ITU: 52.
Synonym: *Commiphora lindensis* in the sense of ITU: 54.
Synonym: *Commiphora subsessilifolia* Engl.; ITU: 54.

Shrub or tree to 4(–6) m high. Bushed grassland, dry bushland, often in rocky places; 500–1900 m.

Acholi, Karamoja, Toro, Bunyoro. Widespread in Africa.

Here assessed preliminarily as Least Concern (LC) because of its wide distribution and habitat range.

Commiphora holtziana Engl. subsp. *holtziana*

Literature: Gillett, FTEA 1991: 80.

Synonym: *Commiphora hildebrandtii* in the sense of ITU: 53.

Tree 3–6(–10) m high. Dry bushland; 500–1100 m.

Karamoja. Widespread in Africa.

Here assessed preliminarily as Least Concern (LC) because of its wide distribution and common habitat.

Commiphora samharensis Schweinf. subsp. *terebinthina* (Vollesen) J. B. Gillett

Literature: Gillett, FTEA 1991: 42.

Synonym: *Commiphora campestris* in the sense of ITU: 53.

Tree to 6(–9) m high. Dry bushland, bushed grassland; 800–1700 m.

Karamoja. Also in Kenya, S Ethiopia, N Tanzania.

Here assessed preliminarily as Least Concern (LC) because of its wide distribution and habitat range.

Commiphora schimperi (O. Berg) Engl.

Literature: Gillett, FTEA 1991: 43.

Synonym: *Commiphora sp. nov.* of ITU: 54

Shrub or tree 2–6 m high. Bushed grassland, dry bushland; 400–1900 m.

Acholi, Karamoja. Widespread in Africa.

Here assessed preliminarily as Least Concern (LC) because of its wide distribution and habitat range.

CANELLACEAE

Warburgia ugandensis Sprague subsp. *ugandensis*

Literature: Verdcourt, FTEA 1956: 3, fig. 1. ITU: 71, plate 5. UFT: 141.

Tree to 42 m high. Moist and dry forest and derived vegetation; 1100–2200 m.

Toro, Ankole, Bunyoro, Karamoja, Busoga. Also in Kenya, N Tanzania, South Africa.

Here assessed preliminarily as Least Concern (LC) because of its wide distribution and habitat range.

CAPPARACEAE

Boscia angustifolia A. Rich.

LITERATURE: Elffers, Graham & DeWolf, FTEA 1964: 55, fig. 9.7.

SYNONYM: *Boscia dawei* Sprague & M. L. Green; ITU: 72.
SYNONYM: *Boscia sp. aff. B. fischeri* of ITU: 72.

Tree to 10 m high. Woodland, bushland, scattered tree grassland; 500–1950 m.

Karamoja, Ankole. Widespread in Africa and Arabia.

Here assessed preliminarily as Least Concern (LC) because of its wide distribution and habitat range.

Boscia coriacea Pax

LITERATURE: Elffers, Graham & DeWolf, FTEA 1964: 56, fig. 9.1–3.

Shrub or tree to 7 m high. Dry bushland, semi-desert scrub, scattered tree grassland; 150–1500 m.

Karamoja. Widespread in E and NE Africa.

Here assessed preliminarily as Least Concern (LC) because of its wide distribution and habitat range.

Boscia salicifolia Oliv.

LITERATURE: Elffers, Graham & DeWolf, FTEA 1964: 52. ITU: 72.

Tree to 14 m high. Woodland, bushland, bamboo thicket, scattered tree grassland; 300–1800 m.

West Nile, Acholi, Karamoja, Madi, Ankole. Widespread in Africa.

Here assessed preliminarily as Least Concern (LC) because of its wide distribution and habitat range.

Cadaba farinosa Forssk. subsp. *farinosa* & subsp. *adenotricha* (Gilg & Gilg-Ben.) R. A. Graham

LITERATURE: Elffers, Graham & DeWolf, FTEA 1964: 75.

Shrub or rarely tree to 8 m high. Dry bushland, scattered tree grassland, semi-desert scrub, riverine; 500–1700 m.

Acholi, Bunyoro, Mbale, Ankole, Busoga, Masaka, Karamoja, West Nile, Teso. Widespread in Africa and Arabia.

Assessed as Least Concern (LC) by Sacande, Sanou & Beentje in *Guide de terrain des arbres du Burkina Faso* (in press).

Capparis tomentosa Lam.

LITERATURE: Elffers, Graham & DeWolf, FTEA 1964: 62.

Scrambling shrub or tree to 10 m high. Dry bushland, thicket, scattered tree grassland, riverine; 500–2500 m.

Acholi, Toro, Bunyoro, Ankole, Busoga, Kigezi, Karamoja. Widespread in Africa.

Here assessed preliminarily as Least Concern (LC) because of its wide distribution and habitat range.

Crateva adansonii DC.

LITERATURE: Elffers, Graham & DeWolf, FTEA 1964: 20, fig. 4. ITU: 73, photo 11, fig. 18.

Shrub or tree 6–15 m high. Scattered tree grassland, riverine forest, old termite mounds; 600–1400 m.

West Nile, Karamoja, Mubende, Bunyoro, Madi, Acholi, Lango, Teso, Mbale, Busoga, Toro. Widespread in Africa.

Here assessed preliminarily as Least Concern (LC) because of its wide distribution and habitat range.

Euadenia eminens Hook. f.

Literature: Elffers, Graham & DeWolf, FTEA 1964: 68, fig. 11. UFT: 196.

Synonym: *Euadenia alimensis* Hua; ITU: 75.

Shrub or tree to 5 m high. Rainforest; 1300–1650 m.

Bunyoro, Toro, Mengo. Widespread in W and central Africa.

Here assessed preliminarily as Least Concern (LC) because of its wide distribution.

Maerua angolensis DC.

Literature: Elffers, Graham & DeWolf, FTEA 1964: 28. ITU: 75.

Shrub or tree to 10 m high. Woodland, bushland, scattered tree grassland, forest edges; 500–1500 m.

Karamoja, Bunyoro, Ankole, West Nile, Madi, Acholi, Lango, Teso, Busoga, Toro. Widespread in Africa.

Here assessed preliminarily as Least Concern (LC) because of its wide distribution and habitat range.

Maerua bussei (Gilg & Gilg-Ben.) R. Wilczek

Literature: Elffers, Graham & DeWolf, FTEA 1964: 31.

Synonym: *Ritchiea sp. near macrocarpa* of ITU: 78.

Shrub or tree to 5 m high. Dry bushland, woodland, dry evergreen forest, riverine; 250–1100 m.

Acholi, Busoga. Also in Tanzania, Zambia, DRC.

Here assessed preliminarily as Least Concern (LC) because of its wide distribution and habitat range.

Maerua crassifolia Forssk.

Literature: Elffers, Graham & DeWolf, FTEA 1964: 40. ITU: 75.

Tree to 9 m high. Dry bushland, thicket, semi-desert scrub; 150–1650 m.

Karamoja. Widespread in Africa and Arabia.

Here assessed preliminarily as Least Concern (LC) because of its wide distribution and habitat range.

Maerua duchesnei (De Wild.) F. White

Literature: Elffers, Graham & DeWolf, FTEA 1964: 29. UFT: 152.

Synonym: *Capparis afzelii* Pax; ITU: 73.

Shrub or tree to 8 m high. Rainforest and forest edges; 1050–1350 m.

Bunyoro, Toro, Mengo, Masaka, Madi, Busoga. Widespread in Africa.

Here assessed preliminarily as Least Concern (LC) because of its wide distribution and habitat range.

Maerua triphylla A. Rich.

Literature: Elffers, Graham & DeWolf, FTEA 1964: 43.

Synonym: *Maerua hoehnelii* Gilg & Gilg-Ben.; ITU: 77.
Synonym: *Maerua sphaerocarpa* Gilg; ITU: 77.

Shrub or tree 5–7.5 m high. Dry or evergreen bushland, scattered tree grassland, thicket edges, sometimes on termite mounds, moist forest edges; 500–2300 m.

Karamoja, Ankole, Kigezi, Toro, Teso, Masaka, Ankole, Mengo, Busoga, Bunyoro, Acholi, West Nile. Widespread in Africa.

Here assessed preliminarily as Least Concern (LC) because of its wide distribution and habitat range.

Ritchiea albersii Gilg

Literature: Elffers, Graham & DeWolf, FTEA 1964: 21, fig. 5. ITU: 77. UFT: 196.

Shrub or tree to 20 m high. Rainforest, evergreen thicket; 1100–2400 m.

Acholi, Toro, Mengo, Masaka, Kigezi, Bunyoro, Karamoja. Widespread in Africa.

Here assessed preliminarily as Least Concern (LC) because of its wide distribution and habitat range.

CARDIOPTERIDACEAE

Leptaulus daphnoides Benth.

Literature: Lucas, FTEA 1968: 2, fig. 1. ITU: 161. UFT: 154.

Shrub or tree to 12 m high. Rainforest, riverine forest; 1050–1250 m.

Bunyoro, Masaka, Mengo, Kigezi. Widespread in Africa.

Here assessed preliminarily as Least Concern (LC) because of its wide distribution.

Leptaulus holstii (Engl.) Engl.

Literature: Lucas, FTEA 1968: 2, fig. 1.

Shrub or tree to 5 m high. Rainforest; 700–1200 m.

Mengo. Also in Angola and DRC.

Here assessed preliminarily as Least Concern (LC) because of its wide distribution.

CECROPIACEAE

Musanga cecropioides Tedlie

Literature: Berg & Hijman, FTEA 1989: 90, fig. 23. ITU: 263, photo 44, partly. UFT: 201.

Tree to 30 m high. Secondary and swamp forest; 750–900 m.

Toro, Mengo, Ankole, Kigezi, Bunyoro. Widespread in Africa.

Here assessed preliminarily as Least Concern (LC) because of its wide distribution.

Musanga leo-errerae Hauman & J. Léonard

Literature: Berg & Hijman, FTEA 1989: 90. UFT: 202.

Synonym: *Musanga cecropioides* of ITU: 263, partly.

Tree to 30 m high. Forest regrowth; 1350 m.

Ankole, Kigezi, limited to Bwindi, Kalinzu and Kasyoha-Kitomi; also in DRC, Rwanda, Burundi.

Here assessed preliminarily as Least Concern (LC) because of its wide distribution (it has an AOO of at least 6,296 km^2) and its habitat: it is a common pioneer in forest gaps.

Myrianthus arboreus P. Beauv.

Literature: Berg & Hijman, FTEA 1989: 87. ITU: 264. UFT: 201.

Tree to 10 m high or shrub. Rainforest, riverine, lakeside and swamp forest, clearings and regrowth; 700–1200 m.

Mengo, Masaka, Mubende, Bunyoro. Widespread in Africa.

Here assessed preliminarily as Least Concern (LC) because of its wide distribution and habitat range.

Myrianthus holstii Engl.

Literature: Berg & Hijman, FTEA 1989: 88, fig. 22. ITU: 264. UFT: 201.

Tree to 20 m high. Rainforest, riverine and regrowth; 900–2100 m.

Toro, Ankole, Kigezi. Widespread in Africa.

Here assessed preliminarily as Least Concern (LC) because of its wide distribution and habitat range.

CELASTRACEAE

Catha edulis Forssk.

Literature: Robson, Hallé, Mathew & Blakelock, FTEA 1994: 25, fig. 5. ITU: 78. UFT: 189.

Tree 2–15(–25) m high. Evergreen forest and forest edges, woodland; 1100–1800 m.

Karamoja, Kigezi, Mbale, Ankole (but certainly more widespread as it is cultivated e.g. in Mengo, Bukedi, etc.). Widespread in Africa.

Here assessed preliminarily as Least Concern (LC) because of its wide distribution and habitat range.

Elaeodendron buchananii (Loes.) Loes.

Literature: Robson, Hallé, Mathew & Blakelock, FTEA 1994: 32. ITU: 79.

Synonym: *Cassine buchananii* Loes.; UFT: 188.

Synonym: *Elaeodendron sp. near keniense* of ITU: 79.

Synonym: *Elaeodendron sp.* of ITU: 80.

Shrub or tree 4.5–30 m high. Forest and riverine forest, woodland, grassland; 1000–2250 m.

Acholi, Toro, Masaka, Mengo, Ankole, West Nile, Madi, Mbale, Teso. Widespread in Africa.

Here assessed preliminarily as Least Concern (LC) because of its wide distribution and habitat range.

Gymnosporia arbutifolia (A. Rich.) Loes.

Literature: Jordaan & van Wyk in *Taxon* 55: 521 (2006)

Synonym: *Maytenus arbutifolia* (A. Rich.) R. Wilczek; Robson, Hallé, Mathew & Blakelock, FTEA 1994: 12.

Synonym: *Gymnosporia atkaio* (A. Rich.) Loes. of ITU: 81.

Shrub or tree 1–8 m high, spiny. Forest edges, riverine forest, thicket; 1350–2400 m.

Kigezi, Mbale, Mubende, Acholi, Karamoja. Fairly widespread in Africa.

Here assessed preliminarily as Least Concern (LC) because of its wide distribution and habitat range.

Gymnosporia buchananii Loes.

LITERATURE: Jordaan & van Wyk in *Taxon* 55: 521 (2006).

SYNONYM: *Maytenus buchananii* (Loes.) R. Wilczek; Robson, Hallé, Mathew & Blakelock, FTEA 1994: 10.

Shrub or tree 2–8 m high. Dry evergreen forest and forest edges, riverine forest; 50–2650 m.

West Nile, Acholi, Ankole. Widespread in Africa.

Here assessed preliminarily as Least Concern (LC) because of its wide distribution and habitat range.

Gymnosporia gracilipes (Oliv.) Loes.

LITERATURE: ITU: 81; Jordaan & van Wyk in *Taxon* 55: 521 (2006).

SYNONYM: *Maytenus gracilipes* (Oliv.) Exell; Robson, Hallé, Mathew & Blakelock, FTEA 1994: 5.

Shrub or tree 0.5–5(–9) m high. Forest and forest edges, riverine, thicket; 1100–2300 m.

Acholi, Toro, Mengo, Ankole, Kigezi, Bunyoro, Mbale. Fairly widespread in Africa.

Here assessed preliminarily as Least Concern (LC) because of its wide distribution and habitat range.

Gymnosporia mossambicensis (Klotzsch) Loes.

SYNONYM: *Maytenus mossambicensis* (Klotzsch) Blakelock; Robson, Hallé, Mathew & Blakelock, FTEA 1994: 8.

Shrub or tree 1–8 m high. Forest, woodland; 600–2900 m.

Ankole. Fairly widespread in Africa.

Here assessed preliminarily as Least Concern (LC) because of its wide distribution and habitat range.

Gymnosporia obscura (A. Rich.) Loes.

LITERATURE: Jordaan & van Wyk in *Taxon* 55: 521 (2006).

SYNONYM: *Maytenus obscura* (A. Rich.) Cufod.; Robson, Hallé, Mathew & Blakelock, FTEA 1994: 11.

Shrub or tree 2–10 m high. Forest and forest edges, scrub, grassland; 2100–2550 m.

Karamoja. Also in Rwanda, Burundi, Kenya, N Tanzania.

Here assessed preliminarily as Least Concern (LC) because of its wide distribution and habitat range.

Gymnosporia senegalensis (Lam.) Loes.

LITERATURE: ITU: 82.

SYNONYM: *Maytenus senegalensis* (Lam.) Exell; Robson, Hallé, Mathew & Blakelock, FTEA 1994: 17.

Shrub or tree 1–9 m high. Woodland, thicket, wooded grassland, scrub; 500–2300 m.

Karamoja, Bunyoro, Ankole, Mengo, Masaka, Kigezi, Mubende, West Nile, Madi, Acholi, Lango, Teso, Mbale, Busoga, Toro. Widespread in Africa.

Here assessed preliminarily as Least Concern (LC) because of its wide distribution and habitat range.

Maytenus acuminata (L. f.) Loes.

Literature: Robson, Hallé, Mathew & Blakelock, FTEA 1994: 19.

Synonym: *Gymnosporia acuminata* (L. f.) Szyszyl.; ITU: 81.

Shrub or tree 1–20 m high. Forest and forest edges; 1050–3300 m.

Toro, Ankole, Masaka. Widespread in Africa.

Assessed as Least Concern (LC) by Sacande, Sanou & Beentje in *Guide de terrain des arbres du Burkina Faso* (in press).

Maytenus heterophylla (Eckl. & Zeyh.) N. Robson

Literature: Robson, Hallé, Mathew & Blakelock, FTEA 1994: 14, fig. 2. UFT: 134.

Synonym: *Gymnosporia buxifolia* (L.) Szyszyl.; ITU: 81.

Shrub or tree 1–7 m high. Forest edges, riverine forest, woodland thicket; 500–2700 m.

Ankole, Kigezi, Masaka, Mengo, West Nile, Acholi, Teso, Karamoja, Mbale, Busoga, Toro. Widespread in Africa.

Here assessed preliminarily as Least Concern (LC) because of its wide distribution and habitat range.

Maytenus undata (Thunb.) Blakelock

Literature: Robson, Hallé, Mathew & Blakelock, FTEA 1994: 18. UFT: 134.

Synonym: *Gymnosporia lancifolia* (Thonn.) Loes. of ITU: 82.
Synonym: *G. fasciculata* Loes.; ITU: 81.
Synonym: *G. luteola* (Delile) Loes; ITU: 82.

Shrub or tree 1–12 m high. Forest and riverine forest, woodland, bushland; 500–3150 m.

Karamoja, Ankole, Masaka, Toro. Widespread in Africa.

Here assessed preliminarily as Least Concern (LC) because of its wide distribution and habitat range.

Mystroxylon aethiopicum (Thunb.) Loes.

Literature: Robson, Hallé, Mathew & Blakelock, FTEA 1994: 21, fig. 3. ITU: 84.

Synonym: *Cassine aethiopica* in the sense of UFT: 134 & of L&S.

Shrub or tree 2–12 m high. Forest and forest edges, woodland; 500–2550 m.

Karamoja, Toro, Mubende, Ankole, Kigezi, Bunyoro, Madi, Acholi, Lango, Teso, Mbale, Busoga. Widespread in Africa.

Here assessed preliminarily as Least Concern (LC) because of its wide distribution and habitat range.

Pleurostylia africana Loes.

Literature: Robson, Hallé, Mathew & Blakelock, FTEA 1994: 27, fig. 6.

Synonym: *Pleurostylia capensis* in the sense of ITU: 84.

Shrub or tree 1–20 m high. Forest edges, woodland; 500–1650 m.

Teso, Madi, Acholi, Karamoja. Widespread in Africa.

Here assessed preliminarily as Least Concern (LC) because of its wide distribution and habitat range.

CHRYSOBALANACEAE

Parinari curatellifolia Benth.

LITERATURE: Graham, FTEA Rosaceae 1960: 50. ITU: 333, fig. 70.

Tree to 15 m high. Woodland, scattered tree grassland; 500–2070 m.

Ankole, Masaka, West Nile, Madi. Not common in Uganda but widespread in Africa.

Here assessed preliminarily as Least Concern (LC) because of its wide distribution and habitat and altitude range.

Parinari excelsa Sabine

LITERATURE: Graham, FTEA Rosaceae 1960: 49. UFT: 141.

SYNONYM: *Parinari holstii* (Engl.) R. Graham; ITU: 333.

Tree to 45 m high. Rainforest, riverine forest; 1000–2100 m.

Bunyoro, Kigezi, Masaka, Mengo, Ankole, Toro, Bunyoro, West Nile, Madi. Widespread in Africa.

Here assessed preliminarily as Least Concern (LC) because of its wide distribution and habitat range.

CLUSIACEAE

Allanblackia kimbiliensis Spirlet

See restricted list, page 168.

Garcinia buchananii Baker

LITERATURE: Bamps, Robson & Verdcourt, FTEA 1978: 24. ITU: 153.

SYNONYM: *Garcinia huillensis* in the sense of UFT: 168?

Shrub or tree 1–15(–25) m high. Forest, riverine forest, thicket, wooded grassland; 50–1800 m.

Mbale, Masaka, Mengo, Madi, Acholi, Teso, Karamoja, Busoga. Widespread in Africa.

Here assessed preliminarily as Least Concern (LC) because of its wide distribution and habitat/altitude range.

Garcinia livingstonei T. Anderson

LITERATURE: Bamps, Robson & Verdcourt, FTEA 1978: 16, fig. 5.

SYNONYM: *Garcinia sp. aff. livingstonei* of ITU: 153.

Tree 3–18 m high. Woodland, thicket, grassland, usually by streams; 500–1650 m.

Karamoja. Widespread in Africa.

Assessed as Least Concern (LC) by Sacande, Sanou & Beentje in *Guide de terrain des arbres du Burkina Faso* (in press).

Garcinia ovalifolia Oliv.

LITERATURE: Bamps, Robson & Verdcourt, FTEA 1978: 22.

SYNONYM: *Garcinia buchananii* in the sense of ITU: 153.

Shrub or tree 2–25 m high. Streamside forest?; 900 m.

West Nile. Widespread in Africa.

Here assessed preliminarily as Least Concern (LC) because of its wide distribution.

Harungana madagascariensis Poir.

Literature: Milne-Redhead, FTEA 1953: 19, fig. 5. ITU: 156. UFT: 168.

Shrub or tree to 12 m high. Rainforest; 500–1800 m.

Ankole, Mbale, Mengo, Masaka, Kigezi, Toro, Bunyoro, Madi, Acholi, Busoga. Widespread in Africa.

Here assessed preliminarily as Least Concern (LC) because of its wide distribution.

Mammea africana Sabine

Literature: Bamps, Robson & Verdcourt, FTEA 1978: 11. ITU: 153, fig. 32. UFT: 169.

Tree 15–45 m high. Rain forest; 1050 m.

Bunyoro. Widespread in Africa.

Here assessed preliminarily as Least Concern (LC) because of its wide distribution.

Psorospermum corymbiferum Hochr.

Literature: Milne-Redhead, FTEA 1953: 16.

Synonym: *Psorospermum febrifugum* in the sense of ITU: 159, partly, fig. 34; & in Bamps, Robson & Verdcourt, FTEA 1978: 32.

Shrub or tree 2–5 m high. Wooded grassland; 1050 m.

West Nile, Masaka, Ankole, Bunyoro, Madi, Lango, Busoga. Widespread in Africa.

Here assessed preliminarily as Least Concern (LC) because of its wide distribution.

Psorospermum febrifugum Spach

Literature: Milne-Redhead, FTEA 1953: 16, fig. 4. ITU: 159.

Synonym: *Psorospermum campestre* Engl.; ITU: 159.

Shrub or tree 1–6 m high. Woodland, wooded grassland; 50–1950 m.

Karamoja, Teso, Mengo, Bunyoro, Ankole, Masaka, West Nile, Acholi, Mbale, Toro. Widespread in Africa.

In Uganda, used commercially for herbal jellies and medicine and may get rare and threatened in future, here assessed preliminarily as Least Concern (LC) because of its wide distribution and altitude range.

Symphonia globulifera L. f.

Literature: Bamps, Robson & Verdcourt, FTEA 1978: 5, fig. 2. ITU: 155. UFT: 168.

Tree 15–25(–40) m high. Rain forest, swamp forest; 800–2550 m.

Ankole, Masaka, Mengo, Mengo, Kigezi, Toro, Bunyoro. Widespread in Africa and America.

Here assessed preliminarily as Least Concern (LC) because of its wide distribution and altitude range.

COMBRETACEAE

Combretum adenogonium A. Rich.

Literature: Lebrun & Stork (1991): 160.

Synonym: *Combretum fragrans* F. Hoffm.; Wickens, FTEA 1973: 29.
Synonym: *Combretum ghasalense* Engl. & Diels; ITU: 86.
Synonym: *Combretum reticulatum* in the sense of ITU: 85.

Tree to 10 m high. Woodland, wooded grassland; 50–1700 m.

West Nile, Karamoja, Teso, Madi, Acholi, Lango, Mbale, Busoga, Mengo, Bunyoro. Widespread in Africa.

Assessed as Least Concern (LC) by Sacande, Sanou & Beentje in *Guide de terrain des arbres du Burkina Faso* (in press).

Combretum collinum Fresen.

LITERATURE: Wickens, FTEA 1973: 24.

SYNONYM: *Combretum binderanum* Kotschy; ITU: 86, fig. 21.
SYNONYM: *Combretum verticillatum* Engl. of ITU: 89.

Shrub or tree to 12 m high. Wooded grassland; 50–2300 m.

West Nile, Teso, Mbale, Lango, Bunyoro, Busoga, Mengo, Mubende, Madi, Acholi, Karamoja. Widespread in Africa.

Assessed as Least Concern (LC) by Sacande, Sanou & Beentje in *Guide de terrain des arbres du Burkina Faso* (in press).

Combretum hereroense Schinz subsp. *grotei* (Exell) Wickens & subsp. *volkensii* (Engl.) Wickens

LITERATURE: Wickens, FTEA 1973: 41.

SYNONYM: *Combretum volkensii* in the sense of ITU: 89.

Shrub or tree to 8 m high. Wooded grassland, dry bushland; 50–2700 m.

Karamoja, Mbale. Widespread in Africa.

Here assessed preliminarily as Least Concern (LC) because of its wide distribution and habitat range.

Combretum molle G. Don

LITERATURE: Wickens, FTEA 1973: 33. ITU: 88.

SYNONYM: *Combretum gueinzii* Sond.; ITU: 88.
SYNONYM: *Combretum sp. near microlepidotum* of ITU: 89.

Shrub or tree to 17 m high. Wooded grassland, bushland; 500–2300 m.

Kigezi, Mbale, Masaka, Ankole, Toro, Bunyoro, Mengo, Mubende, West Nile, Madi, Acholi, Karamoja. Widespread in Africa.

Assessed as Least Concern (LC) by Sacande, Sanou & Beentje in *Guide de terrain des arbres du Burkina Faso* (in press).

Bark and fruits of *Combretum molle* (photo: James Kalema).

Combretum mossambicense (Klotzsch) Engl.
Literature: Wickens, FTEA 1973: 63.

Shrub, liana or tree, 'small'. Riverine forest, woodland, bushland, wooded grassland; 700–1600 m.

Acholi, Bunyoro. Widespread in Africa.

Here assessed preliminarily as Least Concern (LC) because of its wide distribution and habitat range.

Combretum glutinosum DC.
Literature: Keay, FWTA 1954: 271.

Synonym: *Combretum schweinfurthii* Engl. & Diels; Wickens, FTEA 1973: 30. ITU: 88.
Shrub or tree to 3 m high.

Wooded grassland, woodland; 750–1500 m.

West Nile, Madi. Rare in Uganda; elsewhere widespread, occurring all the way to Senegal and South Sudan, and here assessed preliminarily as Least Concern (LC)

Combretum umbricola Engl.
Literature: Wickens, FTEA 1973: 20.

Shrub, liana or 'small' tree. Riverine forest; 650–1050 m.

Toro, Mubende. Also DRC, Kenya and Tanzania.

Rare in Uganda. Elsewhere fairly widespread from 4° N to 11° S and 17° to 39° E; EOO 2,097,040 km^2; AOO 570,316 km^2; and here assessed preliminarily as Least Concern (LC).

Terminalia brownii Fresen.
Literature: Wickens, FTEA 1973: 90. ITU: 90.

Tree to 25 m high. Wooded grassland and woodland, bushland, riverine forest; 700–2000 m.

Karamoja, Toro, Mbale, Bunyoro, West Nile, Madi, Acholi. Widespread in Africa.

Here assessed preliminarily as Least Concern (LC) because of its wide distribution and habitat range.

Terminalia glaucescens Benth.
Literature: Wickens, FTEA 1973: 89.

Synonym: *Terminalia velutina* Rolfe; ITU: 93, fig. 22.
Synonym: *Terminalia sp. aff. schweinfurthii* of ITU: 94.

Tree to 13 m high. Wooded grassland; 1000–1800 m.

West Nile, Teso, Mengo, Acholi, Bunyoro, Mubende, Karamoja. Widespread in Africa.

Here assessed preliminarily as Least Concern (LC) because of its wide distribution.

Terminalia laxiflora Engl. & Diels
Literature: Wickens, FTEA 1973: 87.

Synonym: *Terminalia sp. aff. schweinfurthii* of ITU: 94.

Tree to 10 m high. Wooded grassland; 1800–2200 m.

West Nile, Karamoja. Widespread in Africa.

Here assessed preliminarily as Least Concern (LC) because of its wide distribution.

Terminalia macroptera Guill. & Perr.

LITERATURE: Wickens, FTEA 1973: 87. ITU: 90.

Tree to 13 m high. Woodland, wooded grassland; 750–1400 m.

West Nile, Acholi, Teso, Madi, Lango, Mbale. Widespread in Africa.

Here assessed preliminarily as Least Concern (LC) because of its wide distribution and habitat range.

Terminalia mollis M. A. Lawson

LITERATURE: Wickens, FTEA 1973: 88.

SYNONYM: *Combretum torulosa* F. Hoffm.; ITU: 93.

Tree to 20 m high. Wooded grassland and woodland; 900–2200 m.

West Nile, Acholi, Mengo, Mubende, Madi, Acholi, Lango, Teso, Karamoja. Widespread in Africa.

Here assessed preliminarily as Least Concern (LC) because of its wide distribution and habitat range.

Terminalia spinosa Engl.

LITERATURE: Wickens, FTEA 1973: 82. ITU: 93.

Tree to 20 m high. Bushland, woodland, wooded grassland; 500–1800 m.

Karamoja. Also in South Sudan, Somalia, Kenya, Tanzania.

Here assessed preliminarily as Least Concern (LC) because of its wide distribution and habitat range.

CONNARACEAE

Agelaea pentagyna (Lam.) Baill. of Breteler (ed.) 1989: 144.

SYNONYM: *Connarus pentagyna* (Lam.) Baill.; Lebrun & Stork 1992: 231.
SYNONYM: *Agelaea heterophylla* Gilg; Hemsley, FTEA 1956: 11.
SYNONYM: *Agelaea ugandensis* Schellenb.; Hemsley, FTEA 1956: 12.

Shrub, liana or 'small' tree. Rainforest, especially where regenerating; 900–2100 m.

Bunyoro, Kigezi, Mengo. Widespread in Africa.

Here assessed preliminarily as Least Concern (LC) because of its wide distribution and habitat range.

Cnestis mildbraedii Gilg

See restricted list, page 169.

Connarus longistipitatus Gilg

LITERATURE: Hemsley, FTEA 1956: 24, fig. 9. ITU: 100. UFT: 226.

Shrub, liana or 'small' tree. In thicket, forest or swamp forest; 1100–1800 m.

Ankole, Bunyoro, Mengo, Kigezi. Also in DRC, N Tanzania.

Distribution area from 02° S to 03° N and from 28° to 32°30′ E; EOO 91,889 km^2; AOO 24,438 km^2. Because of wide distribution and habitat and altitude range, here assessed preliminarily as Least Concern (LC).

Rourea thomsonii (Baker) Jongkind

LITERATURE: Lebrun & Stork 1992: 232. Breteler (ed.) 1989: 359.

SYNONYM: *Jaundea pinnata* (P. Beauv.) Schellenb.; Hemsley, FTEA 1956: 21, fig. 7.

Shrub, liana or 'small' tree. Rainforest, forest remnants, in cultivations as a relict; 500–2500 m.

Kigezi, Mengo. Widespread in Africa.

Here assessed preliminarily as Least Concern (LC) because of its wide distribution and habitat range.

CORNACEAE

Afrocrania volkensii (Harms) Hutch.

LITERATURE: Verdcourt, FTEA 1958: 1, fig. 1. ITU: 101. UFT: 172.

Tree to 24 m high. Rainforest, often by streams; 1200–3000 m.

Toro, Kigezi, Mbale, Karamoja. Widespread in central and E Africa.

Here assessed preliminarily as Least Concern (LC) because of its wide distribution.

CUPRESSACEAE

Juniperus procera Endl.

LITERATURE: Melville, FTEA 1958: 16, fig. 5. ITU: 101, fig. 24, photo 14. UFT: 75.

Tree to 40 m high. Upland dry forest; 1350–3100 m.

Karamoja, Mbale. Widespread in Africa.

Here assessed preliminarily as Least Concern (LC) because of its wide distribution and habitat range. IUCN Red list: LR (nt), assessed by WCMC.

CYATHEACEAE

Cyathea camerooniana Hook.

LITERATURE: Edwards, FTEA 2005: 6. UFT: 74.

Tree fern to 3 m high. Moist forest; 1200–1400 m.

Kigezi, Mengo. Widespread in Africa.

Here assessed preliminarily as Least Concern (LC) because of its wide distribution, while the varieties are left unassessed owing to their taxonomic uncertainty [var. *ugandensis* Holttum was assessed in FTEA as Data Deficient (DD)].

Cyathea dregei Kunze

LITERATURE: Edwards, FTEA 2005: 8. ITU: 103. UFT: 74.

Tree fern to 5 m high. Along streams in grassland, swamp edges, riverine forest, forest margins; 1050–2200 m.

Toro, Mbale, Masaka, Ankole. Widespread in Africa.

Here assessed preliminarily as Least Concern (LC) because of its wide distribution and habitat range.

Cyathea manniana Hook.

LITERATURE: Edwards, FTEA 2005: 9. UFT: 74.

SYNONYM: *Cyathea deckenii* Kuhn; ITU: 103, photo 15.

Tree fern to 9 m high. Forest along streams or near springs; 850–2700 m.

Kigezi, Toro, Mbale, Masaka, Ankole. Widespread in Africa.

Here assessed preliminarily as Least Concern (LC) because of its wide distribution and habitat range.

Cyathea mildbraedii (Brause) Domin.

See restricted list, page 170.

DICHAPETALACEAE

Dichapetalum madagascariense Poir.

LITERATURE: Breteler, FTEA 1988: 10.

Shrub, liana or tree to 10 m high. Rainforest, riverine forest, bushland; 500–1700 m.

Toro, Ankole, Kigezi. Widespread in Africa.

Here assessed preliminarily as Least Concern (LC) because of its wide distribution and habitat range.

Tapura fischeri Engl.

LITERATURE: Breteler, FTEA 1988: 16, fig. 4. ITU: 84. UFT: 153.

Shrub or tree to 24 m high. Forest, riverine forest; 500–1200 m.

West Nile, Bunyoro, Busoga, Toro, West Nile, Madi, Acholi, Kigezi. Widespread in Africa.

Here assessed preliminarily as Least Concern (LC) because of its wide distribution.

DRACAENACEAE

Dracaena afromontana Mildbr.

LITERATURE: Mwachala & Mbugua, FTEA 2007: 6. ITU: 2. UFT: 78.

Shrub or tree 2–12 m high. Rain- or bamboo forest; 1600–2700 m.

Karamoja, Mbale, Kigezi, Toro. Widespread in East and central Africa.

Here assessed preliminarily as Least Concern (LC) because of its wide distribution.

Dracaena ellenbeckiana Engl.

LITERATURE: Mwachala & Mbugua, FTEA 2007: 4.

SYNONYM: *Dracaena sp. nov.* of ITU: 3?

Shrub or tree 2–8 m high. Dry forest, evergreen bushland; 1050–2100 m.

Karamoja. Also in Kenya, Ethiopia, South Sudan.

Here assessed preliminarily as Least Concern (LC) because of its wide distribution and habitat range.

Dracaena fragrans (L.) Ker-Gawl.

LITERATURE: Mwachala & Mbugua, FTEA 2007: 8. UFT: 78.

Shrub or tree 1–15 m high. Moist forest; 600–2250 m.

Ankole, Toro, Mengo. Widespread in Africa.

Here assessed preliminarily as Least Concern (LC) because of its wide distribution. Commonly planted as a live hedge and land boundary marker.

Dracaena steudneri Engl.

Literature: Mwachala & Mbugua, FTEA 2007: 8. ITU: 3. UFT: 76.

Tree 3–25 m high. Moist or dry forest, sometimes planted; 850–2300 m.

Kigezi, Mengo, Ankole, Toro, Bunyoro, Acholi, Lango, Mbale, Karamoja, Masaka. Widespread in East and central Africa.

Here assessed preliminarily as Least Concern (LC) because of its wide distribution and habitat range.

EBENACEAE

Diospyros abyssinica (Hiern) F. White

Literature: White & Verdcourt, FTEA 1996: 21. UFT: 145.

Synonym: *Maba abyssinica* Hiern; ITU: 107.

Shrub or tree 2–36 m high. Forest, riverine forest, woodland; 500–2300 m.

Acholi, Bunyoro, Masaka, Mengo, Toro, Ankole. Widespread in Africa.

Here assessed preliminarily as Least Concern (LC) because of its wide distribution and habitat range.

Diospyros bipindensis Gurke

Literature: White & Verdcourt, FTEA 1996: 30.

Shrub or tree 2–10 m high. Forest; ± 700 m.

Toro. Widespread in central Africa.

Here assessed preliminarily as Least Concern (LC) because of its wide distribution.

Diospyros katendei Verdc.

See restricted list, page 171.

Diospyros mespiliformis A. DC.

Literature: White & Verdcourt, FTEA 1996: 40. ITU: 105, fig. 25.

Shrub or tree 1–35 m high. Forest, riverine forest, thicket, scattered tree grassland, woodland; 500–1500 m.

West Nile, Karamoja, Bunyoro, Mengo, Madi, Acholi, Teso, Mbale. Widespread in Africa.

Here assessed preliminarily as Least Concern (LC) because of its wide distribution and habitat range.

Diospyros natalensis (Harv.) Brenan

Literature: White & Verdcourt, FTEA 1996: 13.

Synonym: *Maba natalensis* Harv.; ITU: 108.

Shrub or tree 5–25 m high. Forest, riverine forest, thicket; 500–1950 m.

Toro, Madi. Widespread in Africa.

Here assessed preliminarily as Least Concern (LC) because of its wide distribution and habitat range.

Diospyros scabra (Chiov.) Cufod.

LITERATURE: White & Verdcourt, FTEA 1996: 18.

SYNONYM: *Maba buxifolia* in the sense of ITU: 108.

Shrub or tree 1–11 m high. Dry bushland, open woodland; 400–1500 m.

In Uganda only known from Karamoja & Toro (Sempaya). Also in N Kenya, Ethiopia, South Sudan.

Here assessed preliminarily as Least Concern (LC) because it has a fairly wide distribution area, stretching from 0° to 5° N and from 30 to 37° E; EOO 199,935 km^2; AOO 42,269 km^2; and it occurs in a fairly common habitat with no specific threats.

Diospyros sp. of ITU: 105 is probably one of the *Diospyros* species mentioned here. Kalinzu Forest, Ishasha Gorge.

Euclea divinorum Hiern

LITERATURE: White & Verdcourt, FTEA 1996: 47. ITU: 105.

Shrub or tree 2–9 m high. Scattered tree grassland, dry bushland, woodland, forest remnants, riverine; 500–2100 m.

Karamoja, Kigezi, Mengo, Masaka, Ankole, West Nile, Teso, Mbale, Busoga, Toro. Widespread in Africa.

Here assessed preliminarily as Least Concern (LC) because of its wide distribution and habitat range.

Euclea racemosa Murray

LITERATURE: White & Verdcourt, FTEA 1996: 46.

SYNONYM: *Euclea urijiensis* Hiern; ITU: 107.
SYNONYM: *Euclea latidens* Stapf; ITU: 107. UFT: 160.

Shrub or tree to 12 m high. Scattered tree grassland, thicket, dry bushland, woodland, forest edges; 500–2700 m.

West Nile, Kigezi, Masaka, Toro, Mengo, Ankole, Teso, Karamoja, Acholi, Mbale. Widespread in Africa.

Here assessed preliminarily as Least Concern (LC) because of its wide distribution and habitat range.

ERICACEAE

Agarista salicifolia (Lam.) G. Don

LITERATURE: Beentje, FTEA 2006: 2.

SYNONYM: *Agauria salicifolia* (Lam.) Oliv.; ITU: 109. UFT: 159.

Shrub or tree 1–20 m. Forest and forest margins, evergreen bushland, woodland, wooded grassland, heath zone; 1600–3500 m.

Karamoja, Kigezi, Mbale, Ankole. Widespread in Africa.

Here assessed preliminarily as Least Concern (LC) because of its wide distribution and habitat range.

Erica arborea L.

LITERATURE: Beentje, FTEA 2006: 10. ITU: 109, photo 16. UFT: 80.

Shrub or tree 0.3–8 m. Afroalpine heath zone, forest, grassland, moorland; 2000–4500 m.

Karamoja, Kigezi, Mbale Toro. Widespread in Africa.

Here assessed preliminarily as Least Concern (LC) because of its wide distribution and habitat range.

Erica benguelensis (Engl.) E. G. H. Oliv.

Literature: Beentje, FTEA 2006: 10.

Synonym: *Philippia benguelensis* (Engl.) Britten; ITU: 111; UFT: 80.

Shrub or tree 0.3–8 m. Hillside grassland or bushland, dry forest edges, woodland; 1500–2500 m.

Toro, Kigezi, Ankole, Mbale; also in DRC, Tanzania, and South to Angola and Zimbabwe.

Here assessed preliminarily as Least Concern (LC) because of its wide distribution and habitat range.

Erica kingaensis Engl. subsp. *bequaertii* (De Wild.) R. Ross

Literature: Beentje, FTEA: 2006: 19.

Synonym: *Erica bequaertii* De Wild., *Erica ruwenzoriensis* Alm & T. C. E. Fr.; ITU: 110.

Shrub or tree 3–12 m high. Drier parts of *Sphagnum* bogs, heath vegetation on ridges; 2400–3500 m.

Toro, endemic to Ruwenzori; also in DRC.

Due to fairly common habitat type in a protected area, with a wide altitude range, here assessed preliminarily as Least Concern (LC)

Erica rossii Dorr

Literature: Beentje, FTEA 2006: 15.

Synonym: *Philippia excelsa* Alm & T. C. E. Fr.; ITU: 111. UFT: 80.

Synonym: *Philippia johnstonii* Engl.; ITU: 111 UFT: 80.

Shrub or tree 0.3–15 m high. Moorland, heath zone, upper forest edges and clearings; 2300–4050 m.

Mbale. Also in Kenya, Tanzania, DRC, Rwanda.

Fairly widespread on high mountains in East Africa, with a wide habitat and altitude range, here assessed preliminarily as Least Concern (LC).

Erica trimera (Engl.) Beentje subsp. *trimera*

Literature: Beentje, FTEA 2006: 12.

Synonym: *Philippia trimera* Engl.; ITU: 111. UFT: 80.

Shrub or tree 1–12 m high. Moorland, open forest; 3000–3800 m.

Toro, endemic to Ruwenzori; also in DRC.

Despite its fairly restricted distribution (EOO 195.77 km^2; AOO 39.2 km^2; 8 locations), here assessed preliminarily as Least Concern, due to its occurrence in protected zones of the mountains and common habitat.

subsp. *elgonensis* (Mildbr.) Beentje, endemic to Elgon; also in Kenya.

Despite its fairly restricted distribution, here assessed preliminarily as Least Concern, due to its occurrence in a common habitat (heath zone and alpine zone) and its wide (2850–4500 m) altitudinal range.

Philippia mannii of ITU: 111; species does not occur in Uganda, must be misidentification.

ERYTHROXYLACEAE

Erythroxylum fischeri Engl.
Literature: Verdcourt, FTEA 1984: 5. ITU: 112. UFT: 158.

Shrub or tree 1–9 m high. Evergreen forest, riverine, swamp and lakeside forest, thicket; 450–1350 m.

West Nile, Kigezi, Busoga, Mengo, Toro, Madi, Acholi, Teso, Karamoja. Widespread in East Africa.

Here assessed preliminarily as Least Concern (LC) because of its wide distribution and habitat range.

EUPHORBIACEAE

Acalypha acrogyna Pax
Literature: Radcliffe-Smith, FTEA 1987: 195.

Shrub or tree to 5 m high. Forest edges; 850–1250 m.

Toro, Masaka, Mengo. Widespread in Africa.

Here assessed preliminarily as Least Concern (LC) because of its wide distribution.

Acalypha fruticosa Forssk.
Literature: Radcliffe-Smith, FTEA 1987: 206.

Shrub or tree to 4 m high. Bushland, thicket, wooded grassland, riverine; 500–1900 m.

Karamoja, Bunyoro, Busoga, Mbale, Kigezi. Widespread in Africa.

Here assessed preliminarily as Least Concern (LC) because of its wide distribution and habitat range.

Acalypha neptunica Müll. Arg.
Literature: Radcliffe-Smith, FTEA 1987: 210. UFT: 112.

Shrub or tree to 6 m high. Forest, forest edges, bushland, riverine; 500–1700 m.

Karamoja, Bunyoro, Toro, Busoga, Mengo. Widespread in Africa.

Here assessed preliminarily as Least Concern (LC) because of its wide distribution and habitat range.

Alchornea cordifolia (Schumach. & Thonn.) Müll. Arg.
Literature: Radcliffe-Smith, FTEA 1987: 252, fig. 50. ITU: 114. UFT: 111.

Creeper, shrub or tree to 8 m high. Riverine, lakeside and swamp forest edges; 1150–1550 m.

Bunyoro, Toro, Busoga, Masaka, Mengo, Ankole, Mubende, Mbale. Widespread in Africa.

Here assessed preliminarily as Least Concern (LC) because of its wide distribution and habitat range.

Alchornea floribunda Müll. Arg.
Literature: Radcliffe-Smith, FTEA 1987: 253. ITU: 115. UFT: 136.

Shrub or tree to 5 m high. Swamp forest and rainforest clearings; 1150–1200 m.

Bunyoro, Mengo, Masaka. Widespread in Africa.

Here assessed preliminarily as Least Concern (LC) because of its wide distribution.

Alchornea hirtella Benth.

LITERATURE: Radcliffe-Smith, FTEA 1987: 255. ITU: 115. UFT: 136.

Shrub or tree to 15 m high. Evergreen forest and associated bushland; 500–1600 m.

West Nile, Bunyoro, Busoga, Mengo, Masaka, Ankole, Toro, Kigezi. Widespread in Africa.

Here assessed preliminarily as Least Concern (LC) because of its wide distribution and habitat range.

Alchornea laxiflora (Benth.) Pax & K. Hoffm.

LITERATURE: Radcliffe-Smith, FTEA 1987: 257. ITU: 115. UFT: 111.

Shrub, climber or tree to 8 m high. Riverine, lakeside and swamp forest edges; 1150–1550 m.

Bunyoro, Toro, Busoga, Masaka, Mengo, Madi, West Nile, Karamoja. Widespread in Africa.

Here assessed preliminarily as Least Concern (LC) because of its wide distribution and habitat range.

Argomuellera macrophylla Pax

LITERATURE: Radcliffe-Smith, FTEA 1987: 225, fig. 45. UFT: 136.

Herb, shrub or tree to 8 m high. Forest; 700–1300 m.

Karamoja, Bunyoro, Masaka, Ankole, Kigezi, Toro. Widespread in Africa.

Here assessed preliminarily as Least Concern (LC) because of its wide distribution.

Croton dichogamus Pax

LITERATURE: Radcliffe-Smith, FTEA 1987: 139. ITU: 120.

Shrub or tree 2–8 m high. Dry forest, bushland, thicket; 550–2000 m.

Madi, Karamoja, Busoga. Also widespread in Ethiopia, Kenya, Rwanda, Tanzania.

Here assessed preliminarily as Least Concern (LC) because of its wide distribution and habitat range.

Croton gratissimus Burch.

LITERATURE: Radcliffe-Smith in *Kew Bull.* 45: 556 (1990).

SYNONYM: *Croton zambesicus* Müll. Arg.; Radcliffe-Smith, FTEA 1987: 138.

Tree 4–12 m high. Wooded grassland; 750–1600 m.

Karamoja. Widespread in Africa.

Here assessed preliminarily as Least Concern (LC) because of its wide distribution.

Croton macrostachyus Delile

LITERATURE: Radcliffe-Smith, FTEA 1987: 49. ITU: 120, fig. 27. UFT: 112.

Tree to 35 m high. Secondary forest, forest edges, riverine, lakesides; 200–2300 m.

Karamoja, Ankole, Mengo, Kigezi, Toro, Mubende, Bunyoro, West Nile, Madi, Acholi, Teso, Mbale, Busoga. Widespread in Africa.

Here assessed preliminarily as Least Concern (LC) because of its wide distribution and habitat/altitude range.

Croton megalocarpus Hutch.

LITERATURE: Radcliffe-Smith, FTEA 1987: 144. ITU: 122. UFT: 114.

Tree to 35 m high. Evergreen forest; 1200–2400 m.

Bunyoro, Ankole, Toro, Mengo, Kigezi. Widespread in East Africa.

Here assessed preliminarily as Least Concern (LC) because of its wide distribution and altitude range.

Croton sylvaticus Krauss

LITERATURE: Radcliffe-Smith, FTEA 1987: 155. ITU: 123. UFT: 112.

SYNONYM: *Croton bukobensis* Pax; ITU: 120. UFT: 114.
SYNONYM: *Croton oxypetalus* Müll. Arg.; ITU: 122.

Shrub or tree to 25 m high. Secondary forest, forest edges, riverine, lakesides; 50–1800 m.

Ankole, Kigezi, Masaka, Toro, Bunyoro. Widespread in Africa.

Here assessed preliminarily as Least Concern (LC) because of its wide distribution and habitat/altitude range.

Crotonogynopsis usambarica Pax

LITERATURE: Radcliffe-Smith, FTEA 1987: 215, fig. 40.

Shrub or tree to 10 m high. Rainforest along rivers; 750–1700 m.

Bunyoro. Widespread in Africa.

Here assessed preliminarily as Least Concern (LC) because of its wide distribution and altitude range.

Discoclaoxylon hexandrum (Müll. Arg.) Pax & K. Hoffm.

LITERATURE: Radcliffe-Smith, FTEA 1987: 280, fig. 54.

SYNONYM: *Claoxylon hexandrum* Müll. Arg.; UFT: 137.
SYNONYM: *Discoclaoxylon sp. nov.* of ITU: 123.

Tree to 10 m high. Forest; 1200 m.

Bunyoro, Mengo. Widespread in Africa.

Here assessed preliminarily as Least Concern (LC) because of its wide distribution.

Discoglypremna caloneura (Pax) Prain

LITERATURE: Radcliffe-Smith, FTEA 1987: 195, fig. 44. UFT: 114.

Tree to 45 m high. Rainforest; ± 1050 m.

Bunyoro. Widespread in Africa.

Here assessed preliminarily as Least Concern (LC) because of its wide distribution.

Duvigneaudia leonardii-crispi (J. Léonard) Kruijt & Roebers

See restricted list, page 172.

Elaeophorbia drupifera (Thonn.) Stapf

LITERATURE: Carter & Smith, FTEA 1988: 533, fig. 100. ITU: 125, photo 17.

SYNONYM: *Elaeophorbia sp. nov.* of UFT: 81.

Tree to 22 m high. Forest edges; ± 1000 m.

Toro, Kigezi. Widespread in Africa.

Here assessed preliminarily as Least Concern (LC) because of its wide distribution and common habitat.

Erythrococca atrovirens (Pax) Prain
LITERATURE: Radcliffe-Smith, FTEA 1987: 277.

Shrub or tree to 5 m high. Forest, forest edges and associated bushland; 1000–2100 m.

Bunyoro, Kigezi, Masaka, Mengo. Widespread in central Africa.

Here assessed preliminarily as Least Concern (LC) because of its wide distribution and habitat range.

Erythrococca bongensis Pax
LITERATURE: Radcliffe-Smith, FTEA 1987: 267.

Shrub or tree to 6 m high. Forest edges and associated bushland or thicket; 200–2450 m.

Karamoja, Toro, Mengo. Widespread in East Africa.

Here assessed preliminarily as Least Concern (LC) because of its wide distribution and habitat range.

Erythrococca fischeri Pax
LITERATURE: Radcliffe-Smith, FTEA 1987: 269.

Shrub or tree to 6 m high. Forest edges and associated bushland; 1350–2400 m.

Karamoja, Mbale. Fairly widespread in East Africa.

Here assessed preliminarily as Least Concern (LC) because of its wide distribution and habitat range.

Euphorbia amplophylla Pax
LITERATURE: Gilbert in *Kew Bull.* 45: 196 (1990).
SYNONYM: *Euphorbia obovalifolia* A. Rich.; Carter & Smith, FTEA 1988: 483. UFT: 81.

Tree to 30 m high. Forest; 1800–2500 m.

Karamoja, Mbale. Also in Ethiopia, Kenya, Tanzania, Malawi, N Zambia.

Here assessed preliminarily as Least Concern (LC) because of its wide distribution.

Euphorbia breviarticulata Pax
LITERATURE: Carter & Smith, FTEA 1988: 492. ITU: 126.

Shrub or tree to 6 m high. Dry bushland; 50–1200 m.

Karamoja. Also in Kenya, N Tanzania, S Ethiopia, Somalia.

Here assessed preliminarily as Least Concern (LC) because of its wide distribution and common habitat.

Euphorbia bwambensis S. Carter
See restricted list, page 173.

Euphorbia candelabrum Kotschy
LITERATURE: Carter & Smith, FTEA 1988: 485. ITU: 126, photo 18.

Tree to 12(–20) m high. Wooded grassland; 550–2000 m.

Karamoja, Toro, Mengo, Masaka, Ankole, Kigezi, Bunyoro, West Nile, Madi, Acholi, Lango, Teso. Widespread in Africa.

Here assessed preliminarily as Least Concern (LC) because of its wide distribution.

Euphorbia dawei N. E. Br.
LITERATURE: Carter & Smith, FTEA 1988: 483. ITU: 126, photo 19.

Tree to 15(–25) m high. *Cynometra* forest, woodland, open woodland; 800–1300 m.

Toro, Ankole, Mengo. Also in DRC, Rwanda, Burundi, Tanzania.

Here assessed preliminarily as Least Concern (LC) because of its wide distribution.

Euphorbia grantii Oliv.
LITERATURE: Carter & Smith, FTEA 1988: 457. ITU: 127.

Shrub or tree 1.5–9 m high. Open woodland; 800–1600 m.

Teso, Masaka, Mengo, Ankole. Also in Tanzania, Burundi, DRC, N Zambia.

Here assessed preliminarily as Least Concern (LC) because of its wide distribution and altitude range.

Euphorbia magnicapsula S. Carter
See restricted list, page 174.

(*Euphorbia neocymosa* Bruyns = *Synadenium cymosum* N. E.Br. of ITU, is really a shrub)

Euphorbia teke Pax
LITERATURE: Carter & Smith, FTEA 1988: 481. ITU: 127. UFT: 80.

Shrub or tree to 6(–10) m high. Swamp forest; 900–1200 m.

Toro, Masaka, Mengo, Bunyoro. Also in NE DRC, N Tanzania, South Sudan, CAR.

Here assessed preliminarily as Least Concern (LC) because of its wide distribution.

Euphorbia tirucalli L.
LITERATURE: Carter & Smith, FTEA 1988: 471, fig. 89. ITU: 128.

Shrub or tree to 12 m high. Grassland, open woodland, often near human habitations; 500–2000 m.

Karamoja, Teso, Mengo, Ankole, Toro, Bunyoro, West Nile, Madi, Acholi, Lango, Mbale, Busoga. Widespread in Africa and Asia.

Here assessed preliminarily as Least Concern (LC) because of its wide distribution and habitat range; often cultivated.

(*Euphorbia venenifica* Kotschy of ITU is really a shrub)

Jatropha curcas L.
LITERATURE: Radcliffe-Smith, FTEA 1987: 356.

Shrub or tree to 8 m high. Naturalised from planting as a hedge; 500–1700 m.

Toro, Ankole, Mbale (getting more widespread with its commercial cultivation for biofuel). Native of America.

Least Concern (LC).

Macaranga angolensis (Müll. Arg.) Müll. Arg
LITERATURE: Radcliffe-Smith, FTEA 1987: 242. ITU: 130. UFT: 109.

Climber or tree to 10 m high. Forest, along lakes; 1150–1250 m.

Mengo, Toro. Widespread in central Africa.

Here assessed preliminarily as Least Concern (LC) because of its wide distribution.

Macaranga barteri Müll. Arg.

LITERATURE: Radcliffe-Smith, FTEA 1987: 250.

SYNONYM: *Macaranga lancifolia* Pax; ITU: 132. UFT: 110.

Tree to 37 m high. Rainforest and lakeside forest; 1100–1650 m.

Kigezi, Masaka. Widespread in Africa.

Here assessed preliminarily as Least Concern (LC) because of its wide distribution.

Macaranga capensis (Baill.) Sim

LITERATURE: Radcliffe-Smith, FTEA 1987: 243.

SYNONYM: includes *Macaranga kilimandscharica* Pax; Radcliffe-Smith, FTEA 1987: 245. ITU: 132. UFT: 110; which is now a var. *kilimandscharica* Friis & M. G. Gilbert, 1986.

Tree to 40 m high. Evergreen and secondary forest edges, riverine forest; 50–3050 m.

Acholi, Kigezi, Mbale, Ankole, Toro, Acholi. Widespread in East Africa.

Here assessed preliminarily as Least Concern (LC) because of its wide distribution and habitat/altitude range.

Macaranga monandra Müll. Arg.

LITERATURE: Radcliffe-Smith, FTEA 1987: 248. ITU: 132. UFT: 110.

Tree to 25 m high. Groundwater forest, along rivers, lakes and swamps; 1100–1550 m.

Kigezi, Masaka, Mengo, Ankole. Widespread in Africa.

Here assessed preliminarily as Least Concern (LC) because of its wide distribution and habitat range.

Macaranga schweinfurthii Pax

LITERATURE: Radcliffe-Smith, FTEA 1987: 241. ITU: 133. UFT: 109.

Tree to 40 m high. Groundwater forest, along rivers, lakes and swamps; 1050–1300 m.

Toro, Mengo, Masaka, Ankole, Kigezi, Bunyoro, Mbale. Widespread in Africa.

Here assessed preliminarily as Least Concern (LC) because of its wide distribution and habitat range.

Macaranga spinosa Müll. Arg.

LITERATURE: Radcliffe-Smith, FTEA 1987: 249.

SYNONYM: *Macaranga pynaertii* De Wild.; ITU: 133. UFT: 110.

Tree to 18 m high. Groundwater forest, rainforest edges; 1000–1250 m.

Masaka, Mengo, Bunyoro. Widespread in Africa.

Here assessed preliminarily as Least Concern (LC) because of its wide distribution.

Mallotus oppositifolius (Geiseler) Müll. Arg.

LITERATURE: Radcliffe-Smith, FTEA 1987: 236, fig. 48. ITU: 134. UFT: 173.

Shrub or tree to 5 m high. Forest, forest edges, bushland, thicket, riverine; 500–1500 m.

Bunyoro, Mengo. Widespread in Africa.

Here assessed preliminarily as Least Concern (LC) because of its wide distribution and habitat range.

Neoboutonia macrocalyx Pax

LITERATURE: Radcliffe-Smith, FTEA 1987: 232, fig. 47. ITU: 135. UFT: 111.

Tree to 25 m high. Forest edges and regrowth; 1100–2700 m.

Toro, Kigezi, Mbale, Ankole. Widespread in East and central Africa.

Here assessed preliminarily as Least Concern (LC) because of its wide distribution and common habitat.

Neoboutonia melleri (Müll. Arg.) Prain

LITERATURE: Radcliffe-Smith, FTEA 1987: 234. ITU: 135. UFT: 111.

SYNONYM: *Neoboutonia africana* Müll. Arg.; ITU: 135.

Tree to 15 m high. Groundwater forest, along rivers, lakes and swamps; 1100–1850 m.

West Nile, Mbale, Masaka, Mengo, Toro, Acholi. Widespread in Africa.

Here assessed preliminarily as Least Concern (LC) because of its wide distribution and habitat range.

Pseudagrostistachys ugandensis (Hutch.) Pax & K. Hoffm.

See restricted list, page 175.

Ricinodendron heudelotii (Baill.) Heckel

LITERATURE: Radcliffe-Smith, FTEA 1987: 326, fig. 61. ITU: 138, fig. 29, photo 22. UFT: 199.

Tree to 40 m high. Forest and associated bushland; 100–1200 m.

Bunyoro, Toro, Mengo, Madi. Widespread in Africa.

Here assessed preliminarily as Least Concern (LC) because of its wide distribution and habitat/altitude range.

Shirakiopsis elliptica (Hochst.) Esser

LITERATURE: *World Checklist & Bibliography of Euphorbiaceae* 4: 1470.

SYNONYM: *Sapium ellipticum* (Krauss) Pax; Radcliffe-Smith, FTEA 1987: 390. ITU: 140, fig. 30, photo 23. UFT: 137.

Shrub or tree to 30 m high. Forest edges and associated bushland, riverine; 1000–2150 m.

Acholi, Kigezi, Mbale, Mengo, Masaka, Ankole, Toro, Mubende, Bunyoro, West Nile, Madi, Teso, Busoga. Widespread in Africa.

Here assessed preliminarily as Least Concern (LC) because of its wide distribution and habitat/altitude range.

Suregada procera (Prain) Croizat

LITERATURE: Radcliffe-Smith, FTEA 1987: 378, fig. 71. UFT: 137.

SYNONYM: *Gelonium procerum* Prain; ITU: 129.

SYNONYM: *Gelonium sp.* of ITU: 129.

Shrub or tree 2–24 m high. Forest edges and associated bushland, riverine; 850–1850 m.

West Nile, Toro, Mengo, Masaka, Karamoja. Widespread in Africa.

Here assessed preliminarily as Least Concern (LC) because of its wide distribution and habitat/altitude range.

Synadenium grantii Hook. f.

LITERATURE: Carter & Smith, FTEA 1988: 535. ITU: 142.

Shrub or tree to 10 m high. Open woodland; 950–2100 m.

Ankole, Mbale, Masaka, Mengo, Kigezi, Acholi, Bunyoro. Widespread in East Africa.

Here assessed preliminarily as Least Concern (LC) because of its wide distribution.

Tetrorchidium didymostemon (Baill.) Pax & K. Hoffm.
LITERATURE: Radcliffe-Smith, FTEA 1987: 374. ITU: 142. UFT: 154.

Shrub, liana or tree 4–12 m high. Forest, forest edges and associated bushland, riverine; 1100–1700 m.

Kigezi, Masaka, Mengo. Widespread in Africa.

Here assessed preliminarily as Least Concern (LC) because of its wide distribution and habitat range.

FABACEAE/LEGUMINOSAE, SUBFAM. CAESALPINIOIDEAE

Afzelia africana Pers.
LITERATURE: Brenan, FTEA 1967: 125. ITU: 55, fig. 12.

Tree 6–30 m high. Wooded grassland; 1200–1400 m.

West Nile, Acholi, Bunyoro, Madi. Widespread in Africa.

Assessed as Least Concern (LC) by Sacande, Sanou & Beentje in *Guide de terrain des arbres du Burkina Faso* (in press).

Afzelia bipindensis Harms
LITERATURE: Brenan, FTEA 1967: 128. UFT: 231.
SYNONYM: *Afzelia bella* in the sense of ITU: 57.

Tree 18–40 m high. Rainforest; 900 m.

Toro. Widespread in Africa.

Here assessed preliminarily as Least Concern (LC) because of its wide distribution.

Afzelia africana at Murchison Falls National Park (photo: James Kalema).

Baikiaea insignis Benth.

LITERATURE: Brenan, FTEA 1967: 109. UFT: 231.

SYNONYM: *Baikiaea minor* Oliv.; ITU: 57.

Tree 5–34 m high. Rainforest; 1100–1250 m.

Ankole, Busoga, Masaka, Mengo, Kigezi, Mbale. Widespread in Africa.

Here assessed preliminarily as Least Concern (LC) because of its wide distribution.

Burkea africana Hook.

LITERATURE: Brenan, FTEA 1967: 21, fig. 2. ITU: 58.

Tree 4–20 m high. Woodland; 250–1300 m.

West Nile, Madi. Widespread in Africa.

Here assessed preliminarily as Least Concern (LC) because of its wide distribution.

Cassia mannii Oliv.

LITERATURE: Brenan, FTEA 1967: 59. ITU: 59, photo 7. UFT: 229.

Tree 15–25 m high. Rainforest; 850–1400 m.

West Nile, Bunyoro, Toro, Madi. Widespread in Africa.

Here assessed preliminarily as Least Concern (LC) because of its wide distribution.

Cassia sieberiana DC.

LITERATURE: Brenan, FTEA 1967: 61. ITU: 60, fig. 13.

Shrub or tree 2–12 m high. Wooded grassland; 1070 m.

West Nile, Acholi. Widespread in Africa.

Assessed as Least Concern (LC) by Sacande, Sanou & Beentje in *Guide de terrain des arbres du Burkina Faso* (in press).

Cassia spectabilis DC.

LITERATURE: –

SYNONYM: *Senna spectabilis* (DC.) H. S. Irwin & Barneby

Tree 7–10 m high. Naturalised and now fairly widespread.

Bunyoro, Toro, Ankole, Busoga,Mengo, Masaka, Mubende, Lango (and cultivated in Teso).

Not assessed here.

Cordyla richardii Milne-Redh.

See restricted list, page 176.

Cynometra alexandri C. H. Wright

LITERATURE: Brenan, FTEA 1967: 114. ITU: 63, fig. 14, photo 8. UFT: 231.

Tree 10–50 m high. Rainforest; 700–1250 m.

Toro, Kigezi, Mengo, Ankole, Bunyoro; also in N Tanzania and quite common in DRC.

Here assessed preliminarily as Least Concern because of fairly wide distribution (EOO 49,553 km^2; AOO 7,074 km^2) and being quite common within its habitat.

Daniellia oliveri (Rolfe) Hutch. & Dalziel

LITERATURE: Brenan, FTEA 1967: 132, fig. 24. ITU: 63, fig. 15, photo 9.

Tree 9–25(–45) m high. Wooded grassland; 1050–1550 m.

West Nile, Madi. Widespread in Africa.

Assessed as Least Concern (LC) by Sacande, Sanou & Beentje in *Guide de terrain des arbres du Burkina Faso* (in press).

Delonix elata (L.) Gamble

Literature: Brenan, FTEA 1967: 23, fig. 3. ITU: 65.

Tree 2–15 m high. Dry bushland, thicket; 400–1400 m.

Karamoja. Widespread in Africa and Asia.

Here assessed preliminarily as Least Concern (LC) because of its wide distribution.

Dialium excelsum Steyaert

Literature: Brenan, FTEA 1967: 104. UFT: 231.

Synonym: *Dialium bipindense* in the sense of ITU: 65.

Tree 18–45 m high. Rainforest; 750–900 m.

Bunyoro, Toro. Widespread in Africa and Asia.

Here assessed preliminarily as Least Concern (LC) because of its wide distribution.

Erythrophloeum suaveolens (Guill. & Perr.) Brenan

Literature: Brenan, FTEA 1967: 18, fig. 1. UFT: 233.

Synonym: *Erythrophloeum guineense* G. Don; ITU: 66.

Tree 9–30 m high. Riverine forest, rainforest; 500–1150 m.

West Nile, Bunyoro, Masaka, Mengo, Madi, Acholi, Mbale. Widespread in Africa.

Here assessed preliminarily as Least Concern (LC) because of its wide distribution.

Isoberlinia doka Craib & Stapf

Literature: Brenan, FTEA 1967: 142. ITU: 66, photo 10.

Tree 10–18 m high. Woodland; 1220 m.

West Nile. Widespread in Africa.

Assessed as Least Concern (LC) by Sacande, Sanou & Beentje in *Guide de terrain des arbres du Burkina Faso* (in press).

Piliostigma thonningii (Schumach.) Milne-Redh.

Literature: Brenan, FTEA 1967: 206, fig. 46. ITU: 67, fig. 16.

Tree 3–10 m high. Woodland, wooded grassland, bushland; 500–1850 m.

West Nile, Teso, Mengo, Mubende, Bunyoro, Toro, Madi, Acholi, Lango, Busoga, Ankole. Widespread in Africa.

Here assessed preliminarily as Least Concern (LC) because of its wide distribution and altitude range.

Senna longiracemosa (Vatke) Lock

Literature: Lock 1989: 38.

Synonym: *Cassia longiracemosa* Vatke; Brenan, FTEA 1967: 69.

Shrub or tree to 7 m high. Dry bushland, semi-desert scrub; 300–1650 m.

Karamoja. Also in Ethiopia, Somalia, Kenya, N Tanzania.

Here assessed preliminarily as Least Concern (LC) because of its wide distribution and altitude range.

Senna petersiana (Bolle) Lock

Literature: Lock 1989: 39.

Synonym: *Cassia petersiana* Bolle; Brenan, FTEA 1967: 72. ITU: 60. UFT: 231.

Shrub or tree 0.6–12 m high. Rainforest, riverine forest, woodland, wooded grassland; 500–2150 m.

West Nile, Mbale, Mengo, Mubende, Bunyoro, Lango, Busoga. Widespread in Africa.

Here assessed preliminarily as Least Concern (LC) because of its wide distribution and habitat/altitude range.

Senna septemtrionalis (Viv.) H. S. Irwin & Barneby

Literature: fide African Plant Database

Synonym: *Cassia floribunda* Cav.; Brenan, FTEA 1967: 70. UFT: 229.

Shrub or tree to 3(–4.5) m high. Dry forest, wooded grassland, grassland, river-banks, swamps, ruderal sites; 900–3200 m.

Toro, Ankole. Widespread in the tropics.

Here assessed preliminarily as Least Concern (LC) because of its wide distribution.

Senna singueana (Delile) Lock

Literature: Lock 1989: 39.

Synonym: *Cassia singueana* Delile; Brenan, FTEA 1967: 73, fig. 13. ITU: 60, plate 4.

Shrub or tree 1–15 m high. Woodland, wooded grassland, bushland; 500–2150 m.

Acholi, Karamoja, Teso, West Nile, Madi, Lango, Mbale, Busoga, Ankole. Widespread in Africa.

Assessed as Least Concern (LC) by Sacande, Sanou & Beentje in *Guide de terrain des arbres du Burkina Faso* (in press).

Tamarindus indica L.

Literature: Brenan, FTEA 1967: 153, fig. 32. ITU: 69, fig. 17.

Tree 3–24 m high. Woodland, wooded grassland, dry bushland; 500–1550 m.

Karamoja, Bunyoro, Busoga, Mengo, Toro, West Nile, Madi, Acholi, Lango, Teso, Mbale, Kigezi. Widespread in Africa and Asia.

Assessed as Least Concern (LC) by Sacande, Sanou & Beentje in *Guide de terrain des arbres du Burkina Faso* (in press).

FABACEAE/LEGUMINOSAE, SUBFAM. MIMOSOIDEAE

Acacia abyssinica Benth.

Literature: Brenan, FTEA 1959: 112.

Synonym: *Acacia xiphocarpa* in the sense of ITU: 215, photo 39.

Tree 6–15 m high. Woodland, wooded grassland; 1500–2300 m.

Karamoja, Kigezi, Mbale, Ankole, Teso. Widespread in Africa.

Here assessed preliminarily as Least Concern (LC) because of its wide distribution and habitat range.

Acacia amythethophylla A. Rich.
LITERATURE: Lock (1989): 62.
SYNONYM: *Acacia macrothyrsa* Harms; Brenan, FTEA 1959: 101. ITU: 210.

Shrub or tree 2–12(–15) m high. Woodland, wooded grassland; 600–1850 m.

Lango, Teso, Mbale, Mengo, West Nile. Widespread in Africa.

Assessed as Least Concern (LC) by Sacande, Sanou & Beentje in *Guide de terrain des arbres du Burkina Faso* (in press).

Acacia brevispica Harms
LITERATURE: Brenan, FTEA 1959: 96.
SYNONYM: *Acacia pennata* in the sense of ITU: 212, t. 9.

Shrub or tree 1–7 m high. Bushland, thickets, scrub; 150–1850 m.

Karamoja, Kigezi, Teso, Mengo, Ankole, Toro, Bunyoro, Acholi, Busoga. Widespread in Africa.

Here assessed preliminarily as Least Concern (LC) because of its wide distribution and habitat/altitude range.

Acacia dolichocephala Harms
LITERATURE: Brenan, FTEA 1959: 79.

Tree 4–8 m high. Woodland, wooded grassland; 1150–1550 m, to 1710 m elsewhere.

Karamoja, Mbale. Also in South Sudan, Ethiopia, NE Tanzania.

Here assessed preliminarily as Least Concern (LC) because of its wide distribution (EOO: 381,218 km^2; AOO 124,778 km^2; 10 locations), as far apart as 08°25 N and 04°35 S.

Acacia drepanolobium B. Y. Sjöstedt
LITERATURE: Brenan, FTEA 1959: 121. ITU: 207.
SYNONYM: *Vachellia drepanolobium* (B. Y. Sjöstedt) P. J. H. Hurter of Mabberley's *Plant Book*.

Shrub or tree 1–5(–7.5) m high. Dwarf tree grassland, often on hard-pan soils; 600–2700 m.

Karamoja, Mbale, Teso, Acholi, Madi, West Nile, Bunyoro. Widespread in E and NE Africa.

Here assessed preliminarily as Least Concern (LC) because of its wide distribution and altitude range.

Acacia elatior Brenan
LITERATURE: Brenan, FTEA 1959: 113.
SYNONYM: *Acacia etbaica* in the sense of ITU: 209.

Tree to 25 m high. Riverine woodland; 150–1100 m.

Karamoja, Acholi. Also in South Sudan, Kenya, where reasonably widespread.

Here assessed preliminarily as Least Concern (LC) because of its wide distribution.

Acacia etbaica Schweinf.
LITERATURE: Brenan, FTEA 1959: 114. ITU: 209.

Tree 2–12 m high. Wooded grassland; 250–1550 m.

Karamoja. Widespread in E and NE Africa.

Here assessed preliminarily as Least Concern (LC) because of its wide distribution.

Acacia gerrardii Benth.

LITERATURE: Brenan, FTEA 1959: 86.

SYNONYM: *Acacia hebecladoides* Harms; ITU: 209; *Vachellia gerrardii* (Benth.) P. J. H. Hurter of Mabberley's *Plant Book*.

Shrub or tree 3–15 m high. Woodland, wooded grassland; 900–2150 m.

Karamoja, Ankole, Masaka, Mengo, Kigezi, Toro, Bunyoro, Acholi, Lango, Mbale. Widespread in Africa.

Here assessed preliminarily as Least Concern (LC) because of its wide distribution and altitude range.

Acacia hecatophylla A. Rich.

LITERATURE: Brenan, FTEA 1959: 87. ITU: 210.

Tree 4–8 m high. Scattered tree grassland; 600–1400 m.

West Nile, Madi, Acholi. Also in South Sudan, Ethiopia, Eritrea, DRC.

Here assessed preliminarily as Least Concern (LC) because of its wide distribution and altitude range.

Acacia hockii De Wild.

LITERATURE: Brenan, FTEA 1959: 104.

SYNONYM: *Acacia seyal* Delile var. *multijuga* Baker f.; ITU: 213.

Shrub or tree 2–6(–12) m high. Woodland, wooded grassland, bushland; 500–2300 m.

Ankole, Teso, Masaka, Mengo, Toro, Mubende, Bunyoro, West Nile, Madi, Acholi, Karamoja, Mbale, Busoga. Widespread in Africa.

Assessed as Least Concern (LC) by Sacande, Sanou & Beentje in *Guide de terrain des arbres du Burkina Faso* (in press).

Acacia kirkii Oliv.

LITERATURE: Brenan, FTEA 1959: 106. UFT: 235.

SYNONYM: *Acacia mildbraedii* Harms; ITU: 211, photo 37.

Tree 2–15 m high. Riverine and ground-water forest; 1100–2000 m.

Acholi, Kigezi, Masaka, Mengo, Mubende, Ankole, Toro, Bunyoro, Acholi, Teso, Busoga. Widespread in Africa.

Here assessed preliminarily as Least Concern (LC) because of its wide distribution.

Acacia lahai Benth.

LITERATURE: Brenan, FTEA 1959: 79. ITU: 210.

Tree 3–15 m high. Woodland, wooded grassland; 1800–2450 m.

Karamoja, Mbale. Also in Ethiopia, Eritrea, Kenya and N Tanzania.

Here assessed preliminarily as Least Concern (LC) because of its wide distribution.

Acacia mearnsii De Wild.

LITERATURE: Brenan, FTEA 1959: 95.

Tree 2–15 m high. Roadsides, ruderal sites; 1600–1800 m.

Ankole, Kigezi. Native of Australia, naturalised.

Here assessed preliminarily as Least Concern (LC) because of its distribution and habitat range.

Acacia mellifera (Vahl) Benth.

Literature: Brenan, FTEA 1959: 84. ITU: 211.

Shrub or tree 1–6(–9) m high. Dry bushland, semi-desert scrub; 300–1700 m.

Karamoja, Mbale, Acholi. Widespread in Africa.

Here assessed preliminarily as Least Concern (LC) because of its wide distribution and habitat range.

(*Acacia misera* Vatke of ITU, now a synonym of *Acacia reficiens* Wawra, is really a shrub)

Acacia nilotica (L.) Delile

Literature: Brenan, FTEA 1959: 109.

Synonym: *Acacia subalata* Vatke; ITU: 215; *Vachellia nilotica* (L.) P. J. H. Hurter & Mabb. of Mabberley's *Plant Book*.

Tree 2.5–14 m high. Wooded grassland, bushland, dry scrub; 500–1850 m.

Acholi, Karamoja, Mbale. Widespread in Africa.

Assessed as Least Concern (LC) by Sacande, Sanou & Beentje in *Guide de terrain des arbres du Burkina Faso* (in press).

(*Acacia orfota* in the sense of ITU = *Acacia nubica*, really a shrub)

(*Acacia pennata* in the sense of ITU = *Acacia kamerunensis* Gand., really a climbing shrub)

Acacia persiciflora Pax

Literature: Brenan, FTEA 1959: 86.

Synonym: *Acacia eggelingii* Baker f.; ITU: 209.

Tree 4–9(–15) m high. Woodland, wooded grassland; 1200–2150 m.

West Nile, Mbale, Mubende, Madi, Acholi, Teso, Karamoja. Also in South Sudan, Ethiopia, Kenya, DRC.

Here assessed preliminarily as Least Concern (LC) because of its wide distribution and habitat range.

Acacia polyacantha Willd.

Literature: Brenan, FTEA 1959: 87.

Synonym: *Acacia campylacantha* A. Rich.; ITU: 207.

Tree to 21 m high. Woodland, wooded grassland, bushland, riverine forest; 500–1850 m.

Acholi, Ankole, Teso, Mengo, Toro, Mubende, Bunyoro, West Nile, Madi, Lango, Mbale, Busoga. Also in South Sudan, Ethiopia, Kenya, DRC.

Assessed as Least Concern (LC) by Sacande, Sanou & Beentje in *Guide de terrain des arbres du Burkina Faso* (in press).

Acacia senegal (L.) Willd.

Literature: Brenan, FTEA 1959: 92. ITU: 212, plate 10.

Shrub or tree to 12 m high. Wooded grassland, bushland, dry scrub; 100–1700 m.

Acholi, Mbale, Bunyoro, Mengo, Ankole, Toro, West Nile, Madi, Lango, Teso, Karamoja, Busoga. Widespread in Africa.

Assessed as Least Concern (LC) by Sacande, Sanou & Beentje in *Guide de terrain des arbres du Burkina Faso* (in press).

Acacia seyal Delile

Literature: Brenan, FTEA 1959: 103. ITU: 213.

Synonym: *Vachellia seyal* (Delile) P. J. H. Hurter of Mabberley's *Plant Book*.

Tree 3–12 m high. Wooded grassland, especially on black cotton soil; 600–1850 m.

Lango, Karamoja, Teso, Mengo, Ankole, Bunyoro, Madi, Acholi, Mbale, Busoga. Widespread in Africa.

Assessed as Least Concern (LC) by Sacande, Sanou & Beentje in *Guide de terrain des arbres du Burkina Faso* (in press).

Acacia sieberiana DC.

Literature: Brenan, FTEA 1959: 127. ITU: 214.

Tree 5–18 m high. Woodland, wooded grassland, riverine forest; 900–2150 m.

West Nile, Ankole, Teso, Mengo, Toro, West Nile, Bunyoro, Madi, Acholi, Lango, Karamoja, Mbale, Busoga. Widespread in Africa.

Assessed as Least Concern (LC) by Sacande, Sanou & Beentje in *Guide de terrain des arbres du Burkina Faso* (in press).

Acacia tortilis (Forssk.) Hayne

Literature: Brenan, FTEA 1959: 117. ITU: 215, photo 38.

Synonym: *Vachellia tortilis* (Forssk.) P. J. H. Hurter of Mabberley's *Plant Book*.

Tree 4–21 m high. Woodland, wooded grassland, dry bushland, semi-desert scrub; 600–1500 m.

Karamoja, Acholi. Widespread in Africa.

Assessed as Least Concern (LC) by Sacande, Sanou & Beentje in *Guide de terrain des arbres du Burkina Faso* (in press).

Adenanthera pavonina L.

Literature: Brenan, FTEA 1959: 30

Tree 4–20 m high. Thicket; 1310 m.

Mengo. Native of Asia, naturalised.

Here assessed preliminarily as Least Concern (LC) because of its distribution.

Albizia adianthifolia (Schumach.) W. Wight

Literature: Brenan, FTEA 1959: 160. ITU: 217. UFT: 238.

Tree 4–30 m high. Moist forest, woodland, wooded grassland, upland grassland; 500–1700 m.

Ankole, Masaka, Kigezi. Widespread in Africa.

Here assessed preliminarily as Least Concern (LC) because of its wide distribution and habitat range.

Albizia amara (Roxb.) Boivin

Literature: Brenan, FTEA 1959: 151.

Synonym: *Albizia sericocephala* Benth.; ITU: 222.

Tree to 1.5–12 m high. Wooded grassland, thicket, scrub; 450–1900 m.

Acholi, Karamoja, Teso, Ankole, Lango, Mbale. Widespread in Africa.

Here assessed preliminarily as Least Concern (LC) because of its wide distribution and habitat range.

Albizia anthelmintica Brongn.

Literature: Brenan, FTEA 1959: 148. ITU: 217.

Shrub or tree 3–9(–12) m high. Bushland and dry scrub, especially along luggas; 50–1550 m.

Acholi, Karamoja. Widespread in Africa.

Here assessed preliminarily as Least Concern (LC) because of its wide distribution and habitat/altitude range.

Albizia coriaria Oliv.

Literature: Brenan, FTEA 1959: 143. ITU: 217. UFT: 238.

Tree 6–36 m high. Riverine forest, wooded grassland; 850–1700 m.

Karamoja, Teso, Mengo. All Districts. Widespread in Africa.

Here assessed preliminarily as Least Concern (LC) because of its wide distribution and habitat range.

Albizia ferruginea (Guill. & Perr.) Benth.

Literature: Brenan, FTEA 1959: 144. ITU: 219. UFT: 236.

Tree 6–45 m high. Rainforest; 750–1250 m.

West Nile, Bunyoro, Mengo, Toro, Madi. Widespread in Africa.

Here assessed preliminarily as Least Concern (LC) because of its wide distribution and habitat range.

Albizia glaberrima (Schumach. & Thonn.) Benth.

Literature: Brenan, FTEA 1959: 156. UFT: 238.

Synonym: *Albizia eggelingii* Baker f.; ITU: 219.

Tree 9–24 m high. Forest; 750–1400 m.

Toro, Acholi, West Nile, Mengo, Bunyoro, Madi, Mbale, Ankole. Widespread in Africa.

Here assessed preliminarily as Least Concern (LC) because of its wide distribution.

Albizia grandibracteata Taub.

Literature: Brenan, FTEA 1959: 161. ITU: 220, photo 40. UFT: 238.

Synonym: *Albizia maranguensis* Taub. of ITU: 221.

Tree 6–30 m high. Moist forest, riverine forest, grassland; 1150–2150 m.

Toro, Mengo, Mubende, West Nile, Acholi, Karamoja, Ankole. Also in South Sudan, DRC, W Kenya, N Tanzania.

EOO 163,214 km^2; AOO 82,588 km^2, here assessed preliminarily as Least Concern (LC) because of its wide distribution and habitat range.

Albizia gummifera (J. F. Gmel.) C. A. Sm.

Literature: Brenan, FTEA 1959: 157. ITU: 220. UFT: 238.

Tree to 30 m high. Moist forest, riverine forest, open habitats near forest; 500–2450 m.

Toro, Mbale, Mengo, Masaka, Kigezi, Ankole, Mubende, Bunyoro, West Nile, Acholi, Karamoja, Busoga. Widespread in Africa.

Here assessed preliminarily as Least Concern (LC) because of its wide distribution and habitat range.

Albizia isenbergiana (A. Rich.) E. Fourn.
LITERATURE: Brenan, FTEA 1959: 152. ITU: 221.

Tree to 12 m high. Habitat unknown for Uganda; 1650–1850 m.

Karamoja. Widespread in East Africa.

Here assessed preliminarily as Least Concern (LC) because of its wide distribution.

Albizia lebbeck (L.) Benth.
LITERATURE: Brenan, FTEA 1959: 147.

Tree 2–15 m high. Naturalised near human habitations; 500–1300 m.

West Nile. Pantropical.

Here assessed preliminarily as Least Concern (LC) because of its wide distribution.

Albizia malacophylla (A. Rich.) Walp.
LITERATURE: Brenan, FTEA 1959: 145. ITU: 221.

Tree to 6(–12) m high. Wooded grassland; 1100–1350 m.

West Nile, Teso, Bunyoro, Madi, Acholi, Lango, Mbale, Karamoja. Widespread in Africa.

Here assessed preliminarily as Least Concern (LC) because of its wide distribution.

Albizia petersiana (Bolle) Oliv.
LITERATURE: Brenan, FTEA 1959: 162.

SYNONYM: *Albizia maranguensis* Taub.; ITU: 221.

Tree 3–21 m high. Dry forest, moist forest, riverine forest, evergreen bushland; 1100–2250 m.

Karamoja, Acholi, West Nile. Widespread in Africa.

Here assessed preliminarily as Least Concern (LC) because of its wide distribution and habitat range.

Albizia schimperiana Oliv.
LITERATURE: Brenan, FTEA 1959: 154.

SYNONYM: *Albizia brachycalyx* Oliv.; ITU: 217.

Tree 5–23(–30) m high. Ground-water forest, riverine forest, woodland; 350–1700 m.

Ankole. Widespread in Africa.

Here assessed preliminarily as Least Concern (LC) because of its wide distribution and habitat/altitude range.

Albizia versicolor Oliv.
LITERATURE: Brenan, FTEA 1959: 146. ITU: 222.

Tree to 5–15 m high. Wooded grassland, woodland, bushland; 500–1700 m.

Ankole. Widespread in Africa.

Here assessed preliminarily as Least Concern (LC) because of its wide distribution and habitat range.

Albizia zygia (DC.) J. F. Macbr.
LITERATURE: Brenan, FTEA 1959: 161. ITU: 222. UFT: 239.

Tree 4–30 m high. Moist forest, riverine forest, wooded grassland; 900–1400 m.

Acholi, Bunyoro, Teso, Mengo, Toro, Mubende, West Nile, Madi, Bukedi, Busoga. Widespread in Africa.

Here assessed preliminarily as Least Concern (LC) because of its wide distribution and habitat range.

Amblygonocarpus andongensis (Oliv.) Exell & Torre

Literature: Brenan, FTEA 1959: 34, fig. 9. ITU: 223.

Synonym: *Amblygonocarpus obtusangulus* (Oliv.) Harms; ITU: 223.

Tree 6–25 m high. ?Woodland, wooded grassland; 500–1400 m.

West Nile, Madi. Widespread in Africa.

Here assessed preliminarily as Least Concern (LC) because of its wide distribution.

Cathormion altissimum (Hook. f.) Hutch. & Dandy

Literature: Brenan, FTEA 1959: 166, fig. 23. ITU: 223. UFT: 235.

Synonym: *Albizia altissima* Hook. f.; Lock 1989: 80.

Shrub or tree 5–35 m high. Swamp forest; altitude range unknown.

Bunyoro, West Nile. Widespread in Africa.

Here assessed preliminarily as Least Concern (LC) because of its wide distribution.

Dichrostachys cinerea (L.) Wight & Arn.

Literature: Brenan, FTEA 1959: 37, fig. 11.

Synonym: *Dichrostachys glomerata* (Forssk.) Chiov.; ITU: 224. UFT: 235.
Synonym: *Dichrostachys nyassana* Taub. of ITU: 225.

Shrub or tree 1–8(–12) m high. Dry bushland, wooded grassland, woodland, open forest, secondary bush; 500–1750 m.

Ankole, Karamoja, Teso, Toro, Bunyoro, West Nile, Acholi, Lango, Mbale, Busoga, Mengo, Mubende. Widespread in Africa and Asia.

Assessed as Least Concern (LC) by Sacande, Sanou & Beentje in *Guide de terrain des arbres du Burkina Faso* (in press).

Entada abyssinica A. Rich.

Literature: Brenan, FTEA 1959: 13, fig. 2. ITU: 225.

Shrub or tree 3–10 m high. Woodland and wooded grassland; 400–2300 m.

West Nile, Mbale, Mengo, Masaka, Ankole, Kigezi, Toro, Mubende, Bunyoro, Acholi, Karamoja, Busoga. Widespread in Africa.

Here assessed preliminarily as Least Concern (LC) because of its wide distribution and altitude range.

Entada africana Guill. & Perr.

Literature: Brenan, FTEA 1959: 12.

Synonym: *Entada sudanica* Schweinf.; ITU: 225.

Shrub or tree 1–10 m high. Habitat and altitude in Uganda uncertain.

Acholi, West Nile, Lango, Karamoja. Widespread in Africa.

Here assessed preliminarily as Least Concern (LC) because of its wide distribution.

Faidherbia albida (Delile) A. Chev.

Literature: Name change accepted generally.

Synonym: *Acacia albida* Delile; Brenan, FTEA 1959: 78. ITU: 205.

Tree 6–30 m high. Riverine and ground-water forest, woodland; 600–1850 m.

Karamoja, Ankole, Acholi, Mbale, Mengo. Widespread in Africa.

Assessed as Least Concern (LC) by Sacande, Sanou & Beentje in *Guide de terrain des arbres du Burkina Faso* (in press).

Leucaena leucocephala (L.) De Wit

Literature: Lock 1989: 93.

Synonym: *Leucaena glauca* (L.) Benth.; Brenan, FTEA 1959: 48.

Shrub or tree 0.6–9 m high. Ruderal, often open sites; 500–1300 m.

Ankole, Mengo. Originally from Central America, now widespread in the tropics and subtropics.

Here assessed preliminarily as Least Concern (LC) because of its wide distribution and habitat range.

Newtonia buchananii (Baker) G. C. C. Gilbert & Boutique

Literature: Brenan, FTEA 1959: 23, fig. 6. UFT: 235.

Synonym: *Piptadenia buchananii* Baker; ITU: 230.

Tree 10–40 m high. Moist forest, usually by streams, riverine forest; 600–2150 m.

Ankole, Masaka, Mengo, Kigezi, Toro, West Nile. Widespread in Africa.

Here assessed preliminarily as Least Concern (LC) because of its wide distribution and altitude range.

Parkia filicoidea Oliv.

Literature: Brenan, FTEA 1959: 7, fig. 1. ITU: 227. UFT: 236.

Tree 8–30 m high. Moist forest, riverine forest; 250–1400 m.

Bunyoro, Mengo, Masaka, Kigezi, Toro, Madi. Widespread in Africa.

Here assessed preliminarily as Least Concern (LC) because of its wide distribution.

Piptadeniastrum africanum (Hook. f.) Brenan

Literature: Brenan, FTEA 1959: 21, fig. 5. UFT: 235.

Synonym: *Piptadenia africana* Hook. f.; ITU: 228, fig. 51.

Tree to 50 m high. Rainforest, riverine forest; 1100–1250 m.

Bunyoro, Masaka, Mengo, Toro, Ankole. Widespread in Africa.

Here assessed preliminarily as Least Concern (LC) because of its wide distribution.

Prosopis africana (Guill. & Perr.) Taub.

Literature: Brenan, FTEA 1959: 36, fig. 10. ITU: 230.

Tree 4–12(–21) m high. Wooded grassland; 900–1250 m.

West Nile, Acholi, Teso, Madi, Lango, Bunyoro. Widespread in Africa.

Here assessed preliminarily as Least Concern (LC) because of its wide distribution.

Tetrapleura tetraptera (Schumach. & Thonn.) Taub.
Literature: Brenan, FTEA 1959: 32, fig. 8. ITU: 231. UFT: 236.

Tree 6–30 m high. Moist forest; 50–1250 m.

Bunyoro, Mengo, Toro, Madi. Widespread in Africa.

Here assessed preliminarily as Least Concern (LC) because of its wide distribution and altitude range.

FABACEAE/LEGUMINOSAE, SUBFAM. PAPILIONOIDEAE

Aeschynomene elaphroxylon (Guill. & Perr.) Taub.
Literature: Gillett, Polhill & Verdcourt, FTEA 1971: 375. ITU: 295.

Shrub or tree 2–12 m high. Lake-sides, swamps, usually in standing water; 500–1350 m.

Kigezi, Masaka, Mengo, Ankole, Toro, Bunyoro, West Nile, Madi, Acholi, Lango, Busoga, Bukedi. Widespread in Africa.

Here assessed preliminarily as Least Concern (LC) because of its wide distribution.

Andira inermis (W. Wright) DC.
Literature: Gillett, Polhill & Verdcourt, FTEA 1971: 63.

Synonym: *Millettia sp. nov. aff. macrophylla* in the sense of ITU: 308.

Tree 7–10(–30) m high. Riverine forest; 1200 m.

Karamoja, Madi. Widespread in Africa.

Assessed as Least Concern (LC) by Sacande, Sanou & Beentje in *Guide de terrain des arbres du Burkina Faso* (in press).

Baphia capparidifolia Baker
Literature: Gillett, Polhill & Verdcourt, FTEA 1971: 53. ITU: 297. UFT: 152.

Synonym: *Baphia multiflora* Harms; ITU: 297.

Climbing shrub or tree to 4.5 m high. Rainforest, riverine forest, secondary grassland; 650–1200 m.

Toro, Kigezi. Widespread in Africa.

Here assessed preliminarily as Least Concern (LC) because of its wide distribution and habitat range.

Baphia wollastonii Baker f.
See restricted list, page 177.

Baphiopsis parviflora Baker
Literature: Brenan, FTEA 1967: 218, fig. 217. UFT: 152.

Synonym: *Baphiopsis stuhlmannii* Taub.; ITU: 297.

Shrub or tree 2–15 m high. Swamp forest, rainforest; 1100–1350 m.

Kigezi, Busoga, Mengo. Widespread in Africa.

Here assessed preliminarily as Least Concern (LC) because of its wide distribution.

Calpurnia aurea (Aiton) Benth.
Literature: Gillett, Polhill & Verdcourt, FTEA 1971: 47, fig. 9.

Synonym: *Calpurnia subdecandra* (L'Hér.) Schweick.; ITU: 298.

Shrub or tree 2–10 m high. Rainforest edges and clearings, riverine forest; 1300–2250 m.

Karamoja. Widespread in Africa.

Here assessed preliminarily as Least Concern (LC) because of its wide distribution and habitat/altitude range.

Craibia brownii Dunn

Literature: Gillett, Polhill & Verdcourt, FTEA 1971: 152. ITU: 299. UFT: 231.

Tree 6–20 m high. Forest and riverine; 1100–2200 m.

Bunyoro, Toro, Ankole, Busoga. Also in NE DRC, Kenya, N Tanzania.

Here assessed preliminarily as Least Concern (LC) because of its wide distribution and habitat/altitude range.

Craibia laurentii (De Wild.) De Wild.

Literature: Gillett, Polhill & Verdcourt, FTEA 1971: 149. ITU: 299.

Tree to 18 m high. Dry forest; 600–1350 m.

West Nile, Teso, Mbale, Karamoja. Widespread in Africa.

Here assessed preliminarily as Least Concern (LC) because of its wide distribution.

Dalbergia lactea Vatke

Literature: Gillett, Polhill & Verdcourt, FTEA 1971: 111.

Shrub, climber or tree to 9 m high. Rainforest and its margins, riverine and swamp forest, evergreen bushland, grassland; 500–2400 m.

West Nile, Karamoja, Kigezi. Widespread in Africa.

Here assessed preliminarily as Least Concern (LC) because of its wide distribution and habitat/altitude range.

Dalbergia melanoxylon Guill. & Perr.

Literature: Gillett, Polhill & Verdcourt, FTEA 1971: 100, fig. 21. ITU: 300, photo 50, fig. 64.

Shrub or tree to 12(–30) m high. Woodland, bushland, wooded grassland; 500–1350 m.

Acholi, Bunyoro, Mbale, West Nile, Madi, Karamoja. Widespread in Africa.

IUCN Red List rating: assessed by WCMC in 1997 as Lower Risk/Near Threatened (LR/NT) due to the high levels of exploitation of mature trees for their treasured hardwood and international trade therein, with the resultant risk for erosion of genetic diversity, here assessed preliminarily as Near Threatened (NT, A3) because of its exploitation throughout its range of distribution even though the species is still widespread.

Dalbergia nitidula Baker

Literature: Gillett, Polhill & Verdcourt, FTEA 1971: 109. ITU: 300.

Shrub or tree to 12 m high. Woodland, bushland, thicket, wooded grassland; 350–1650 m.

Ankole. Widespread in Africa.

Here assessed preliminarily as Least Concern (LC) because of its wide distribution and habitat/altitude range.

Erythrina abyssinica DC.

Literature: Gillett, Polhill & Verdcourt, FTEA 1971: 555. ITU: 302, plate 15.

Tree 2–15 m high. Scattered tree grassland, open woodland, bushland, forest edges; 200–2100 m.

Karamoja, Teso, Mengo. All Districts. Widespread in Africa.

Here assessed preliminarily as Least Concern (LC) because of its wide distribution and habitat/ altitude range.

Erythrina droogmansiana De Wild. & T. Durand

LITERATURE: – Lock 1989: 410

SYNONYM: *Erythrina sp. C* of FTEA: 558 & ITU: 304; *Erythrina sp. D* of FTEA: 560; *Erythrina sp.* of UFT: 194.

Tree 12–27 m high.Rainforest; 1100–1500 m.

Bunyoro, Toro; also in Cameroon, DRC, Angola.

Here assessed preliminarily as Least Concern (LC) because of its wide distribution.

Erythrina excelsa Baker

LITERATURE: Gillett, Polhill & Verdcourt, FTEA 1971: 559. ITU: 303. UFT: 194.

SYNONYM: *Erythrina bagshawei* Baker f.; ITU: 303.

Tree 9–30 m high. Swamp and riverine/lakeside forest; 1100–1500 m.

West Nile, Mengo, Masaka, Ankole. Widespread in Africa.

Here assessed preliminarily as Least Concern (LC) because of its wide distribution.

Erythrina mildbraedii Harms

LITERATURE: Gillett, Polhill & Verdcourt, FTEA 1971: 558. ITU: 303.

Tree 12–30 m high. Rainforest; 1500–1600 m.

Bunyoro, Toro, Mengo. Widespread in Africa.

Here assessed preliminarily as Least Concern (LC) because of its wide distribution.

Erythrina × lanigera P. A. Duvign. & Rochez

LITERATURE: Lock 1989: 410.

SYNONYM: *Erythrina sp. B* of Gillett, Polhill & Verdcourt, FTEA 1971: 557.

Tree ± 12 m high. Scattered tree grassland; 1100–1200 m.

Busoga, Mengo. Also in DRC.

Thought to be a hybrid between *E. abyssinica* and *E. mildbraedii*.

Mildbraediodendron excelsum Harms

LITERATURE: Brenan, FTEA 1967: 225, fig. 52. ITU: 306. UFT: 229.

Tree 20–50 m high. Rainforest; 750–1000 m.

West Nile, Bunyoro, Mengo, Kigezi, Toro, Madi. Widespread in Africa.

Here assessed preliminarily as Least Concern (LC) because of its wide distribution.

Millettia dura Dunn

LITERATURE: Gillett, Polhill & Verdcourt, FTEA 1971: 144, fig. 25. ITU: 307, plate 16. UFT: 233.

SYNONYM: *Millettia drastica* in the sense of ITU: 307.

Shrub or tree to 13 m high. Forest and forest margins; 1200–1650 m.

Ankole, Kigezi, Toro, Madi. Also in DRC, Rwanda, Burundi, Kenya, Tanzania.

Here assessed preliminarily as Least Concern (LC) because of its wide distribution and habitat.

Millettia eetveldeana (Micheli) Hauman

Literature: Gillett, Polhill & Verdcourt, FTEA 1971: 141. UFT: 233.

Synonym: *Millettia sp.* of ITU: 308.

Shrub or tree to 17 m high. Forest and forest margins; 100–900 m.

Toro. Only known from Bwamba. Widespread in Africa.

Here assessed preliminarily as Least Concern (LC) because of its wide distribution and altitude range.

Millettia psilopetala Harms

Literature: Gillett, Polhill & Verdcourt, FTEA 1971: 127. ITU: 307. UFT: 233.

Shrub, liana or tree to 10 m high. Rainforest; 750–1800 m.

Toro, Ankole, Kigezi. Also in DRC, where it is widespread.

Here assessed preliminarily as Least Concern (LC) because of its wide distribution and altitude range.

(*Millettia* sp. near *M. lucens* in the sense of ITU is *M. lacus-alberti* J. B. Gillett, a liana or shrub)

Mundulea sericea (Willd.) A. Chev.

Literature: Gillett, Polhill & Verdcourt, FTEA 1971: 155, fig. 28. ITU: 309.

Shrub or tree 2–7 m high. Evergreen forest and bushland; 500–1950 m.

Acholi. Widespread in Africa and Asia.

Here assessed preliminarily as Least Concern (LC) because of its wide distribution and habitat/altitude range.

Ormocarpum trachycarpum (Taub.) Harms

Literature: Gillett, Polhill & Verdcourt, FTEA 1971: 359, fig. 53. ITU: 309.

Synonym: *Ormocarpum sp. near O. trachycarpum* of ITU: 309.

Shrub or tree 1–6 m high. Woodland, grassland, dry bushland; 950–1800 m.

Karamoja, Ankole, Kigezi, West Nile, Madi, Mbale. Also in Ethiopia, Kenya, Tanzania and fairly widespread.

Here assessed preliminarily as Least Concern (LC) because of its wide distribution and habitat/altitude range.

Philenoptera laxiflora (Guill. & Perr.) Roberty

Literature: Schrire B. D. 2000. A synopsis of the genus *Philenoptera* (Leguminosae–Millettieae) from Africa and Madagascar. *Kew Bull.* 55: 81–94.

Synonym: *Lonchocarpus laxiflorus* Guill. & Perr.; Gillett, Polhill & Verdcourt, FTEA 1971: 67. ITU: 304, fig. 65.

Tree 3–12 m high. Wooded grassland; 1000–1900 m.

Karamoja, Busoga, Mengo, Bunyoro, Acholi, West Nile. Widespread in Africa.

Assessed as Least Concern (LC) by Sacande, Sanou & Beentje in *Guide de terrain des arbres du Burkina Faso* (in press).

Pterocarpus lucens Guill. & Perr.

Literature: Gillett, Polhill & Verdcourt, FTEA 1971: 82.

Synonym: *Pterocarpus abyssinicus* Hochst.; ITU: 310.

Tree 7–18 m high. Wooded grassland; 1050–1200 m.

West Nile, Bunyoro, Madi, Acholi. Widespread in Africa.

Assessed as Least Concern (LC) by Sacande, Sanou & Beentje in *Guide de terrain des arbres du Burkina Faso* (in press).

(*Sesbania bispinosa* of ITU is really a shrub)

Sesbania dummeri E. Phillips & Hutch.
See restricted list, page 178.

Sesbania macrantha E. Phillips & Hutch.
LITERATURE: Gillett, Polhill & Verdcourt, FTEA 1971: 341. ITU: 311.

Tree 1–6 m high. Swamps, streamsides, lake edges; 1100–2000 m.

Acholi, Ankole, Mubende, Mengo, Bunyoro, Karamoja. Widespread in Africa.

Here assessed preliminarily as Least Concern (LC) because of its wide distribution and habitat as well as altitude range.

(*Sesbania pubescens* of ITU, now a synonym of *Sesbania sericea*, is really a woody herb)

Sesbania sesban (L.) Merr.
LITERATURE: Gillett, Polhill & Verdcourt, FTEA 1971: 339. ITU: 311.

Tree 1–7 m high. Streamsides, lake edges; 100–2200 m.

Karamoja, Toro, Kigezi, Mengo, Masaka, Mubende, Acholi, Mbale, Busoga, Teso. Widespread in Africa.

Lake Bunyonyi — cultivation right to the edge of the lake (photo: James Kalema).

Assessed as Least Concern (LC) by Sacande, Sanou & Beentje in *Guide de terrain des arbres du Burkina Faso* (in press).

(*Smithia kotschyi* of ITU, a synonym of *Kotschya africana*, is really a shrub)

HAMAMELIDACEAE

Trichocladus ellipticus Eckl. & Zeyh.

LITERATURE: Verdcourt, FTEA 1971: 2, fig. 1. UFT: 153.

SYNONYM: *Trichocladus malosanus* Baker; ITU: 155, plate 6.

Shrub or tree 5–10 m high. Rainforest along streams, riverine forest, swamp forest, woodland; 1100–2700 m.

Acholi, Mbale, Mengo, Masaka, Karamoja, Busoga. Widespread in Africa.

Here assessed preliminarily as Least Concern (LC) because of its wide distribution and habitat range.

HYPERICACEAE

Hypericum bequaertii De Wild.

See restricted list, page 179.

Hypericum revolutum Vahl

LITERATURE: UFT: 169.

SYNONYM: *Hypericum lanceolatum* Lam.; Milne-Redhead, FTEA 1953: 4.
SYNONYM: *Hypericum leucoptychodes* A. Rich.; ITU: 157, fig. 33.

Shrub or tree 1–12 m high. Dry forest, evergreen bushland, grassland, often by streams; 1800–3400 m.

Kigezi, Toro, Mbale. Widespread in Africa.

Here assessed preliminarily as Least Concern (LC) because of its wide distribution and habitat/altitude range.

subsp. *keniense* (Schweinf.) N. Robson

SYNONYM: *Hypericum keniense* Schweinf.; Milne-Redhead, FTEA 1953: 5. UFT: 169.
SYNONYM: *Hypericum ruwenzoriense* De Wild.; ITU: 159.

Shrub or tree 1–10 m high. Dry forest, evergreen bushland; 2400–3550 m.

Toro, Karamoja, Mbale. Also in DRC (Ruwenzori, Mt Kahuzi), Kenya and Tanzania.

Here assessed preliminarily as Least Concern (LC) because of its wide distribution and habitat range.

Hypericum roeperanum A. Rich.

LITERATURE: Milne-Redhead, FTEA 1953: 3. ITU: 159. UFT: 170.

Shrub or tree 0.6–5 m high. Dry forest, bamboo thicket, evergreen bushland, grassland, often by streams; 1500–2900 m.

Karamoja, Mbale, Toro. Also in South Sudan, Ethiopia, Kenya and Tanzania.

Here assessed preliminarily as Least Concern (LC) because of its wide distribution and habitat range.

ICACINACEAE

Apodytes dimidiata Arn.
LITERATURE: Lucas, FTEA 1968: 4, fig. 2. ITU: 161, plate. 7. UFT: 153.

Shrub or tree to 25 m high. Rainforest, other types of forest; 1000–2500 m.

West Nile, Masaka, Toro, Karamoja, Busoga. Widespread in Africa.

Here assessed preliminarily as Least Concern (LC) because of its distribution and habitat/altitude range.

IRVINGIACEAE

Irvingia gabonensis (O'Rorke) Baill.
LITERATURE: Verdcourt, FTEA 1984: 7, fig. 3. ITU: 409, fig. 85. UFT: 144.

Tree 18–36(–50) m high. Evergreen forest; 900–1200 m.

Bunyoro, Toro, Mengo. Widespread in Africa.

Here assessed preliminarily as Least Concern (LC) because of its wide distribution.

Klainedoxa gabonensis Engl.
LITERATURE: Verdcourt, FTEA 1984: 4, fig. 2. ITU: 411, fig. 86. UFT: 144.

Tree 12–45 m high. Rainforest, swamp forest; 900–1200 m.

Bunyoro, Masaka, Mengo, Toro. Widespread in Africa.

Here assessed preliminarily as Least Concern (LC) because of its wide distribution.

LAMIACEAE/LABIATAE

(note: these have been transferred from the family Verbenaceae to the Labiatae/Lamiaceae. The FTEA references below stand for the FTEA treatment of Verbenaceae.)

Clerodendrum formicarum Gürke
LITERATURE: Verdcourt, FTEA 1992: 116, fig. 17.

Shrub, climber or tree 1–8 m high. Forest, lakeside forest, dry bushland; 350–1800 m.

Kigezi, Masaka, Mengo. Widespread in Africa.

Here assessed preliminarily as Least Concern (LC) because of its wide distribution and habitat/altitude range.

Clerodendrum johnstonii Oliv.
LITERATURE: Verdcourt, FTEA 1992: 118.

Shrub, climber or tree 1–5 m high. Forest, thicket, ruderal sites; 1200–2550 m.

Kigezi, Mbale, Teso, Ankole, West Nile. Widespread in central and East Africa.

Here assessed preliminarily as Least Concern (LC) because of its wide distribution and habitat/altitude range.

Clerodendrum rotundifolium Oliv.
LITERATURE: Verdcourt, FTEA 1992: 99.

Usually a shrub, rarely a 'small' tree. Wooded grassland, bushland; 350–2150 m.

Karamoja, Ankole, Mengo. Widespread in East Africa.

Here assessed preliminarily as Least Concern (LC) because of its wide distribution and habitat/altitude range.

Clerodendrum tanganyikense Baker

LITERATURE: Verdcourt, FTEA 1992: 111.

Shrub or tree 1–4.5 m high. Forest, hillside scrub, woodland; 1200–2100 m.

Ankole, Kigezi. Widespread in central Africa.

Here assessed preliminarily as Least Concern (LC) because of its wide distribution and habitat/altitude range.

Premna angolensis Gürke

LITERATURE: Verdcourt, FTEA 1992: 70. ITU: 442. UFT: 160.

Tree 4–27 m high. Forest, forest edges, bushland, grassland; 1150–1650 m.

Kigezi, Mbale, Mengo, Ankole, Toro, Mubende, Bunyoro, Madi, Acholi, Busoga. Widespread in Africa.

Here assessed preliminarily as Least Concern (LC) because of its wide distribution and habitat range.

Premna resinosa (Hochst.) Schauer

LITERATURE: Verdcourt, FTEA 1992: 71.

Shrub or tree 1–4 m high. Dry bushland, thicket, woodland; 500–1200 m.

Karamoja. Widespread in East Africa.

Here assessed preliminarily as Least Concern (LC) because of its wide distribution and habitat range.

Vitex doniana Sweet

LITERATURE: Verdcourt, FTEA 1992: 62. ITU: 443, fig. 93.

Tree 4–12(–24) m high. Wooded grassland, woodland, elephant grassland, forest; 500–1950 m.

Vitex doniana in fruit, North Mengo (photo: James Kalema).

West Nile, Bunyoro, Teso, Mengo, Mubende, Madi, Acholi, Lango, Karamoja, Mbale, Busoga, Toro. Widespread in Africa.

Here assessed preliminarily as Least Concern (LC) because of its wide distribution and habitat/ altitude range.

Vitex ferruginea Schumach. & Thonn.

LITERATURE: Verdcourt, FTEA 1992: 66.

SYNONYM: *Vitex amboniensis* in the sense of ITU: 442. UFT: 202.

Shrub or tree 1–14 m high. Forest, riverine and lakeside forest; 1100–1200 m.

Bunyoro, Mengo, Ankole, Toro, Madi. Widespread in Africa.

Here assessed preliminarily as Least Concern (LC) because of its wide distribution and habitat.

Vitex fischeri Gürke

LITERATURE: Verdcourt, FTEA 1992: 59. ITU: 445.

Shrub or tree 2–9(–15) m high. Forest, rich bushland, wooded grassland, thicket; 950–2100 m.

Teso, Busoga, Mengo, Masaka, Ankole, Madi, Acholi, Lango, Mbale. Widespread in Africa.

Here assessed preliminarily as Least Concern (LC) because of its wide distribution and habitat/ altitude range.

Vitex madiensis Oliv.

LITERATURE: Verdcourt, FTEA 1992: 60. ITU: 445.

SYNONYM: *Vitex simplicifolia* Oliv.; ITU: 442.

Shrub or tree 1–8 m high. Bushland, woodland; 600–1350 m.

West Nile, Teso, Mubende, Acholi, Lango. Widespread in Africa.

Here assessed preliminarily as Least Concern (LC) because of its wide distribution and habitat type.

LAURACEAE

Beilschmiedia ugandensis Rendle

LITERATURE: Verdcourt, FTEA 1996: 4. UFT: 142.

SYNONYM: *Tylostemon ugandensis* (Rendle) Stapf; ITU: 163.

Shrub or tree 3–27 m high. Riverine and lakeside forest; 900–1500 m.

Toro, Masaka, Mengo, Ankole, West Nile, Madi. Also in South Sudan, Tanzania, DRC.

Here assessed preliminarily as Least Concern (LC) because of its wide distribution.

Ocotea kenyensis (Chiov.) Robyns & R. Wilczek

LITERATURE: Verdcourt, FTEA 1996: 11. UFT: 144.

SYNONYM: *Ocotea sp. nov.* of ITU: 162.

Tree 10–30 m high. Rainforest; 900–2400 m.

Acholi, Kigezi. Widespread in Africa.

Here assessed preliminarily as Least Concern (LC) because of its wide distribution and altitude range.

Ocotea usambarensis Engl.
Literature: Verdcourt, FTEA 1996: 10, fig. 3. ITU: 162. UFT: 142.

Tree 3–45 m high. Moist forest; 900–2400 m.

Toro, Kigezi. Widespread in Africa.

Here assessed preliminarily as Least Concern (LC) because of its wide distribution and altitude range.

LECYTHIDACEAE

Brazzeia longipedicellata Verdc.
See restricted list, page 180.

LINACEAE

Hugonia platysepala Oliv.
Literature: Smith, FTEA 1966: 2. FPU: 62. UFT: 132.

Shrub, climber or tree 2–20 m high. Riverine forest, secondary growth; 1050–1250 m.

Bunyoro, Masaka, Mengo. Widespread in Africa.

Here assessed preliminarily as Least Concern (LC) because of its wide distribution and habitat.

LOBELIACEAE

Lobelia giberroa, *L. stuhlmannii*, *L. wollastonii* and the like may be up to 10 m high, but cannot really be called trees — they are shrubs.

LOGANIACEAE

Anthocleista grandiflora Gilg
Literature: Leeuwenberg 1961: 11.

Synonym: *Anthocleista zambesiaca* Baker; Bruce & Lewis, FTEA 1960: 10, fig. 2. UFT: 170.
Synonym: *Anthocleista pulcherrima* Gilg; ITU: 165.

Tree 7–27 m high. Rainforest, riverine forest, forest remnants; 500–2300 m.

Toro, Mbale, Ankole, Kigezi, Acholi. Widespread in Africa.

Here assessed preliminarily as Least Concern (LC) because of its wide distribution and habitat/altitude range.

Anthocleista schweinfurthii Gilg
Literature: Bruce & Lewis, FTEA 1960: 11. ITU: 165, photo 28. UFT: 170.

Synonym: *Anthocleista inermis* in the sense of ITU: 164.
Synonym: *Anthocleista insulana* S. Moore of ITU: 164.

Tree 6–20 m high. Rainforest, secondary evergreen bushland; 1200–1400 m.

Masaka. Ankole, Mengo. Widespread in Africa.

Here assessed preliminarily as Least Concern (LC) because of its wide distribution and habitat range.

Anthocleista vogelii Planch.
LITERATURE: Bruce & Lewis, FTEA 1960: 8. UFT: 170.
SYNONYM: *Anthocleista nobilis* in the sense of ITU: 164.
Tree 5–18 m high. Swamp forest; 1200–1400 m.
West Nile, Ankole, Masaka, Kigezi, West Nile. Widespread in Africa.
Here assessed preliminarily as Least Concern (LC) because of its wide distribution.

Nuxia congesta Fresen.
LITERATURE: Bruce & Lewis, FTEA 1960: 44. UFT: 160.
SYNONYM: *Lachnopylis congesta* (Fresen.) C.A. Sm.; ITU: 166.
SYNONYM: *Lachnopylis compacta* C.A. Sm. of ITU: 166.
Tree to 25 m high. Rainforest; 1800–2700 m.
Acholi, Kigezi, Mbale, Ankole, Teso, Karamoja, Toro. Widespread in Africa.
Here assessed preliminarily as Least Concern (LC) because of its wide distribution and altitude range.

Nuxia floribunda Benth.
LITERATURE: Bruce & Lewis, FTEA 1960: 43, fig. 8. UFT: 160.
SYNONYM: *Lachnopylis floribunda* (Benth.) C. A. Sm.; ITU: 166.
Shrub or tree to 20 m high. Rainforest edges and -remnants; 1200–2000 m.
Kigezi. Widespread in Africa.
Here assessed preliminarily as Least Concern (LC) because of its wide distribution and habitat.

Nuxia oppositifolia (Hochst.) Benth.
LITERATURE: Bruce & Lewis, FTEA 1960: 43.
SYNONYM: *Lachnopylis oppositifolia* Hochst.; ITU: 166.
Shrub or tree to 12 m high. Riverine forest; 750–2000 m.
Karamoja, Acholi, Teso, Mbale. Widespread in Africa.
Here assessed preliminarily as Least Concern (LC) because of its wide distribution and altitude range.

Strychnos congolana Gilg
LITERATURE: Bruce & Lewis, FTEA 1960: 16.
'Small' tree or climber. Rainforest; 1300 m.
Mengo. Widespread in Africa.
Here assessed preliminarily as Least Concern (LC) because of its wide distribution.

Strychnos henningsii Gilg
LITERATURE: Bruce & Lewis, FTEA 1960: 32.
SYNONYM: *Strychnos myrcioides* S. Moore; ITU: 169.
Shrub or tree 2–10 m high. Rainforest, moist bushland, dry forest, riverine forest; 300–2000 m.
Bunyoro, Busoga, Mengo, West Nile, Karamoja, Busoga, Toro. Widespread in Africa.
Here assessed preliminarily as Least Concern (LC) because of its wide distribution and habitat/altitude range.

Strychnos innocua Delile

LITERATURE: Bruce & Lewis, FTEA 1960: 25. ITU: 167, photo 29, fig. 35.

Shrub or tree 2–12 m high. Woodland; 500–1400 m.

West Nile, Acholi, Lango, Teso, Mengo, Mubende, Bunyoro, Madi, Karamoja, Mbale, Busoga. Widespread in Africa.

Here assessed preliminarily as Least Concern (LC) because of its wide distribution and altitude range.

Strychnos mitis S. Moore

LITERATURE: Bruce & Lewis, FTEA 1960: 21. ITU: 167. UFT: 170.

Tree 10–35 m high. Rainforest, riverine forest; 500–2300 m.

Karamoja, Toro, Mengo, Masaka, Ankole, Bunyoro, Madi, Mbale, Busoga. Also in South Sudan, Kenya, Tanzania and Zimbabwe.

Here assessed preliminarily as Least Concern (LC) because of its wide distribution and altitude range.

Strychnos spinosa Lam.

LITERATURE: Bruce & Lewis, FTEA 1960: 17. ITU: 169

Shrub or tree to 6.5 m high. Scattered tree grassland, woodland; 400–2200 m.

West Nile, Acholi, Teso, Madi, Lango, Karamoja. Widespread in Africa.

Assessed as Least Concern (LC) by Sacande, Sanou & Beentje in *Guide de terrain des arbres du Burkina Faso* (in press).

Strychnos usambarensis Gilg

LITERATURE: Bruce & Lewis, FTEA 1960: 34.

SYNONYM: *Strychnos sp. aff. S. cerasifera* of ITU: 170.

Shrub or tree 2–10 m high. Rainforest, moist bushland; 50–2000 m.

Karamoja, Toro, Busoga. Widespread in Africa.

Here assessed preliminarily as Least Concern (LC) because of its wide distribution and altitude range.

LYTHRACEAE

Lawsonia inermis L.

LITERATURE: Verdcourt, FTEA 1994: 10, fig. 4.

Shrub or tree 1–7 m high. Luggas, riverine thicket; 01–350 m.

Karamoja. Widespread in Africa and Asia.

Assessed as Least Concern (LC) by Sacande, Sanou & Beentje in *Guide de terrain des arbres du Burkina Faso* (in press).

Woodfordia uniflora (A. Rich.) Koehne

LITERATURE: Verdcourt, FTEA 1994: 4.

Shrub or tree 0.8–5 m high. Rocky slope bushland, stream banks, gullies; 1200–2250 m.

Karamoja, Mbale. Widespread in Africa.

Here assessed preliminarily as Least Concern (LC) because of its wide distribution and altitude range.

MALVACEAE

Bombax buonopozense P. Beauv.

Literature: Beentje, FTEA 1989: 5. UFT: 201.

Synonym: *Bombax reflexum* Sprague; ITU: 44, fig. 10.

Tree to 40 m high. Rainforest; 900–1200 m.

Bunyoro, Toro, Mengo, Kigezi. Widespread in Africa.

Here assessed preliminarily as Least Concern (LC) because of its wide distribution.

Ceiba pentandra (L.) Gaertn.

Literature: Beentje, FTEA 1989: 9, fig. 4.

Tree to 40 m high. Moist forest, riverine or lakeside forest; 800–1200 m.

Bunyoro. Widespread in Africa, America and Asia.

Here assessed preliminarily as Least Concern (LC) because of its wide distribution.

Cola congolana De Wild. & T. Durand

Literature: Cheek & Dorr, FTEA 2007: 34.

Synonym: *Cola bracteata* De Wild.; ITU: 415.

Tree 3–10 m high. Evergreen forest; ± 1500 m.

Toro, Ankole. Also in DRC.

EOO 40,809 km^2; AOO 6,697 km^2; 14 collections from 7 locations. Although known in East Africa only from four specimens, *C. congolana* is relatively widespread and abundant in the forests of DRC and so is considered Least Concern (LC), fide FTEA (Cheek 2006). IUCN (Red List 2000, www.redlist.org) list this species under its synonym *C. bracteata* De Wild., as VU B1+2c on the fallacious basis that it is endemic to Uganda.

Cola gigantea A. Chev. of UFT: 117.

Literature: Cheek & Dorr, FTEA 2007: 24.

Synonym: *Cola cordifolia* (Cav.) R. Br.; ITU: 415, photo 36.

Tree to 36 m. Forest; 950–1500 m.

Budongo, Mengo, Ankole, Toro, Bunyoro, Madi, West Nile. Also in Ghana, Togo, Benin, Nigeria, Cameroon, Central African Republic, DRC, South Sudan.

EOO 18,398 km^2; AOO 5,883 km^2; 13 collections from 5 locations, here assessed preliminarily as Least Concern (LC) because of its wide distribution. In agreement with Cheek in FTEA.

Cola pierlotii R. Germ.

Literature: Cheek & Dorr, FTEA 2007: 24.

Tree to 15 m. Forest; 1350–1550 m.

Kigezi. Also in DRC. Known from only two collections in Uganda, both from Ishasha Gorge, but well-distributed in the eastern part of DRC.

It is a restricted range species but we have not been able to get the specimen geodata to prepare its distribution map, here assessed preliminarily as Least Concern (LC) because of its numbers in DRC, in agreement with Cheek in FTEA.

Desplatsia chrysochlamys (Mildbr. & Burret) Mildbr. & Burret

Literature: Whitehouse, Cheek, Andrews & Verdcourt, FTEA 2001: 64. UFT: 122.

Synonym: *Ledermannia chrysochlamys* Mildbr. & Burret; ITU: 429.

Shrub or tree 4–9 m high. Moist forest; 750–1100 m.

Toro, Bunyoro, Mengo. Widespread in Africa.

Here assessed preliminarily as Least Concern (LC) because of its wide distribution.

Desplatsia dewevrei (De Wild. & T. Durand) Burret

Literature: Whitehouse, Cheek, Andrews & Verdcourt, FTEA 2001: 64, fig. 10. UFT: 122.

Synonym: *Desplatsia lutea* Hutch. & Dalziel; ITU: 426, partly.

Tree 4–20 m high. Moist forest; 750–1150 m.

Toro, Bunyoro. Widespread in Africa.

Here assessed preliminarily as Least Concern (LC) because of its wide distribution.

Desplatsia mildbraedii Burret

See restricted list, page 181.

[*Dombeya bagshawei* Baker f. of ITU: 416 (now *Dombeya buettneri* K. Schum.) is really a shrub. Seyani, Dombeya in Africa (1991): 75.]

[*Dombeya dawei* Sprague of ITU (now *Dombeya burgessiae* Harv.) is really a shrub. Seyani, Dombeya in Africa (1991): 66.]

[*Dombeya nairobiensis* Engl. of ITU (now *Dombeya burgessiae* Harv.) is really a shrub. Seyani, Dombeya in Africa (1991): 66.]

Dombeya kirkii Mast.

Literature: Seyani 1991: 112. Cheek & Dorr, FTEA 2007: 70.

Synonym: *Dombeya mukole* Sprague; ITU: 419. UFT: 118.

Tree to 15 m high. Forest edges, secondary forest; 500–2000 m.

Mengo, Ankole, Masaka, Toro, Bunyoro, Busoga, Mbale. Widespread in Africa.

Here assessed preliminarily as Least Concern (LC) because of its wide distribution and habitat and altitude range. In agreement with Cheek in FTEA.

Dombeya quinqueseta (Delile) Exell

Literature: Seyani 1991: 105. Cheek & Dorr, FTEA 2007: 68.

Shrub or tree to 6 m high. Woodland, wooded grassland; 500–2200 m.

Karamoja. Widespread in East and central Africa.

Here assessed preliminarily as Least Concern (LC) because of its wide distribution and habitat type and altitude range. In agreement with Cheek in FTEA.

Dombeya rotundifolia (Hochst.) Planch.

Literature: Seyani 1991: 96. ITU: 419. Cheek & Dorr, FTEA 2007: 69.

Tree to 9 m high. Woodland, wooded grassland; 1000–2400 m.

Bunyoro, Ankole, West Nile, Madi, Acholi, Lango, Teso, Karamoja, Mbale. Widespread in Africa.

Here assessed preliminarily as Least Concern (LC) because of its wide distribution, common habitat and altitude range. In agreement with Cheek in FTEA.

Dombeya torrida (J. F. Gmel.) Bamps

LITERATURE: Seyani 1991: 80. Cheek & Dorr, FTEA 2007: 66.

SYNONYM: *Dombeya goetzenii* K. Schum.; ITU: 418. UFT: 118.

Tree 7–20 m high. Secondary and open forest; 2100–3050 m.

Toro, Kigezi, Acholi, Mbale. Widespread in Africa and Arabia.

Here assessed preliminarily as Least Concern (LC) because of its wide distribution and common habitat type. In agreement with Cheek in FTEA.

Glyphaea brevis (Spreng.) Monach.

LITERATURE: Whitehouse, Cheek, Andrews & Verdcourt, FTEA 2001: 99, fig. 17. UFT: 122.

SYNONYM: *Glyphaea lateriflora* (G. Don.) Hutch. & Dalziel; ITU: 426.

Shrub or tree 1.5–20 m high. Forest, forest edges, secondary bushland; 650–1500 m.

Bunyoro, Busoga, Mengo, Masaka, Toro. Widespread in Africa.

Here assessed preliminarily as Least Concern (LC) because of its wide distribution and habitat/ altitude range.

Grewia arborea (Forssk.) Lam.

LITERATURE: Whitehouse, Cheek, Andrews & Verdcourt, FTEA 2001: 53, fig. 8.

Shrub or tree to 9 m high. Woodland, scattered shrub grassland, bushland and thickets, on rocky outcrops; 450–1700 m.

Karamoja. Widespread in East Africa.

Here assessed preliminarily as Least Concern (LC) because of its wide distribution and habitat/ altitude range.

Grewia bicolor Juss.

LITERATURE: Whitehouse, Cheek, Andrews & Verdcourt, FTEA 2001: 42. ITU: 427.

Shrub or tree to 6 m high. Woodland, grassland, dry bushland and thickets; 650–1650 m.

Karamoja, Acholi, Mbale. Widespread in Africa.

Here assessed preliminarily as Least Concern (LC) because of its wide distribution and habitat/ altitude range.

Grewia flavescens Juss.

LITERATURE: Whitehouse, Cheek, Andrews & Verdcourt, FTEA 2001: 33.

Shrub, liana or tree to 8 m high. Woodland, forest edges, bushland and thickets, riverine; 600–1450 m.

Karamoja. Widespread in Africa.

Here assessed preliminarily as Least Concern (LC) because of its wide distribution and habitat/ altitude range.

Grewia mildbraedii Burret

LITERATURE: Whitehouse, Cheek, Andrews & Verdcourt, FTEA 2001: 58. ITU: 428. UFT: 122.

Tree to 12 m high. Moist forest, often near river; 850–2050 m.

Kigezi, almost exclusively from Ishasha; forest obligate. Also in DRC (including the SW), where it is widespread, Rwanda, Burundi and Tanzania.

EOO 971,517 km^2; AOO 528,962 km^2; 40 collections, some 21 locations. Eilu *et al.* (2004) described it as rare in the Ugandan Albertine Rift area; here assessed preliminarily as Least Concern (LC) because of its wide distribution and altitude range.

Grewia mollis Juss.

Literature: Whitehouse, Cheek, Andrews & Verdcourt, FTEA 2001: 44. ITU: 428.

Shrub or tree to 11 m high. Forest, woodland, wooded grassland; 700–1550 m.

West Nile, Bunyoro, Mbale, Mengo, Masaka, Ankole, Toro, Mubende, Madi, Acholi, Teso, Karamoja, Busoga. Widespread in Africa.

Here assessed preliminarily as Least Concern (LC) because of its wide distribution and habitat/altitude range.

Grewia pubescens P. Beauv.

Literature: Whitehouse, Cheek, Andrews & Verdcourt, FTEA 2001: 25. UFT: 122.

Shrub, liana or tree to 8 m high. Moist forest; 1050–1250 m.

Bunyoro, Toro, Mengo. Widespread in Africa.

Here assessed preliminarily as Least Concern (LC) because of its wide distribution.

Grewia rugosifolia De Wild.

See restricted list, page 182

Grewia seretii De Wild.

Literature: Whitehouse, Cheek, Andrews & Verdcourt, FTEA 2001: 59.

Shrub, liana or tree to 3(–6) m high. Forest and forest edges, woodland, swamp forest; 900–1250 m.

Bunyoro, Mengo. Widespread in central Africa.

Here assessed preliminarily as Least Concern (LC) because of its wide distribution and habitat range.

Grewia similis K. Schum.

Literature: Whitehouse, Cheek, Andrews & Verdcourt, FTEA 2001: 13.

Shrub, liana or rarely a tree to 10 m high. Woodland, grassland, bushland and thickets; 600–2250 m.

Toro, Mengo, Ankole, Bunyoro, Acholi, Kigezi. Widespread in East Africa.

Here assessed preliminarily as Least Concern (LC) because of its wide distribution and habitat/altitude range.

Grewia stolzii Ulbr.

Literature: Whitehouse, Cheek, Andrews & Verdcourt, FTEA 2001: 22.

Shrub, liana or tree to 4 m high. Riverine forest, woodland and thicket; 1200–2150 m.

Karamoja, Toro, Ankole. Widespread in Africa.

Here assessed preliminarily as Least Concern (LC) because of its wide distribution and habitat/altitude range.

Grewia trichocarpa A. Rich.

Literature: Whitehouse, Cheek, Andrews & Verdcourt, FTEA 2001: 45. ITU: 428.

Shrub or tree to 6 m high. Riverine forest, scattered tree grassland, bushland; 900–2150 m.

Karamoja, Teso, Mengo, Masaka, Ankole, Bunyoro, West Nile, Acholi, Mbale, Busoga. Widespread in Africa.

Here assessed preliminarily as Least Concern (LC) because of its wide distribution and habitat/altitude range.

Grewia ugandensis Sprague
See restricted list, page 183.

Grewia velutina (Forssk.) Lam.
LITERATURE: Whitehouse, Cheek, Andrews & Verdcourt, FTEA 2001: 44.

Shrub or tree to 6 m high. Rocky hillside, bushland; 450–1550 m.

Acholi, Karamoja. Widespread in NE Africa.

Here assessed preliminarily as Least Concern (LC) because of its wide distribution and habitat type and altitude range.

Grewia sp. *A* of FTEA
See restricted list, page 184.

Hibiscus diversifolius Jacq. has been called a small tree but in Uganda has always been described as a shrub.

Leptonychia mildbraedii Engl.
See restricted list, page 185.

Nesogordonia kabingaensis (K. Schum.) Capuron
LITERATURE: ITU: 414. Cheek & Dorr, FTEA 2007: 93, fig. 16.

SYNONYM: *Cistanthera kabingaensis* K. Schum.

Tree to 27 m high. Riverine forest edge; ± 750 m.

Kigezi, Toro. A widespread and variable species of W Africa and the Congo Basin.

Here assessed preliminarily as Least Concern (LC) because of its wide distribution. Cheek in FTEA rates this as Near Threatened (NT) on the basis of reports that the Congo basin "is to be made accessible for the export of timber".

Pterygota mildbraedii Engl.
LITERATURE: UFT: 117. Cheek & Dorr, FTEA 2007: 49.

SYNONYM: *Pterygota sp. nov.* of ITU: 422.

Tree 30–45 m high. Evergreen forest; 750–1550 m.

Kigezi, Ankole, Toro, Mubende, Bunyoro, West Nile. Also in Cameroon, DRC, Rwanda, Burundi, Zambia.

Here assessed preliminarily as Least Concern (LC) because of its wide distribution. EOO 30,808 km^2; AOO 9,525 km^2; 13 collections from 10 locations. In agreement with Cheek in FTEA.

Sterculia dawei Sprague
LITERATURE: ITU: 422. UFT: 118. Cheek & Dorr, FTEA 2007: 10.

Tree to 27 m high. Forest; 900–1500 m.

Mengo, Toro, Bunyoro, Mbale, Busoga, Ankole. Also in Cameroon, Gabon, DRC, Angola

Here assessed preliminarily as Least Concern (LC) because of its wide distribution. In agreement with Cheek in FTEA.

Sterculia setigera Delile

LITERATURE: ITU: 423. Cheek & Dorr, FTEA 2007: 14.

Tree 4–12 m high. Wooded grassland; 750–1200 m.

Mengo, Bunyoro, West Nile, Madi, Acholi, Lango, Karamoja, Teso. Also in Tanzania, Senegal, Gambia, Ghana, Togo, Benin, Nigeria, Cameroon, Central African Republic, DRC, South Sudan, Ethiopia, Angola

Here assessed preliminarily as Least Concern (LC) because of its wide distribution and habitat type. In agreement with Cheek in FTEA.

Sterculia stenocarpa H. J. P. Winkl.

LITERATURE: Cheek & Dorr, FTEA 2007: 16.

SYNONYM: *Sterculia rhynchocarpa* in the sense of ITU: 423.

Tree 4–12 m high. Dry bushland with *Commiphora*, *Terminalia*, *Combretum* and *Acacia*; 300–1700 m.

Karamoja. Also in Kenya, Tanzania, South Sudan, Ethiopia, Somalia.

Here assessed preliminarily as Least Concern (LC) because of its wide distribution and habitat type and altitude range. In agreement with Cheek in FTEA.

MELASTOMATACEAE

Dichaetanthera corymbosa (Cogn.) Jacq.-Fél.

LITERATURE: Wickens, FTEA 1975: 14. UFT: 172.

SYNONYM: *Sakersia laurentii* Cogn.; ITU: 171.

Tree 9–15 m high. Rainforest; 1450–1950 m.

Ankole, Kigezi, Toro. Also in Cameroon, DRC, Zambia and Angola.

Here assessed preliminarily as Least Concern (LC) because of its wide distribution.

Lijndenia bequaertii (De Wild.) Borhidi

See restricted list, page 186.

Lijndenia jasminioides (Gilg) Borhidi

LITERATURE: for name change, see Borhidi in *Opera Botanica* 121: 149 (1993).

SYNONYM: *Memecylon jasminioides* Gilg; Wickens, FTEA 1975: 83. ITU: 170. UFT: 172; *Warneckea jasminioides* (Gilg) Jacq.-Fél. in *Adansonia* 18(2): 231 (1978).

Shrub or tree to 15 m high. Rainforest, swamp forest; 1050–1200 m.

Bunyoro, Masaka, Mengo, Ankole; also in Cameroon, CAR, DRC, Tanzania.

Here assessed preliminarily as Least Concern (LC) because of its wide distribution (EOO > 47,887 km^2; AOO >11,536 km^2; >> 10 locations).

Memecylon myrianthum Gilg

LITERATURE: Wickens, FTEA 1975: 87. ITU: 170. UFT: 192.

Tree to 9 m high. Rainforest; 1100–1200 m.

Masaka, Mengo. Widespread in central Africa.

Here assessed preliminarily on a global scale as Least Concern (LC) because of its wide distribution.

MELIACEAE

Carapa procera DC.

LITERATURE: Styles & White, FTEA 1991: 61, fig. 20.

SYNONYM: *Carapa grandiflora* Sprague; ITU: 172, fig. 36. UFT: 216.

Tree to 25 m high. Lake-shore, riverine and rainforest; 1100–1850 m.

Toro, Ankole, Mengo, Masaka, Kigezi. Widespread in central Africa.

Assessed as Least Concern (LC) by Sacande, Sanou & Beentje in *Guide de terrain des arbres du Burkina Faso* (in press).

Ekebergia capensis Sparrm.

LITERATURE: Styles & White, FTEA 1991: 38. UFT: 215.

SYNONYM: *Ekebergia complanata* Baker f. of ITU: 172.
SYNONYM: *Ekebergia rueppelliana* (Fresen.) A. Rich.in of ITU: 174.
SYNONYM: *Ekebergia senegalensis* A. Juss. of ITU: 174. UFT: 213; *Ekebergia sp.* of ITU: 175.

Tree to 30 m high. Forest and riverine forest, sometimes in woodland and wooded grassland; 600–2650 m.

Karamoja, Bunyoro, Mbale, Mengo, Kigezi, Acholi, Teso. Widespread in Africa.

Here assessed preliminarily on a global scale as Least Concern (LC) because of its wide distribution and habitat & altitude range.

Entandrophragma angolense (Welw.) C. DC.

LITERATURE: Styles & White, FTEA 1991: 56, fig. 17. ITU: 176, photo 30, t. 37. UFT: 217.

Tree to 50 m high. Rainforest; 1100–1850 m.

Bunyoro, Mbale, Mengo, Toro, Madi. Widespread in Africa.

Used for firewood, charcoal, timber (furniture). The species has been overharvested in Uganda, resulting in a very small remnant population size (Katende *et al.* 1995) and is accordingly on the country's 'Reserved Species' list. It is however, here assessed preliminarily on a global scale as Least Concern (LC) because the species is widespread in Africa.

Entandrophragma cylindricum (Sprague) Sprague

LITERATURE: Styles & White, FTEA 1991: 50, fig. 15. ITU: 178, photo 31, t. 38. UFT: 217.

Tree to 60 m high. Rainforest; 1100–1500 m.

Bunyoro, Mengo, Kigezi. Widespread in Africa.

Used for firewood, charcoal, timber (furniture), veneer; a valuable export commodity in West Africa. It is overharvested in Uganda (Katende *et al.* 1995) and is on the country's 'Reserved Species' list. It is however, here assessed preliminarily on a global scale as Least Concern (LC) because the species is widespread in Africa.

Entandrophragma excelsum (Dawe & Sprague) Sprague

LITERATURE: Styles & White, FTEA 1991: 54, fig. 17. ITU: 180, t. 39. UFT: 219.

SYNONYM: *Entandrophragma sp.* of ITU: 183.

Tree to 60 m high. Rainforest, riverine forest; 1250–2150 m.

Toro, Ankole, Mbale, Kigezi, Mengo. Also DRC, Tanzania, Malawi.

Here assessed preliminarily as Least Concern (LC) because of its wide distribution and altitude range. The species is on Uganda's 'Reserved Species' list but the timber has little utility as it tends to twist badly if used unseasoned (Styles & White 1991).

Entandrophragma utile (Dawe & Sprague) Sprague
LITERATURE: Styles & White, FTEA 1991: 52, fig. 16. ITU: 180, photo 32, t. 40. UFT: 217.

Tree to 50 m high. Rainforest; 1100–1400 m.

Acholi, Bunyoro, Mengo, Toro, Madi. Widespread in Africa.

Its timber is exported largely from Ivory Coast and Ghana. It was common in Budongo and Mabira forests but is rare elsewhere in Uganda. It is used for firewood, charcoal, timber, veneer. It is one of the rarest of all the *Entandrophragma* spp. and indiscriminate harvesting during the years of political unrest in Uganda brought it close to extinction (Katende *et al.* 1995). The species is on the country's 'Reserved Species' list. Here assessed preliminarily on a global scale as Least Concern (LC) because of its wide distribution in Africa.

Guarea cedrata (A. Chev.) Pellegr.
LITERATURE: Styles & White, FTEA 1991: 43. ITU: 184. UFT: 219.

Tree to 45 m high. Rainforest; ± 1100 m.

Bunyoro, Toro, Mengo, Masaka, Kigezi. Widespread in Africa.

It is not easily distinguishable from *Entandrophragma angolense* and it has been harvested under the name of the latter (Katende *et al.* 1995). Commercially exploited elsewhere. Here assessed preliminarily as Least Concern (LC) because of its wide distribution.

Guarea mayombensis Pellegr.
LITERATURE: Styles & White, FTEA 1991: 43, fig. 12. ITU: 184.

SYNONYM: *Leplaea mayombensis* (Pellegr.) Staner; UFT: 216.

Tree to 25 m high. Rainforest; 1200–1600 m.

Kigezi. Also in Gabon and DRC.

Here assessed preliminarily as Least Concern (LC) because of its wide distribution.

Khaya anthotheca (Welw.) C. DC.
LITERATURE: Styles & White, FTEA 1991: 47, fig. 14. ITU: 185, fig. 41, photo 33. UFT: 219.

Tree to 60 m high. Rainforest and riverine forest; 100–1550 m.

Bunyoro, Toro. Widespread in Africa.

One of the most valuable trees in Uganda (Styles & White 1991) and its high quality timber is heavily traded in East Africa (Katende *et al.* 1995) and the Great Lakes region generally. It was formerly the source of nearly half the converted timber from Budongo (Styles & White 1991). The species is on Uganda's 'Reserved Tree Species' list. Here assessed preliminarily on a global scale as Least Concern (LC) because of its wide distribution and altitude range.

Khaya grandifoliola C. DC.
LITERATURE: Styles & White, FTEA 1991: 49. ITU: 187, fig. 42, photo 34. UFT: 219.

Tree to 30 m high. Riverine forest; 750–1250 m.

West Nile, Acholi, Bunyoro, Madi. Widespread in Africa.

Used for firewood, charcoal and timber but because of low branching (Styles & White 1991), exploitation for timber is very limited. The species is on Uganda's 'Reserved Tree Species' list. Here assessed preliminarily on a global scale as Least Concern (LC) because of its wide distribution.

Khaya senegalensis (Desr.) A. Juss.

LITERATURE: Styles & White, FTEA 1991: 49. ITU: 189, fig. 43, photo 35.

Tree to 20 m high. Woodland and riverine; 900–1550 m.

West Nile, Acholi, Madi. Widespread in Africa.

Assessed as Least Concern (LC) by Sacande, Sanou & Beentje in *Guide de terrain des arbres du Burkina Faso* (in press).

Lepidotrichilia volkensii (Gürke) J. F. Leroy

LITERATURE: Styles & White, FTEA 1991: 37, fig. 10.

SYNONYM: *Trichilia volkensii* Gürke; ITU: 198.

Tree 5–20 m high. Forest; 1550–2600 m.

Acholi, Ankole, Mbale, Mengo, Kigezi, Toro, West Nile. Widespread in central Africa.

Here assessed preliminarily as Least Concern (LC) because of its wide distribution and altitude range.

Lovoa swynnertonii Baker f.

LITERATURE: Styles & White, FTEA 1991: 59. ITU: 193. UFT: 220.

Tree to 50 m high. Rainforest; 150–1550 m.

Toro, Ankole, Mengo. Widespread in central, eastern and southern Africa.

Used for high grade timber, firewood and charcoal. Much exploited in Zimbabwe and elsewhere. The species is on Uganda's list of 'Reserved Tree Species'. Here assessed preliminarily as Least Concern (LC) because of its wide distribution and altitude range.

Lovoa trichilioides Harms

LITERATURE: Styles & White, FTEA 1991: 59, fig. 19. UFT: 220.

SYNONYM: *Lovoa brownii* Sprague; ITU: 191, fig. 44, photo 36.

Tree to 40 m high. Rainforest, lake-shore forest; 1100–1200 m.

Bunyoro, Mengo, Masaka, Toro. Widespread in central Africa.

The species is heavily exploited in Uganda for timber and is on the country's list of 'Reserved Tree Species' but is still common in its areas of occurrence. Here assessed preliminarily as Least Concern (LC) because of its wide distribution, though its Ugandan altitude range seems restricted.

Melia azedarach L.

LITERATURE: Styles & White, FTEA 1991: 22, fig. 4.

Tree to 15 m high. Planted and locally naturalised; 500–2150 m.

Acholi, Bunyoro, Mengo. Originally from Asia.

Here assessed preliminarily as Least Concern (LC) because of its distribution and altitude range.

Pseudocedrela kotschyi (Schweinf.) Harms

LITERATURE: Styles & White, FTEA 1991: 56, fig. 18. ITU: 193, fig. 45.

Tree to 12 m high. Woodland, wooded grassland; 750–1200 m.

West Nile, Acholi, Mengo, Madi, Lango, Teso, Karamoja, Mbale, Bunyoro. Widespread in Africa.

Assessed as Least Concern (LC) by Sacande, Sanou & Beentje in *Guide de terrain des arbres du Burkina Faso* (in press).

Trichilia dregeana Sond.

LITERATURE: Styles & White, FTEA 1991: 34. UFT: 215.

SYNONYM: *Trichilia megalantha* in the sense of ITU: 197.

SYNONYM: *Trichilia splendida* A. Chev.; ITU: 198

SYNONYM: *Trichilia sp. 2* of ITU: 199.

Tree 20–40 m high. Rainforest, riverine and swamp forest; 750–1550 m.

Bunyoro, Ankole, Mbale, Mengo, Masaka, Toro, West Nile, Busoga. Widespread in Africa.

Here assessed preliminarily as Least Concern (LC) because of its wide distribution and habitat range.

Trichilia emetica Vahl

LITERATURE: Styles & White, FTEA 1991: 34. ITU: 195, fig. 46.

Tree 8–20 m high. Riverine forest and woodland; 500–1300 m.

West Nile, Acholi, Mbale, Ankole, Bunyoro, Madi, Karamoja, Lango, Teso. Widespread in Africa.

Assessed as Least Concern (LC) by Sacande, Sanou & Beentje in *Guide de terrain des arbres du Burkina Faso* (in press).

Trichilia martineaui Aubrév. & Pellegr.

LITERATURE: Styles & White, FTEA 1991: 32. UFT: 215.

SYNONYM: *Trichilia sp. 1* of ITU: 199.

Tree to 35 m high. Rainforest; 1000–1200 m.

Bunyoro, Mengo, Kigezi. Widespread in Africa.

Here assessed preliminarily as Least Concern (LC) because of its wide distribution.

Trichilia prieuriana A. Juss.

LITERATURE: Styles & White, FTEA 1991: 30. ITU: 197. UFT: 215.

Tree 10–15(–25) m high. Rainforest, riverine forest; 900–1250 m.

Lango, Teso, Mengo, Bunyoro, Kigezi, Acholi, Karamoja, Mbale. Widespread in Africa.

Here assessed preliminarily as Least Concern (LC) because of its wide distribution.

Trichilia rubescens Oliv.

LITERATURE: Styles & White, FTEA 1991: 30. ITU: 197. UFT: 216.

Shrub or tree to 10(–18) m high. Rainforest, swamps; 900–1500 m.

Bunyoro, Masaka, Mengo, Ankole, Madi. Widespread in Africa.

Here assessed preliminarily as Least Concern (LC) because of its wide distribution.

Turraea abyssinica A. Rich.

LITERATURE: Styles & White, FTEA 1991: 12.

Tree 2–8(–16) m high. Montane forest, also riverine forest; 1800–2250 m.

Karamoja. Also in Ethiopia, Kenya, N Tanzania, fairly widespread.

Here assessed preliminarily as Least Concern (LC) because of its wide distribution and habitat range.

Turraea fischeri Gürke

LITERATURE: Styles & White, FTEA 1991: 19. ITU: 200.

Shrub or tree to 9 m high. Bushland and scrub forest; 1200–1850 m.

Acholi, Karamoja. Also in Tanzania and Zimbabwe, fairly widespread.

Here assessed preliminarily as Least Concern (LC) because of its wide distribution and habitat range.

Turraea floribunda Hochst.

Literature: Styles & White, FTEA 1991: 21. ITU: 200. UFT: 150.

Shrub or tree to 10(–15) m high. Evergreen forest, riverine forest, and secondary forest; 100–2150 m.

Bunyoro, Mbale, Mengo, Toro, Madi, Acholi. Widespread in Africa.

Here assessed preliminarily as Least Concern (LC) because of its wide distribution and habitat/altitude range.

(*Turraea heterophylla* in the sense of ITU is really a shrub, and now a synonym of *Turraea vogelioides* Bagsh. & Baker f., treated in UFT)

Turraea holstii Gürke

Literature: Styles & White, FTEA 1991: 13. ITU: 200.

Shrub or tree to 18 m high. Forest and forest edges; 700–2500 m.

Acholi. Widespread in Africa.

Here assessed preliminarily as Least Concern (LC) because of its wide distribution and habitat range.

Turraea pellegriana Keay

Literature: Styles & White, FTEA 1991: 12.

Shrub or tree to 8 m high. Rainforest edges; 1050 m.

Bunyoro. Widespread in Africa.

Here assessed preliminarily as Least Concern (LC) because of its wide distribution.

Turraea robusta Gürke

Literature: Styles & White, FTEA 1991: 8. ITU: 200. UFT: 150.

Tree 2–8(–16) m high. Evergreen forest, especially riverine, and secondary forest; 900–1750 m.

Kigezi, Busoga, Mengo, Ankole, Toro, Teso, Mbale. Widespread in central Africa.

Here assessed preliminarily as Least Concern (LC) because of its wide distribution and habitat/altitude range.

(*Turraea vogelii* Hook. f. of ITU and UFT is really a climbing shrub)

Turraeanthus africanus (C. DC.) Pellegr.

Literature: Styles & White, FTEA 1991: 40, fig. 11. UFT: 220.

Synonym: *Turraeanthus sp.* of ITU: 201.

Tree to 20 m high. Rainforest along streams; 750–1550 m.

Toro, Ankole. Widespread in Africa.

Here assessed preliminarily as Least Concern (LC) because of its wide distribution.

MELIANTHACEAE

Bersama abyssinica Fresen.

LITERATURE: Verdcourt, FTEA 1958: 2, fig. 1. ITU: 202. UFT: 206.

Tree to 17 m high. Secondary forest, scattered tree grassland, woodland, scrub; 1100–2550 m.

West Nile, Karamoja, Kigezi, Mengo, Mbale, Ankole, Toro, Busoga, Bunyoro. Widespread in Africa.

Here assessed preliminarily as Least Concern (LC) because of its wide distribution and habitat/altitude range.

MONIMIACEAE

Xymalos monospora (Harv.) Warb.

LITERATURE: Verdcourt, FTEA 1968: 1, fig. 1. ITU: 232. UFT: 188.

Shrub or tree 6–20 m high. Rainforest, forest on isolated hill-tops; 900–2700 m.

Ankole, Kigezi, Mengo, Toro, West Nile, Mbale. Widespread in Africa.

Here assessed preliminarily as Least Concern (LC) because of its wide distribution and altitude range.

MORACEAE

Antiaris toxicaria Lesch.

LITERATURE: Berg & Hijman, FTEA 1989: 10, fig. 7. ITU: 233. UFT: 93.

Tree to 40(–60) m high. Rainforest, riverine forest, wooded grassland; 500–1700 m.

Bunyoro, Acholi, Busoga, Mbale, Mengo. All districts except Karamoja, West Nile. Widespread in Africa.

Assessed here as Least Concern (LC) because of its wide distribution and habitat/altitude range.

Broussonetia papyrifera Vent.

LITERATURE: –

Tree to 15 m high. Naturalising in rainforest margins and gaps, and becoming an invader.

Bunyoro, Mengo. Originally from East Africa.

Not assessed here.

(*Craterogyne kameruniana* (Engl.) Lanj. of ITU, now *Dorstenia kameruniana* Engl., is really a shrub)

Ficus abutilifolia (Miq.) Miq.

LITERATURE: Berg & Hijman, FTEA 1989: 67.

SYNONYM: *Ficus discifera* Warb.; ITU: 248.

Tree to 15 m high. Rock outcrops, riverine, lakesides; 550–1050 m.

Mengo, Karamoja. Widespread in Africa.

Here assessed preliminarily as Least Concern (LC) because of its wide distribution.

Ficus amadiensis De Wild.

LITERATURE: Berg & Hijman, FTEA 1989: 70.

SYNONYM: *Ficus kitubalu* Hutch.; ITU: 252.

Tree to 15 m high. Wooded grassland, lakeside forest and thicket, rarely planted; 950–2100 m.

Acholi, Toro, Mengo, Masaka, Ankole. Fairly widespread in East Africa.

Here assessed preliminarily as Least Concern (LC) because of its wide distribution and habitat/altitude range.

Ficus artocarpoides Warb.

LITERATURE: Berg & Hijman, FTEA 1989: 78. UFT: 102.

SYNONYM: *Ficus kisantuensis* Warb.; ITU: 252.

Tree to 10 m high. Rainforest, riverine forest and secondary associations; 500–1200 m.

Mengo, Masaka. Widespread in Africa.

Here assessed preliminarily as Least Concern (LC) because of its wide distribution and habitat range.

Ficus barteri Sprague

LITERATURE: Berg & Hijman, FTEA 1989: 83.

SYNONYM: *Ficus stipulifera* in the sense of ITU: 258, partly. UFT: 102.

Tree to 10 m high or shrub. Forest, rainforest; 1000–1200 m.

Bunyoro, Mengo. Widespread in Africa.

Here assessed preliminarily as Least Concern (LC) because of its wide distribution.

Ficus bubu Warb.

LITERATURE: Berg & Hijman, FTEA 1989: 81.

Tree to 30 m high. Forest, riverine, lakeside and ground-water forest; 500–1200 m.

Bunyoro, Toro, Mengo. Widespread in Africa.

Here assessed preliminarily as Least Concern (LC) because of its wide distribution.

(*Ficus capreifolia* Delile of ITU is really a shrub)

Ficus conraui Warb.

LITERATURE: Berg & Hijman, FTEA 1989: 83.

'Treelet' or shrub. Forest, rainforest; 1100–1200 m.

Masaka. Widespread in Africa.

Here assessed preliminarily as Least Concern (LC) because of its wide distribution.

Ficus cordata Thunb.

LITERATURE: Berg & Hijman, FTEA 1989: 60.

SYNONYM: *Ficus pretoriae* Burtt Davy; ITU: 257.

Tree to 15(–35) m high. Seasonal streams, rock outcrops, lava or limestone; 950–2400 m.

Acholi, Karamoja, Bunyoro. Widespread in Africa.

Here assessed preliminarily as Least Concern (LC) because of its wide distribution and habitat/altitude range.

Ficus craterostoma Mildbr. & Burrett

LITERATURE: Berg & Hijman, FTEA 1989: 71.

SYNONYM: *Ficus pilosula* De Wild.; ITU: 256. UFT: 102.

Tree to 10 m high or shrub. Forest, especially ground-water and riverine, occasionally planted; 300–2100 m.

Kigezi, Masaka, Ankole. Widespread in Africa.

Here assessed preliminarily as Least Concern (LC) because of its wide distribution and altitude range.

Ficus cyathistipula Warb.

LITERATURE: Berg & Hijman, FTEA 1989: 83. ITU: 245, fig. 58a. UFT: 100.

Tree to 8(–15) m high. Rainforest, riverine, lakeside and ground-water forest; 700–1800 m.

Ankole, Busoga, Mengo, Masaka, Toro, Acholi. Widespread in Africa.

Here assessed preliminarily as Least Concern (LC) because of its wide distribution and habitat/altitude range.

Ficus densistipulata De Wild.

LITERATURE: Berg & Hijman, FTEA 1989: 84.

SYNONYM: *Ficus namalalensis* Hutch.; ITU: 253.

Tree to 10 m high, or shrub. Rainforest, semi-swamp forest; 1100–1300 m.

Bunyoro, Masaka, Mengo. Widespread in central Africa.

Here assessed preliminarily as Least Concern (LC) because of its wide distribution.

Ficus dicranostyla Mildbr.

LITERATURE: Berg & Hijman, FTEA 1989: 59, fig. 19. ITU: 247, fig. 54c.

Tree to 20(–30) m high. Wooded grassland, especially in rocky places; 900–1100 m.

Acholi, Karamoja, Teso, Mengo, Bunyoro, West Nile, Madi, Lango, Mbale, Busoga. Fairly widespread in Africa.

Here assessed preliminarily as Least Concern (LC) because of its wide distribution.

Ficus exasperata Vahl

LITERATURE: Berg & Hijman, FTEA 1989: 52. ITU: 248. UFT: 96.

Tree to 20(–30) m high. Forest, forest edges, riverine; 500–1200 m.

Mengo, Masaka, Toro, Mubende, Mbale, Busoga, Karamoja, West Nile. Widespread in Africa and Asia.

Here assessed preliminarily as Least Concern (LC) because of its wide distribution and habitat range.

Ficus glumosa Delile

LITERATURE: Berg & Hijman, FTEA 1989: 65. ITU: 248, fig. 55a.

SYNONYM: *Ficus sonderi* Miq.; ITU: 258.

Tree to 15 m high. Rock outcrops, lava in wooded grassland and dry bushland; 500–2000 m.

West Nile, Acholi, Teso, Mbale, Busoga, Bunyoro, Karamoja. Widespread in Africa.

Here assessed preliminarily as Least Concern (LC) because of its wide distribution and habitat/altitude range.

Ficus ingens (Miq.) Miq.

LITERATURE: Berg & Hijman, FTEA 1989: 60, fig. 20. ITU: 250, fig. 56b. UFT: 100.

Tree to 18 m high. Rock outcrops, lava and limestone, riverine, also in disturbed forest and wooded grassland; 500–2600 m.

Mengo, Acholi, Teso, Mbale, Karamoja. Widespread in Africa.

Here assessed preliminarily as Least Concern (LC) because of its wide distribution and habitat/altitude range.

Ficus katendei Verdc.

See restricted list, page 187.

Ficus lingua De Wild. & T. Durand

Literature: Berg & Hijman, FTEA 1989: 72.

Synonym: *Ficus depauperata* in the sense of ITU: 247.

Tree to 30 m high. Forest; 1050–1200 m.

Bunyoro, Masaka, Kigezi. Widespread in Africa.

Here assessed preliminarily as Least Concern (LC) because of its wide distribution.

Ficus lutea Vahl

Literature: Berg & Hijman, FTEA 1989: 69.

Tree to 20 m high. Forest and forest edges, riverine, lake-sides, also planted by settlements; 500–1800 m.

Bunyoro, Masaka. Widespread in Africa.

Here assessed preliminarily as Least Concern (LC) because of its wide distribution and habitat/altitude range.

Ficus mucuso Ficalho

Literature: Berg & Hijman, FTEA 1989: 56. ITU: 253, photo 42. UFT: 98.

Tree to 30(–40) m high. Rain forest; 300–1200 m.

Bunyoro, Masaka, Mengo, Toro, Bunyoro, Madi, Busoga, Ankole. Widespread in Africa.

Here assessed preliminarily as Least Concern (LC) because of its wide distribution and altitude range.

Ficus natalensis Hochst.

Literature: Berg & Hijman, FTEA 1989: 71. ITU: 253. UFT: 102.

Tree to 30 m high or shrub. Forest, ground-water forest, riverine, lake-sides, woodland, often planted; 500–2200 m.

Bunyoro, Masaka, Mengo, Ankole, Toro, Mubende, Busoga. Widespread in Africa.

Here assessed preliminarily as Least Concern (LC) because of its wide distribution and habitat/altitude range.

Ficus oreodryadum Mildbr.

Literature: Berg & Hijman, FTEA 1989: 82.

'Large' tree or shrub. Swamp forest, forest; 1300–2500 m.

Kigezi. Widespread in Africa.

Here assessed preliminarily as Least Concern (LC) because of its wide distribution and altitude range.

Ficus ottoniifolia (Miq.) Miq.

Literature: Berg & Hijman, FTEA 1989: 76. UFT: 99.

Synonym: *Ficus lukanda* (properly *lucanda* Ficalho!) of ITU: 252.

Tree to 15 m high, or shrub or liana. Forest, riverine and lakeside forest; 500–1500 m.

Ankole, Kigezi, Masaka, Mengo, Toro. Widespread in Africa.

Here assessed preliminarily as Least Concern (LC) because of its wide distribution and habitat/ altitude range.

Ficus ovata Vahl

Literature: Berg & Hijman, FTEA 1989: 81.

Synonym: *Ficus brachypoda* Hutch.; ITU: 242. UFT: 99.

Tree to 10(–25) m high. Woodland, wooded grassland, riverine, lakeside and often planted; 750–2100 m.

West Nile, Kigezi, Mengo, Masaka, Ankole, Toro, Mubende, Madi, Acholi, Lango, Teso, Busoga. Widespread in Africa.

Here assessed preliminarily as Least Concern (LC) because of its wide distribution and habitat/ altitude range.

Ficus platyphylla Delile

Literature: Berg & Hijman, FTEA 1989: 64. ITU: 256.

Tree to 15 m high. Wooded grassland, rocky places; 950–1200 m.

Acholi, Karamoja, Teso, Mbale, Busoga. Widespread in Africa.

Here assessed preliminarily as Least Concern (LC) because of its wide distribution.

Ficus polita Vahl

Literature: Berg & Hijman, FTEA 1989: 80. UFT: 100.

Tree to 15(–40) m high. Rainforest; 500–1200 m.

Bunyoro, Mengo, Toro. Widespread in Africa.

Here assessed preliminarily as Least Concern (LC) because of its wide distribution and altitude range.

Ficus populifolia Vahl

Literature: Berg & Hijman, FTEA 1989: 68. ITU: 257, fig. 55c.

Tree to 10(–30?) m high. Rock outcrops, riverine, lava and scarps in dry bushland and wooded grassland; 500–1600 m.

Acholi, Karamoja. Widespread in Africa.

Here assessed preliminarily as Least Concern (LC) because of its wide distribution and habitat/ altitude range.

Ficus preussii Warb.

Literature: Berg & Hijman, FTEA 1989: 85.

Tree to 10 m high or shrub. Rainforest; 1150–1200 m.

Mengo. Widespread in Africa.

Here assessed preliminarily as Least Concern (LC) because of its wide distribution.

Ficus pseudomangifera Hutch.

Literature: Berg & Hijman, FTEA 1989: 82. ITU: 257. UFT: 104.

Tree to 10 m high. Rainforest; 1050–1200 m.

Bunyoro, Toro, Masaka, Mengo. Widespread in Africa.

Here assessed preliminarily as Least Concern (LC) because of its wide distribution.

Ficus sansibarica Warb.

Literature: Berg & Hijman, FTEA 1989: 79.

Synonym: *Ficus brachylepis* Hiern.; ITU: 242. UFT: 100.

Tree to 20(–40) m high. Rainforest, riverine, lake-sides; 1050–1200 m.

Bunyoro, Masaka, Toro. Widespread in Africa.

Here assessed preliminarily as Least Concern (LC) because of its wide distribution.

Ficus saussureana DC.

Literature: Berg & Hijman, FTEA 1989: 69.

Synonym: *Ficus dawei* Hutch.; ITU: 245.
Synonym: *Ficus eriobotryoides* Kunth & Bouché of UFT: 100.

Tree to 20 m high. Forest, forest edges, riverine, lake-sides; 900–1600 m.

West Nile, Bunyoro, Masaka, Madi, Toro, Ankole, Mengo. Widespread in Africa.

Here assessed preliminarily as Least Concern (LC) because of its wide distribution and habitat range.

Ficus stuhlmannii Warb.

Literature: Berg & Hijman, FTEA 1989: 66. ITU: 258.

Tree to 18 m high. Wooded grassland, bushland, often on rock outcrops, along lakes or streams, on hardpan soils; 500–1800 m.

Karamoja. Widespread in Africa.

Here assessed preliminarily as Least Concern (LC) because of its wide distribution and habitat/altitude range.

Ficus sur Forssk.

Literature: Berg & Hijman, FTEA 1989: 56, fig. 18.

Synonym: *Ficus capensis* Thunb.; ITU: 243, fig. 53. UFT: 99.

Tree to 30 m high. Forest, wooded grassland, riverine; 500–2300 m.

West Nile, Bunyoro, Mengo, Masaka, Ankole, Toro, Mubende, Acholi, Lango, Teso, Mbale. Widespread in Africa.

Here assessed preliminarily as Least Concern (LC) because of its wide distribution and habitat/altitude range.

Ficus sycomorus L.

Literature: Berg & Hijman, FTEA 1989: 54. ITU: 260.

Synonym: *Ficus gnaphalocarpa* (Miq.) A. Rich.; ITU: 250, fig. 55b

Tree to 20(–30) m high. Forest edges, riverine, lakesides, foot of rocky hills; 500–2200 m.

Karamoja, Teso, Toro, West Nile, Madi, Acholi, Lango, Mbale, Busoga, Bunyoro. Widespread in Africa and Arabia.

Here assessed preliminarily as Least Concern (LC) because of its wide distribution and habitat/altitude range.

Ficus thonningii Blume

Literature: Berg & Hijman, FTEA 1989: 73, fig. 21. ITU: 260. UFT: 102.

Synonym: *Ficus dekdekana* (Miq.) A. Rich.; ITU: 247.
Synonym: *Ficus persicifolia* Warb.; ITU: 256, fig. 57b. UFT: 104.
Synonym: *Ficus rhodesica* Mildbr. & Burret; ITU: 258.

Tree to 15(–30) m high or shrub. Forest, woodland, bushland and wooded grassland, also planted; 500–2500 m.

West Nile, Kigezi, Masaka, Teso, Acholi, Mengo, Bunyoro, Lango, Mbale, Ankole. Widespread in Africa.

Here assessed preliminarily as Least Concern (LC) because of its wide distribution and habitat/altitude range.

Ficus tremula Warb.

LITERATURE: Berg & Hijman, FTEA 1989: 77.

Tree to 10 m high, or shrub or liana. Rainforest; 1650–2300 m.

Kigezi; widespread in Africa, while the subsp. *acuta* (De Wild.) C. C. Berg which occurs in Uganda is otherwise restricted to eastern DRC, Rwanda, Burundi and western Kenya.

Here assessed preliminarily as Least Concern, because of the wide distribution even for the subspecies, with quite a wide altitude range.

Ficus trichopoda Baker

LITERATURE: Berg & Hijman, FTEA 1989: 68.

SYNONYM: *Ficus congensis* Engl.; ITU: 245, fig. 56a. UFT: 98.

Tree to 10(–20) m high or shrub. Ground-water forest, swamp edges, riverine; 50–1200 m.

West Nile, Acholi, Mengo, Mengo, Masaka, Mubende, Toro, West Nile, Madi, Acholi. Widespread in Africa.

Here assessed preliminarily as Least Concern (LC) because of its wide distribution and altitude range.

(*Ficus urceolaris* Hiern of ITU and UFT, now a synonym of *Ficus asperifolia* Miq., is really a shrub)

Ficus vallis-choudae Delile

LITERATURE: Berg & Hijman, FTEA 1989: 58. ITU: 261, fig. 57d. UFT: 98.

Tree to 20 m high. Ground-water forest, riverine, lake-sides; 450–1800 m.

Acholi, Toro, Mengo, Masaka, Kigezi, Mubende, Bunyoro, Madi, Mbale, Busoga. Widespread in Africa.

Here assessed preliminarily as Least Concern (LC) because of its wide distribution and altitude range.

Ficus variifolia Warb.

LITERATURE: Berg & Hijman, FTEA 1989: 59. ITU: 261.

Tree to 35 m high. Forest, pioneer of cleared areas; 900–1200 m.

Bunyoro, Toro, Ankole. Widespread in Africa.

Here assessed preliminarily as Least Concern (LC) because of its wide distribution and habitat.

Ficus vasta Forssk.

LITERATURE: Berg & Hijman, FTEA 1989: 64. ITU: 262.

Tree to 25 m high. Riverine; 1100–1650 m.

Acholi, Karamoja, Bunyoro, Teso. N Kenya, South Sudan, Ethiopia, Somalia and Arabia.

Here assessed preliminarily as Least Concern (LC) because of its wide distribution.

Ficus verruculosa Warb.
LITERATURE: Berg & Hijman, FTEA 1989: 63. ITU: 262.

Shrub or tree to 7 m high. Swamps, lake-sides, riverine, occasionally in wooded grassland; 800–1850 m.

Kigezi, Mbale, Masaka, Mengo, Bunyoro, Teso, Busoga. Widespread in Africa.

Here assessed preliminarily as Least Concern (LC) because of its wide distribution and habitat/altitude range.

Ficus vogeliana (Miq.) Miq.
LITERATURE: Berg & Hijman, FTEA 1989: 58. ITU: 262. UFT: 100.

Tree to 20 m high. Swamp forest; 1200 m.

Mengo, Toro. Widespread in Africa.

Here assessed preliminarily as Least Concern (LC) because of its wide distribution.

Ficus wakefieldii Hutch.
LITERATURE: Berg & Hijman, FTEA 1989: 65.

Tree to 25 m high. Wooded grassland, riverine, lakesides, rocky outcrops or at foot of rocky hills; 200–2000 m.

Acholi, Busoga. Widespread in East Africa.

Here assessed preliminarily as Least Concern (LC) because of its wide distribution and habitat/altitude range.

Ficus wildemaniana De Wild. & T. Durand
LITERATURE: Berg & Hijman, FTEA 1989: 85.

Tree or shrub (size not mentioned). Rainforest; 1100 m.

Bunyoro. Widespread in Central Africa.

Here assessed preliminarily as Least Concern (LC) because of its wide distribution.

Milicia excelsa (Welw.) C. C. Berg
LITERATURE: Berg & Hijman, FTEA 1989: 4, fig. 2.

SYNONYM: *Chlorophora excelsa* (Welw.) Benth. & Hook. f.; ITU: 234, fig. 52, photo 41. UFT: 94.

Tree to 30(–50) m high. Rainforest, evergreen forest, riverine and ground-water forest; 500–1350 m.

Bunyoro, Mengo, Masaka, Mubende, Toro, West Nile, Madi, Acholi, Lango, Teso, Mbale, Busoga. Widespread in Africa.

Logged commercially throughout its range, and used for firewood, charcoal, timber, used especially for quality indoor and outdoor furniture. It is on Uganda's list of 'Reserved Tree Species'. In Kenya, the species is now rare and threatened. Here assessed preliminarily as Least Concern (LC) because of its wide distribution and habitat range.

Morus mesozygia Stapf
LITERATURE: Berg & Hijman, FTEA 1989: 2, fig. 1.

SYNONYM: *Morus lactea* (Sim) Mildbr.; ITU: 263, photo 43. UFT: 94.

Tree to 35 m high. Rainforest, evergreen forest; 450–1600 m.

Bunyoro, Mengo, Kigezi, Toro, Madi, Mbale, Busoga. Widespread in Africa.

Here assessed preliminarily as Least Concern (LC) because of its wide distribution and altitude range.

Treculia africana Trécul

Literature: Berg & Hijman, FTEA 1989: 10, fig. 5 & 6. ITU: 265. UFT: 96.

Tree to 30(–50) m high. Forest, riverine forest; 500–1200 m.

Bunyoro, Mengo, Masaka, Toro. Widespread in Africa.

Here assessed preliminarily as Least Concern (LC) because of its wide distribution.

Trilepisium madagascariensis DC.

Literature: Berg & Hijman, FTEA 1989: 17, fig. 10.

Synonym: *Bosquiea phoberos* Baill.; ITU: 234. UFT: 96.

Tree to 25(–40) m high. Rainforest, evergreen forest, riverine and ground-water forest; 1100–1800 m.

Bunyoro, Kigezi, Mengo. Widespread in Africa.

Here assessed preliminarily as Least Concern (LC) because of its wide distribution and habitat range.

MORINGACEAE

Moringa oleifera Lam.

Literature: Verdcourt, FTEA 1986: 3, fig. 2.

Shrub or tree 2–10 m high. Cultivations, grassland on black clay; 500–1350 m.

West Nile, Mengo. Originally from India, now naturalised.

Here assessed preliminarily as Least Concern (LC) because of its wide distribution and habitat.

(*Moringa peregrina* in the sense of ITU, really *Moringa stenopetala* (Baker f.) Cufod., was included in ITU based on a single collection, which was from Kenya)

MYRICACEAE

Morella kandtiana (Engl.) Verdc. & Polhill

Literature: Polhill & Verdcourt, FTEA 2000: 4.

Synonym: *Myrica kandtiana* Engl.; ITU: 266. UFT: 135.

Shrub or tree 1–6 m high. Swamps, swamp forest edges; 1150–2000 m.

West Nile, Kigezi, Mengo, Ankole, Toro, Masaka. Also in Kenya, N Tanzania, DRC, Rwanda, Burundi.

Here assessed preliminarily as Least Concern (LC) because of its wide distribution (EOO 82,299 km^2; AOO 14,640 km^2; 16 locations) and altitude range.

Morella salicifolia (A. Rich.) Verdc. & Polhill

Literature: Polhill & Verdcourt, FTEA 2000: 5, fig. 1.

Synonym: *Myrica salicifolia* A. Rich.; ITU: 266. UFT: 159.

Shrub or tree 3–12 m high. Rainforest, evergreen bushland, secondary thicket, wooded grassland, heath zone; 1350–3000 m.

Toro, Kigezi, Karamoja, Mbale. Widespread in East Africa.

Here assessed preliminarily as Least Concern (LC) because of its wide distribution and habitat/altitude range.

MYRISTICACEAE

Pycnanthus angolensis (Welw.) Warb.

Literature: Verdcourt, FTEA 1997: 4, fig. 2. ITU: 267. UFT: 142.

Tree 9–38 m high. Rainforest, secondary bushland, riverine forest, open woodland; 750–1200 m.

Mbale, Mengo, Masaka, Toro, Bunyoro, Busoga. Widespread in East Africa.

Here assessed preliminarily as Least Concern (LC) because of its wide distribution and habitat range.

Staudtia kamerunensis Warb.

Literature: Verdcourt, FTEA 1997: 8, fig. 3. ITU: 268. UFT: 142.

Tree 15–35 m high. Evergreen forest; 1050–1350 m.

Bunyoro, Mengo. Widespread in Africa.

Here assessed preliminarily as Least Concern (LC) because of its wide distribution.

MYRTACEAE

Eugenia bukobensis Engl.

Literature: Verdcourt, FTEA 2001: 59. ITU: 272. UFT: 192.

Shrub or tree 1.5–9 m high. Swamp and lakeside forest, thicket, woodland; 1100–1800 m.

Ankole, Kigezi, Mengo, Masaka. Also in DRC, W Kenya, NE Tanzania, and possibly Ethiopia, the Comoros.

Here assessed preliminarily as Least Concern (LC) because of its wide distribution (EOO 25,187 km^2; AOO 23,658 km^2; 19 locations) and habitat/altitude range.

Syzygium congolense Amshoff

Literature: Verdcourt, FTEA 2001: 76.

Synonym: *Syzygium guineense* of ITU: 273, partly. UFT: 192.

Tree 6–30 high. Rainforest at lake edges; 1100–1500 m.

Kigezi, Mengo, Masaka. Widespread in central Africa.

Here assessed preliminarily as Least Concern (LC) because of its wide distribution.

Syzygium cordatum C. Krauss

Literature: Verdcourt, FTEA 2001: 73, fig. 15. ITU: 273. UFT: 192.

Shrub or tree 3–20 m high. Swamp forest, riverine woodland, open woodland; 900–2400 m.

Kigezi, Mbale, Masaka, Mengo, Ankole, Acholi, Karamoja. Widespread in Africa.

Here assessed preliminarily as Least Concern (LC) because of its wide distribution and habitat/altitude range.

Syzygium cumini (L.) Skeels

Literature: Verdcourt, FTEA 2001: 72.

Shrub or tree 6–25 m high. Cultivations; 500–1650 m.

Mengo. Originally from Asia but often naturalised.

Here assessed preliminarily as Least Concern (LC) because of its wide distribution and habitat.

Syzygium guineense (Willd.) DC.
Literature: Verdcourt, FTEA 2001: 77. ITU: 273, plate 12 & fig. 60.

Shrub or tree 0.2–30 m high. Evergreen forest, riverine forest, woodland, scattered tree grassland; 800–2250 m.

Mengo, Mubende, Acholi, Kigezi, Mbale, West Nile, Bunyoro, Busoga, Masaka, Ankole, Karamoja. Widespread in Africa.

Here assessed preliminarily as Least Concern (LC) because of its wide distribution and habitat/altitude range.

Syzygium owariense (P. Beauv.) Benth.
Literature: Verdcourt, FTEA 2001: 76. ITU: 275.

Tree 9–30 m high. Swamp forest, riverine forest; 1100–2000 m.

Masaka, Madi, Acholi, Lango, Karamoja, Mbale. Widespread in Africa.

Here assessed preliminarily as Least Concern (LC) because of its wide distribution.

OCHNACEAE

Gomphia densiflora (De Wild. & T. Durand) Verdc.
Literature: Verdcourt, FTEA 2005: 50.

Synonym: *Ouratea densiflora* De Wild. & T. Durand; ITU: 281. UFT: 131.

Shrub or tree 1–12 m high. Evergreen forest and swamp forest; 800–1500 m.

Bunyoro, Masaka, Mengo, Toro. Widespread in Africa.

Here assessed preliminarily as Least Concern (LC) because of its wide distribution.

Gomphia likimiensis (De Wild.) Verdc.
Literature: Verdcourt, FTEA 2005: 46.

Synonym: *Ouratea bukobensis* of ITU: 281, partly.

Shrub or tree 1–5 m high. Forest edges; 1050–1500 m.

Bunyoro, Mengo. Also in DRC, South Sudan, western Kenya and Tanzania.

Here assessed preliminarily as Least Concern (LC) because of its distribution (EOO > 7,627 km^2; AOO 2,527 km^2) and habitat.

Gomphia mildbraedii (Gilg) Verdc.
See restricted list, page 188.

Gomphia reticulata P. Beauv.
Literature: Verdcourt, FTEA 2005: 45.

Shrub or tree 1–12 m high. Evergreen forest and riverine forest; 500–1500 m.

Kigezi. Widespread in Africa.

Here assessed preliminarily as Least Concern (LC) because of its wide distribution and altitude range.

Gomphia vogelii Hook. f.
Literature: Verdcourt, FTEA 2005: 47.

Synonym: *Ouratea bukobensis* of ITU: 281, partly.
Synonym: *Ouratea hiernii* in the sense of UFT: 132, probably.

Shrub or tree 1–12 m high. Swamp forest, secondary evergreen forest; 1100–1200 m.

Masaka. Widespread in Africa.

Here assessed preliminarily as Least Concern (LC) because of its wide distribution and habitat.

Lophira lanceolata Keay

Literature: Verdcourt, FTEA 2005: 53.

Synonym: *Lophira alata* of ITU: 276, fig. 61.

Tree to 16 m high. Open woodland; 900–1500 m.

West Nile, Acholi, Madi. Widespread in Africa.

Here assessed preliminarily as Least Concern (LC) because of its wide distribution.

Ochna afzelii Oliv.

Literature: Verdcourt, FTEA 2005: 34. ITU: 279. UFT: 132.

Synonym: *Ochna alba* of ITU: 278.

Shrub or tree 3–12 m high. Evergreen forest, grassland, bushland; 1150–1350 m.

West Nile, Mbale, Mengo, Masaka. Widespread in Africa.

Here assessed preliminarily as Least Concern (LC) because of its wide distribution and habitat range.

(*Ochna bracteosa* of UFT is really a shrub)

Ochna holstii Engl.

Literature: Verdcourt, FTEA 2005: 23. ITU: 279 (280?). UFT: 132.

Synonym: *Ochna prunifolia* Engl.; ITU: 280.

Shrub or tree 3–27 m high. Evergreen forest and riverine forest, grassland, thicket; 900–2350 m.

Acholi, Mbale, Karamoja. Widespread in Africa.

Here assessed preliminarily as Least Concern (LC) because of its wide distribution and habitat/altitude range.

Ochna inermis (Forssk.) Schweinf.

Literature: Verdcourt, FTEA 2005: 18. ITU: 279.

Synonym: *Ochna monantha* of ITU: 280, partly.

Synonym: *Ochna ovata* of ITU: 280, partly.

Shrub or tree 1–6 m high. Dry bushland, woodland; 500–1500 m.

Karamoja. Widespread in Africa.

Here assessed preliminarily as Least Concern (LC) because of its wide distribution and habitat/altitude range.

Ochna insculpta Sleumer

Literature: Verdcourt, FTEA 2005: 14.

Synonym: *Ochna sp. near macrocalyx* of ITU: 281.

Shrub or tree 1–9 m high. Evergreen forest and forest margins, riverine forest; 1050–2100 m.

Mengo. Fairly widespread in East Africa.

Here assessed preliminarily as Least Concern (LC) because of its wide distribution and habitat/altitude range.

Ochna leucophloeos A. Rich.
See restricted list, page 189.

Ochna membranacea Oliv.
Literature: Verdcourt, FTEA 2005: 7. ITU: 280. UFT: 132.

Shrub or tree 1–12 m high. Evergreen forest and forest margins; 1050–1250 m.

Toro, Ankole, Mengo, Kigezi. Widespread in Africa.

Here assessed preliminarily as Least Concern (LC) because of its wide distribution and habitat.

Ochna ovata F. Hoffm.
Literature: Verdcourt, FTEA 2005: 19. ITU: 280.

Shrub or tree 2–9 m high. Woodland, dry forest; 500–2100 m.

Karamoja. Widespread in East Africa.

Here assessed preliminarily as Least Concern (LC) because of its wide distribution and habitat/altitude range.

Ochna schweinfurthiana F. Hoffm.
Literature: Verdcourt, FTEA 2005: 33. ITU: 281.

Shrub or tree 2–9 m high. Woodland, wooded grassland, bamboo bushland, riverine forest; 750–2100 m.

West Nile, Teso, Madi, Acholi. Widespread in Africa.

Here assessed preliminarily as Least Concern (LC) because of its wide distribution and habitat/altitude range.

Ochna sp. 40 of FTEA
See restricted list, page 190.

OLACACEAE

Heisteria parvifolia Sm.
Literature: Lucas, FTEA 1968: 3, fig. 1. UFT: 120.

Shrub or tree to 15 m high. Rainforest; ± 1170 m.

Mengo. Widespread in Africa.

Here assessed preliminarily as Least Concern (LC) because of its wide distribution.

Strombosia scheffleri Engl.
Literature: Lucas, FTEA 1968: 11, fig. 4. ITU: 282. UFT: 120.

Tree to 30 m high. Rainforest; 800–2500 m.

West Nile, Mbale, Masaka, Kigezi, Ankole, Toro, Bunyoro, Mbale. Widespread in Africa.

Here assessed preliminarily as Least Concern (LC) because of its wide distribution and altitude range.

Strombosiopsis tetrandra Engl.
Literature: Lucas, FTEA 1968: 13, fig. 5. UFT: 120.

Tree to 30 m high. Rainforest; 1200 m.

Kigezi. Widespread in Africa.

Here assessed preliminarily as Least Concern (LC) because of its wide distribution.

Edible fruits of *Ximenia americana*, Lake Opeta (photo: James Kalema).

Ximenia americana L.

LITERATURE: Lucas, FTEA 1968: 3, fig. 2. ITU: 283, fig. 62.

Shrub or tree to 15 m high. Wooded grassland, deciduous bushland; 50–1950 m.

West Nile, Mbale, Mengo, Masaka, Madi, Acholi, Lango, Teso, Karamoja, Busoga, Ankole, Toro.

Widespread in Africa, Asia and America.

Assessed as Least Concern (LC) by Sacande, Sanou & Beentje in *Guide de terrain des arbres du Burkina Faso* (in press).

Ximenia caffra Sond.

LITERATURE: Lucas, FTEA 1968: 5. ITU: 283.

Shrub or tree to 8 m high. Woodland, wooded grassland; 500–2000 m.

Acholi, Karamoja, Ankole. Widespread in Africa.

Here assessed preliminarily as Least Concern (LC) because of its wide distribution and altitude range.

OLEACEAE

Chionanthus africanus (Knobl.) Stearn

LITERATURE: For name change, see Stearn in *Bot. J. Linn. Soc.* 80: 191 (1980).

SYNONYM: *Linociera johnsonii* Baker; Turrill, FTEA 1952: 12, fig. 4. ITU: 285. UFT: 191.

Shrub or tree 3–18 m high. Rainforest; 1300–3150 m.

Bunyoro, Toro, Mengo, Masaka, Kigezi. Widespread in Africa.

Here assessed preliminarily as Least Concern (LC) because of its wide distribution and altitude range.

Chionanthus mildbraedii (Gilg & Schellenb.) Stearn

Literature: For name change, see Stearn in *Bot. J. Linn. Soc.* 80: 191 (1980).

Synonym: *Linociera latipetala* M. Taylor; ITU: 285.
Synonym: *Olea mildbraedii* (Gilg & Schellenb.) Knobl. of Turrill, FTEA 1952: 7.

Shrub, climber or tree 3–12 m high. Rainforest, swamp forest; 1200–2100 m.

Mengo, Ankole, Toro. Also in Cameroon, Kenya, Tanzania.

Here assessed preliminarily as Least Concern (LC) because of its wide distribution and altitude range.

Chionanthus niloticus (Oliv.) Stearn

Literature: For name change, see Stearn in *Bot. J. Linn. Soc.* 80: 191 (1980).

Synonym: *Linociera nilotica* Oliv.; Turrill, FTEA 1952: 14. ITU: 285.

Shrub or tree 3–18 m high. Riverine forest; 300–1650 m.

West Nile, Madi, Acholi, Karamoja. Widespread in Africa.

Assessed as Least Concern (LC) by Sacande, Sanou & Beentje in *Guide de terrain des arbres du Burkina Faso* (in press).

Olea capensis L.

Literature: Green, 2002: 108.

Synonym: *Olea hochstetteri* Baker; Turrill, FTEA 1952: 10. ITU: 286. UFT: 191.

Tree 5–24 m high. Rainforest, dry forest; 750–2650 m.

Ankole, Bunyoro, Toro, Karamoja, Kigezi. Mbale. Widespread in Africa.

Here assessed preliminarily as Least Concern (LC) because of its wide distribution and habitat/altitude range.

Olea capensis L. subsp. *welwitschii* (Knobl.) Friis & P. S. Green

Synonym: *Olea welwitschii* (Knobl.) Gilg & Schellenb.; Turrill, FTEA 1952: 12, fig. 3. UFT: 190; *Steganthus welwitschii* (Knobl.) Knobl.; ITU: 288.

Tree 12–24 m high. Rainforest, evergreen forest; 750–1950 m.

Bunyoro, Mengo, Masaka, Ankole, Toro, Mbale. Widespread in Africa.

Here assessed preliminarily as Least Concern (LC) because of its wide distribution and altitude range.

Olea europaea L.

Literature: Green, 2002: 93.

Synonym: *Olea africana* of UFT: 191.
Synonym: *Olea chrysophylla* Lam. of Turrill, FTEA 1952: 9, fig. 2. ITU: 286.

Tree 5–20 m high. Rainforest, evergreen woodland or wooded grassland; 1300–3150 m.

Toro, Masaka, Ankole, Karamoja, Mbale (Tororo Rock is also mentioned and this is in what used to be called Bukedi). Widespread in Africa and Asia.

Here assessed preliminarily as Least Concern (LC) because of its wide distribution and habitat/altitude range.

Schrebera alata (Hochst.) Welw.

LITERATURE: Turrill, FTEA 1952: 4, fig. 1. ITU: 287, plate 13. UFT: 206.

Shrub or tree 9–24 m high. Forest, bushland; 1950–2250 m.

Mbale, Karamoja. Widespread in Africa.

Here assessed preliminarily as Least Concern (LC) because of its wide distribution and habitat range.

Schrebera arborea A. Chev.

LITERATURE: Turrill, FTEA 1952: 2. UFT: 190.

SYNONYM: *Schrebera macrantha* Gilg & Schellenb.; ITU: 288, fig. 63.

Tree 18–35 m high. Forest; 750–1200 m.

Bunyoro, West Nile, Mengo, Masaka, Toro, Bunyoro, Madi. Widespread in Africa.

Here assessed preliminarily as Least Concern (LC) because of its wide distribution.

OLINIACEAE

Olinia rochetiana A. Juss.

LITERATURE: Verdcourt, FTEA 1975: 2, fig. 1.

SYNONYM: *Olinia usambarensis* Gilg; ITU: 290, plate 14. UFT: 191.

Shrub or tree 4–16 m high. Moist forest, forest edges and relicts, thicket; 1650–3000.

Karamoja, Kigezi, Mbale, Toro. Widespread in Africa.

Here assessed preliminarily as Least Concern (LC) because of its wide distribution and habitat/altitude range.

PANDACEAE

Microdesmis puberula Planch.

LITERATURE: Carter & Radcliffe-Smith, FTEA 1988: 581, fig. 108. ITU: 134.

Shrub or tree to 6 m. Rainforest; 1100 m.

Bunyoro. From Nigeria to Uganda and S to Angola.

Here assessed preliminarily as Least Concern (LC) because of its wide distribution.

PANDANACEAE

Pandanus chiliocarpus Stapf

See restricted list, page 191.

PASSIFLORACEAE

Barteria nigritana Hook. f.

LITERATURE: de Wilde, FTEA 1975: 3, fig. 1.

SYNONYM: *Barteria fistulosa* Mast.; ITU: 312.

SYNONYM: *Barteria acuminata* Baker f. of ITU: 313. UFT: 153.

Tree to 10(–25) m high, or shrub. Lake-side forest, clearings, forest edges; 1000–1500 m.

subsp. *fistulosa* (Mast.) Sleumer

Masaka, Mengo; also in Tanzania, Cameroon and DRC.

Here assessed preliminarily as Least Concern (LC) because of its wide distribution and habitat range.

Paropsia guineensis Oliv.
LITERATURE: de Wilde, FTEA 1975: 6, fig. 2. ITU: 313. UFT: 140.

Tree to 20 m high, or shrub. Rainforest, riverine forest; 1000–1500 m.

Mengo, Bunyoro. Widespread in Africa.

Here assessed preliminarily as Least Concern (LC) because of its wide distribution and habitat.

PENTAPHYLACACEAE

Balthasaria schliebenii (Melch.) Verdc.
See restricted list, page 192.

Ficalhoa laurifolia Hiern
LITERATURE: Verdcourt, FTEA 1962: 8, fig. 3. ITU: 110. UFT: 140.

Tree 6–2(–36) m high. Rainforest, riverine forest, bushland, secondary growth; 1350–2400 m.

Toro, Kigezi. Widespread in Africa.

Here assessed preliminarily as Least Concern (LC) because of its wide distribution and habitat/altitude range.

PERACEAE

Clutia abyssinica Jaub. & Spach.
LITERATURE: Radcliffe-Smith, FTEA 1987: 333, fig. 63.

Woody herb, shrub or tree to 8 m high. Forest edges and associated bushland, wooded grassland, riverine, thicket; 1000–3700 m.

Karamoja, Kigezi, Mbale, Masaka. Widespread in Africa.

Here assessed preliminarily as Least Concern (LC) because of its wide distribution and habitat range.

PHYLLANTHACEAE

Antidesma laciniatum Müll. Arg.
LITERATURE: Carter & Smith, FTEA 1988: 572. ITU: 116. UFT: 155.

Tree to 12 m high. Forest and forest edges; 950–1200 m.

Bunyoro, Mengo. Widespread in Africa.

Here assessed preliminarily as Least Concern (LC) because of its wide distribution and habitat range.

Antidesma membranaceum Müll. Arg.

LITERATURE: Carter & Smith, FTEA 1988: 574. ITU: 116. UFT: 155.

SYNONYM: *Antidesma meiocarpum* J. Léonard of ITU: 116.

Shrub or tree 1–9 m high. Forest and forest edges, associated bushland, riverine, wooded grassland; 500–1850 m.

Bunyoro, Busoga, Masaka, Mengo, Ankole, Kigezi, Acholi, Madi, Lango, Teso, Mbale. Widespread in Africa.

Here assessed preliminarily as Least Concern (LC) because of its wide distribution and habitat range.

Antidesma venosum Tul.

LITERATURE: Carter & Smith, FTEA 1988: 573.

Shrub or tree 1–9 m high. Forest and forest edges, associated bushland, woodland, wooded grassland; 500–1850 m.

Karamoja, Kigezi, Mbale, West Nile, Masaka. Widespread in Africa.

Here assessed preliminarily as Least Concern (LC) because of its wide distribution and habitat range.

Antidesma vogelianum Müll. Arg.

LITERATURE: Carter & Smith, FTEA 1988: 576, fig. 106.

Shrub or tree 1–9 m high. Forest, associated bushland; 500–1500 m.

Bunyoro, Masaka, Mengo, Busoga, Masaka. Widespread in Africa.

Here assessed preliminarily as Least Concern (LC) because of its wide distribution and habitat range.

Bridelia atroviridis Müll. Arg.

LITERATURE: Radcliffe-Smith, FTEA 1987: 125, fig. 23. ITU: 117.

Shrub or tree 2–12 m high. Forest edges, bushland, thicket; 50–1700 m.

Bunyoro, Busoga, Mengo. Widespread in Africa.

Here assessed preliminarily as Least Concern (LC) because of its wide distribution and habitat range.

Bridelia brideliifolia (Pax) Fedde

LITERATURE: Radcliffe-Smith, FTEA 1987: 126. ITU: 117. UFT: 155.

Shrub or tree to 30 m high. Evergreen forest, riverine, bushland, thicket; 1200–2450 m.

Toro, Ankole, Kigezi, Acholi. Widespread in central Africa.

Here assessed preliminarily as Least Concern (LC) because of its wide distribution and habitat range.

Bridelia micrantha (Hochst.) Baill.

LITERATURE: Radcliffe-Smith, FTEA 1987: 127. ITU: 117. UFT: 155.

Shrub, climber or tree to 18(–27) m high. Evergreen forest, riverine, lakesides, swamps, bushland, thicket; 50–2300 m.

West Nile, Kigezi, Busoga, Mengo, Masaka, Ankole, Toro, Mubende, Lango, Mbale. Widespread in Africa.

Assessed as Least Concern (LC) by Sacande, Sanou & Beentje in *Guide de terrain des arbres du Burkina Faso* (in press).

Bridelia ndellensis Beille

LITERATURE: Radcliffe-Smith, FTEA 1987: 129.

SYNONYM: *Bridelia ferruginea* Benth.; ITU: 117.

Tree 8–14 m high. Evergreen forest, swamps, bushland; 1350–2000 m.

West Nile, Acholi, Bunyoro, Toro, Madi. Widespread in central Africa.

Here assessed preliminarily as Least Concern (LC) because of its wide distribution and habitat range.

Bridelia scleroneura Müll. Arg.

LITERATURE: Radcliffe-Smith, FTEA 1987: 122.

SYNONYM: *Bridelia scleroneuroides* Pax; ITU: 118, fig. 26.

Shrub or tree to 10 m high. Woodland, thicket, wooded grassland; 750–2400 m.

Karamoja, Toro, Mbale, Mengo, Mubende, West Nile, Madi, Acholi, Lango, Teso, Busoga, Bunyoro, Kigezi, Ankole. Widespread in Africa.

Assessed as Least Concern (LC) by Sacande, Sanou & Beentje in *Guide de terrain des arbres du Burkina Faso* (in press).

Cleistanthus polystachyus Planch.

LITERATURE: Radcliffe-Smith, FTEA 1987: 133. ITU: 118. UFT: 156.

Shrub or tree 3–18 m high. Evergreen forest, riverine, semi-swamp forest; 700–1650 m.

West Nile, Ankole, Mengo, Masaka, Toro. Widespread in Africa.

Here assessed preliminarily as Least Concern (LC) because of its wide distribution and habitat range.

Flueggea virosa (Willd.) Voigt

LITERATURE: Radcliffe-Smith, FTEA 1987: 68, fig. 7. ITU: 128.

SYNONYM: *Securinega virosa* Willd.; UFT: 137.

Scandent shrub or tree 2–6 m high. Forest edges, thickets, bushland, ruderal sites; 500–2300 m.

Karamoja, Bunyoro, Mbale, Mengo, Masaka, Ankole, Toro, Mubende, Madi, Acholi, Lango, Teso, Busoga, West Nile, Kigezi. Widespread in Africa and Asia.

Here assessed preliminarily as Least Concern (LC) because of its wide distribution and habitat range.

Heywoodia lucens Sim

LITERATURE: Radcliffe-Smith, FTEA 1987: 86, fig. 12.

SYNONYM: *Heywoodia sp.* of ITU: 129, partly.

Tree to 30 m high. Riverine forest, swamp forest, evergreen forest remnants; 1150–1800 m.

Ankole. Also in Kenya, Tanzania, South Africa.

Here assessed preliminarily as Least Concern (LC) because of its wide distribution and habitat range.

Hymenocardia acida Tul.

LITERATURE: Carter & Radcliffe-Smith, FTEA 1988: 579. ITU: 130, fig. 28.

Shrub or tree to 10 m high. Woodland, wooded grassland, riverine; 600–1700 m.

West Nile, Teso, Busoga, Mengo, Mubende, Bunyoro, Madi, Acholi, Lango, Mbale. Widespread in Africa.

Here assessed preliminarily as Least Concern (LC) because of its wide distribution and habitat range.

Maesobotrya floribunda Benth.

LITERATURE: Radcliffe-Smith, FTEA 1987: 112, fig. 20.

SYNONYM: *Maesobotrya purseglovei* Verdc.; ITU: 133. UFT: 137.

Tree to 20 m high. Moist forest; 1250–1550 m.

Kigezi. Widespread in central Africa.

Here assessed preliminarily as Least Concern (LC) because of its wide distribution.

Margaritaria discoidea (Baill.) G. L. Webster

LITERATURE: Radcliffe-Smith, FTEA 1987: 63, fig. 64.

SYNONYM: *Phyllanthus discoideus* (Baill.) Müll. Arg.; ITU: 136. UFT: 155.

Shrub or tree 1–25 m high. Forest, wooded grassland, bushland, woodland; 500–1900 m.

West Nile, Teso, Mengo, Acholi, Bunyoro, Karamoja, Kigezi, Toro, Madi, Lango, Mbale, Busoga. Widespread in Africa.

Here assessed preliminarily as Least Concern (LC) because of its wide distribution and habitat range.

Phyllanthus inflatus Hutch.

LITERATURE: Radcliffe-Smith, FTEA 1987: 25.

SYNONYM: *Phyllanthus polyanthus* in the sense of ITU: 137.

Shrub or tree to 12 m high. Rainforest edges, riverine forest; 750–1850 m.

Bunyoro, Ankole, Mbale. Widespread in Africa.

Here assessed preliminarily as Least Concern (LC) because of its wide distribution and habitat range.

Phyllanthus müllerianus (Kuntze) Exell

LITERATURE: Radcliffe-Smith, FTEA 1987: 24.

SYNONYM: *Phyllanthus floribundus* (Baill.) Müll. Arg.; ITU: 137.

Shrub or tree 2–12 m high. Wooded grassland, woodland, riverine forest; 250–1550 m.

Acholi, Bunyoro, Mbale, West Nile, Madi Lango. Widespread in Africa.

Here assessed preliminarily as Least Concern (LC) because of its wide distribution and habitat range.

Phyllanthus ovalifolius Forssk.

LITERATURE: Radcliffe-Smith, FTEA 1987: 32, fig. 4.

SYNONYM: *Phyllanthus guineensis* Pax; ITU: 137.

Shrub, climber or tree to 9 m high. Forest edges, riverine and thicket; 350–2000 m.

Karamoja, Ankole, Mengo, Masaka, Lango. Widespread in Africa.

Here assessed preliminarily as Least Concern (LC) because of its wide distribution and habitat range.

Phyllanthus reticulatus Poir.

Literature: Radcliffe-Smith, FTEA 1987: 34. ITU: 137.

Shrub or tree to 4.5(–18) m high. Forest, lake and swamp edges, riverine, bushland, woodland; 500–1300 m.

Toro, Teso, Mengo, Ankole. Widespread in Africa and Asia.

Here assessed preliminarily as Least Concern (LC) because of its wide distribution and habitat range.

Spondianthus preussii Engl.

Literature: Radcliffe-Smith, FTEA 1987: 105, fig. 17. ITU: 140. UFT: 154.

Tree 9–18 m high. Swamp forest, forest; 1100–1250 m.

Bunyoro, Masaka, Mengo, Madi. Widespread in Africa.

Here assessed preliminarily as Least Concern (LC) because of its wide distribution.

Thecacoris lucida (Pax) Hutch.

Literature: Radcliffe-Smith, FTEA 1987: 107, fig. 18. ITU: 143. UFT: 156.

Shrub or tree to 12 m high. Moist forest; 700–1200 m.

Bunyoro, Toro, Mengo, Kigezi. Widespread in Central Africa.

Here assessed preliminarily as Least Concern (LC) because of its wide distribution.

Uapaca paludosa Aubrév. & Leandri

Literature: Carter & Smith, FTEA 1988: 571.

Synonym: *Uapaca guineensis* in the sense of ITU: 143, photo 24. UFT: 154.

Tree to 40 m high. Swamp forest, riverine and lakeside forest, rainforest; 600–1400 m.

Masaka, Mengo, Kigezi. Widespread in Africa.

Here assessed preliminarily as Least Concern (LC) because of its wide distribution and habitat range.

Uapaca sansibarica Pax

Literature: Carter & Smith, FTEA 1988: 568, fig. 105. ITU: 143.

Tree to 15 m high. Woodland, wooded grassland, bushland, riverine forest; 500–1850 m.

West Nile, Acholi, Madi. Widespread in Africa.

Here assessed preliminarily as Least Concern (LC) because of its wide distribution and habitat/altitude range.

PITTOSPORACEAE

Pittosporum abyssinicum Delile

Literature: fide African Plants Database

Synonym: *Pittosporum lanatum* Hutch. & E. A. Bruce; Cufodontis, FTEA 1966: 12. ITU: 313.

Tree 3–19 m high. Rainforest, dry forest, grassland; 2250–2850 m.

Karamoja. Also in Ethiopia, Kenya, Tanzania.

Here assessed preliminarily as Least Concern (LC) because of its wide distribution and habitat range.

Pittosporum viridiflorum Sims

Literature: Cufodontis, FTEA 1966: 3. UFT: 158.

Synonym: *Pittosporum mannii* Hook. f.; Cufodontis, FTEA 1966: 8. UFT: 158.
Synonym: *Pittosporum quartinianum* Cufod. of ITU: 314.
Synonym: *Pittosporum ripicolum* J. Léonard of ITU: 314.
Synonym: *Pittosporum spathicalyx* De Wild. of Cufodontis, FTEA 1966: 9. ITU: 314. UFT: 158.

Shrub or tree to 20 m high. Dry evergreen forest, riverine forest, wooded grassland; 900–2400 m.

Karamoja, Mbale, Busoga, Mengo, Masaka, Ankole, Kigezi. Widespread in Africa.

Here assessed preliminarily as Least Concern (LC) because of its wide distribution and habitat/altitude range.

POACEAE/GRAMINEAE

Oreobambos buchwaldii K. Schum.

Literature: Clayton, FTEA 1970: 13, fig. 4. ITU: 151. UFT: 78.

Bamboo 4–18 m high. Open forest, often by streams; 300–1950 m.

Bunyoro, Mengo, Masaka, Busoga. Also in Tanzania, Zambia, Malawi.

Here assessed preliminarily as Least Concern (LC) because of its wide distribution and altitude range.

Oxytenanthera abyssinica (A. Rich.) Munro

Literature: Clayton, FTEA 1970: 11, fig. 3. ITU: 151, photo 26.

Bamboo 3–10 m high. Wooded grassland, often along streams; 500–2000 m.

West Nile, Mbale, Bunyoro, Acholi, Karamoja. Widespread in Africa.

Here assessed preliminarily as Least Concern (LC) because of its wide distribution and altitude range.

Sinarundinaria alpina (K. Schum.) C. S. Chao & Renvoize

Literature: Name changed by Chao & Renvoize in *Kew Bull.* 44: 349 (1989).

Synonym: *Arundinaria alpina* K. Schum.; Clayton, FTEA 1970: 9, fig. 2. ITU: 151, photo 25. UFT: 78.

Bamboo 2–20 m high. Montane thicket and forest; (1800–)2400–3000 m.

Toro, Kigezi, Mbale. Widespread in Africa.

Here assessed preliminarily as Least Concern (LC) because of its wide distribution and habitat range.

PODOCARPACEAE

Afrocarpus dawei (Stapf) C. N. Page

See restricted list, page 193.

Afrocarpus gracilior (Pilg.) C. N. Page

LITERATURE: Farjon 2001: 251.

SYNONYM: *Podocarpus gracilior* Pilg.; Melville, FTEA 1958: 12. ITU: 315, fig. 66. UFT: 75.

Tree to 30 m high. Upland forest; 1500–2400 m.

Karamoja, Mbale. Also in DRC, South Sudan, Ethiopia, Kenya, Tanzania and down to South Africa.

Here assessed preliminarily as Least Concern (LC) because of its wide distribution and altitude range.

Podocarpus latifolius (Thunb.) Mirb.

SYNONYM: *Podocarpus milanjianus* Rendle; Melville, FTEA 1958: 11. ITU: 317, fig. 66. UFT: 74.

Tree to 35 m high. Upland forest; 900–3150 m.

Acholi, Toro, Mbale, Masaka, Kigezi. Widespread in Africa. Widely distributed and common throughout East Africa and Uganda.

Here assessed preliminarily as Least Concern (LC) because of its wide distribution and altitude range.

POLYGALACEAE

Carpolobia goetzei Gürke

LITERATURE: Paiva, FTEA 2007: 3.

Shrub or tree to 5 m. Forest, riverine forest, woodland; 500–1400 m.

Ankole, Toro, Mengo. Widespread in Africa.

Here assessed preliminarily as Least Concern (LC) because of its wide distribution and habitat range.

Securidaca longipedunculata Fresen.

LITERATURE: Paiva, FTEA 2007: 6. ITU: 318, fig. 67.

Shrub or tree to 6 m high. Wooded grassland; 500–1700 m.

Mengo, Masaka, Mubende, Ankole, Bunyoro, West Nile, Madi, Acholi, Lango, Teso, Mbale, Busoga. Widespread in Africa.

Here assessed preliminarily as Least Concern (LC) because of its wide distribution and altitude range.

PRIMULACEAE

Ardisia staudtii Gilg

LITERATURE: Halliday, FTEA 1984: 17, fig. 5.

Shrub or tree 1–2.4 m high. Rainforest; 1150–1200 m.

Kigezi, Masaka, Mengo. Widespread in central Africa.

Here assessed preliminarily as Least Concern (LC) because of its wide distribution.

(*Embelia nilotica* of ITU is really a climbing shrub, as is *Embelia schimperi* of ITU: 269)

Maesa lanceolata Forssk.

Literature: Halliday, FTEA 1984: 3. ITU: 269, fig. 59. UFT: 134.

Shrub or tree 1–24 m high. Secondary and riverine forest, thicket, bushland; 350–2550 m.

Karamoja, Kigezi, Masaka, Mengo, Bunyoro. Widespread in Africa and Arabia.

Here assessed preliminarily as Least Concern (LC) because of its wide distribution and habitat/altitude range.

Maesa welwitschii Gilg

Literature: Halliday, FTEA 1984: 5, fig. 1.

Shrub, climber or tree 1–2(–6) m high. Forest, secondary woodland edges; 1050–1200 m.

Mbale, Mengo. Widespread in Africa.

Here assessed preliminarily on a global scale as Least Concern (LC) because of its wide distribution.

Myrsine africana L.

Literature: Halliday, FTEA 1984: 6, fig. 2.

Shrub or tree 1–6 m high. Forest edges, wooded grassland, stony hillsides; 1200–3600 m.

Karamoja. Widespread in Africa and Asia.

Here assessed preliminarily as Least Concern (LC) because of its wide distribution and habitat/altitude range.

Rapanea melanophloeos (L.) Mez

Literature: Halliday, FTEA 1984: 8, fig. 3.

Synonym: *Rapanea neurophylla* (Gilg) Mez; ITU: 271.
Synonym: *Rapanea pulchra* Gilg & Schellenb.; ITU: 271.
Synonym: *Rapanea rhododendroides* of ITU: 271, photo. 45. UFT: 159.

Shrub or tree 2–4.5 m high. Evergreen, swamp and riverine forest, thicket, woodland, grassland; 900–3750 m.

Toro, Mbale, Masaka, Acholi, Karamoja, Kigezi. Widespread in Africa.

Here assessed preliminarily as Least Concern (LC) because of its wide distribution and habitat/altitude range.

PROTEACEAE

Faurea arborea Engl.

Literature: Brummitt & Marner, FTEA 1993: 7. ITU: 320.

Shrub or tree to 15 m high. Dry forest edges, grassland, rocky outcrops, heath scrub; 1600–3100 m.

Karamoja. Widespread in Africa.

Here assessed preliminarily as Least Concern (LC) because of its wide distribution and habitat/altitude range.

Faurea rochetiana (A. Rich.) Pic. Serm.

Literature: Brummitt & Marner, FTEA 1993: 3.

Synonym: *Faurea speciosa* Welw.; ITU: 321.

Shrub or tree to 10 m high. Woodland, wooded grassland, rocky slopes; 900–2400 m.

Acholi, Karamoja, Ankole, Toro, West Nile, Mbale. Widespread in Africa.

Here assessed preliminarily as Least Concern (LC) because of its wide distribution and habitat/altitude range.

Faurea wentzeliana Engl.

LITERATURE: Brummitt & Marner, FTEA 1993: 6, fig. 1.

SYNONYM: *Faurea saligna* in the sense of ITU: 320, plate 17. UFT: 159.

Tree to 35 m high. Rainforest; 1400–3000 m.

Toro, Ankole, Kigezi, Mbale, Karamoja. Widespread in central Africa.

Here assessed preliminarily as Least Concern (LC) because of its wide distribution and altitude range.

Protea caffra Meisn.

LITERATURE: Brummitt & Marner, FTEA 1993: 15.

SYNONYM: *Protea kilimandscharica* Engl.; ITU: 321. UFT: 159.

Shrub or tree 1–6 m high. Heath scrub, stream-banks, rocky hillsides, grassland, forest edges; 2300–3700 m.

Karamoja, Mbale. Widespread in Africa.

Here assessed preliminarily as Least Concern (LC) because of its wide distribution and habitat/altitude range.

Protea gaguedi J. F. Gmel.

LITERATURE: Brummitt & Marner, FTEA 1993: 17.

SYNONYM: *Protea abyssinica* Willd.; ITU: 321.

Shrub or tree to 1–8 m high. Grassland, woodland, heathland, rocky sites; 900–2100 m.

Acholi, Karamoja, Mbale. Widespread in Africa.

Here assessed preliminarily as Least Concern (LC) because of its wide distribution and habitat/altitude range.

Protea madiensis Oliv.

LITERATURE: Brummitt & Marner, FTEA 1993: 10, fig. 2. ITU: 322, photo 51.

Shrub or tree 1–5 m high. Grassland, woodland; (500–)1500–2150 m.

Karamoja, Ankole, Masaka, Kigezi, Bunyoro, West Nile, Madi, Acholi, Lango, Teso, Mbale, Busoga. Widespread in Africa.

Here assessed preliminarily as Least Concern (LC) because of its wide distribution and habitat range.

Protea welwitschii Engl.

LITERATURE: Brummitt & Marner, FTEA 1993: 19.

SYNONYM: *Protea melliodora* Engl. & Gilg; ITU: 322.

Shrub or tree to 1–3 m high. Grassland, woodland; 1200–2900 m.

Kigezi. Widespread in Africa.

Here assessed preliminarily as Least Concern (LC) because of its wide distribution and habitat/altitude range.

PUTRANJIVACEAE

Drypetes bipindensis (Pax) Hutch.
Literature: Radcliffe-Smith, FTEA 1987: 96. ITU: 124. UFT: 140.

Shrub or tree to 8(–15) m high. Moist forest; 1200–1550 m.

Ankole, Kigezi. Widespread in central Africa.

Here assessed preliminarily as Least Concern (LC) because of its wide distribution.

Drypetes calvescens Pax & K. Hoffm.
Literature: Radcliffe-Smith, FTEA 1987: 96.

Tree to 8 m high. Moist forest; 1200 m.

Mengo. Also in Cameroun and E DRC.

Here assessed preliminarily as Least Concern (LC) because of its wide distribution.

Drypetes gerrardii Hutch.
Literature: Radcliffe-Smith, FTEA 1987: 97. UFT: 138.

Synonym: *Drypetes sp. nov.* of ITU: 124.
Synonym: *D. sp. aff. battiscombei* of ITU: 124.
Synonym: *D. sp. aff. principum* of ITU: 124.
Synonym: *Drypetes sp.* (*183*) of UFT: 138.

Shrub or tree to 30 m high. Moist and dry forest, riverine forest; 1150–2300 m.

Ankole, Busoga, Kigezi, Mbale, Mengo, Bunyoro, Mbale, Toro. Widespread in Africa.

Here assessed preliminarily as Least Concern (LC) because of its wide distribution and habitat range.

Drypetes ugandensis (Rendle) Hutch.
See restricted list, page 194.

RHAMNACEAE

Berchemia discolor (Klotzsch) Hemsl.
Literature: Johnston, FTEA 1972: 32, fig. 10. ITU: 323.

Shrub or tree to 10(–25) m high. Thicket, semi-desert grassland, wooded grassland; 500–2000 m.

Karamoja, Mengo. Widespread in Africa.

Here assessed preliminarily as Least Concern (LC) because of its wide distribution and habitat/altitude range.

Lasiodiscus pervillei Baill.
Synonym: *Lasiodiscus mildbraedii* of Johnston, FTEA 1972: 5, fig. 2. ITU: 323. UFT: 174 (misapplied name).

Shrub or tree 1–5 m high. Rainforest, semi-swamp forest; 500–1600 m.

Toro, Busoga, Masaka, Mengo, Kigezi, Bunyoro, Madi. Also in Tanzania and Mozambique.

Here assessed preliminarily as Least Concern (LC) because of its wide distribution and altitude range.

Maesopsis eminii Engl.

LITERATURE: Johnston, FTEA 1972: 38, fig. 12. ITU: 323, photo 52, fig. 68. UFT: 140.

Tree 15–25(–42) m high. Rainforest, riverine forest; 800–1200 m.

Bunyoro, Busoga, Mengo, Masaka, Ankole, Kigezi, Mubende, Toro. Widespread in Africa.

Here assessed preliminarily as Least Concern (LC) because of its wide distribution.

Rhamnus prinioides L'Hér.

LITERATURE: Johnston, FTEA 1972: 18, fig. 6. ITU: 325. UFT: 135.

Shrub or tree to 8 m high. Forest, evergreen bushland, thicket; 700–3700 m.

Kigezi, Mbale, Mengo, Masaka, Karamoja, Busoga. Widespread in Africa.

Here assessed preliminarily as Least Concern (LC) because of its wide distribution and habitat/altitude range.

Rhamnus staddo A. Rich.

LITERATURE: Johnston, FTEA 1972: 18.

Shrub or tree 0.5–7 m high. Forest margins, evergreen bushland; 1000–3000 m.

Ankole, Kigezi, Masaka. Widespread in Africa.

Here assessed preliminarily as Least Concern (LC) because of its wide distribution and habitat/altitude range.

Scutia myrtina (Burm. f.) Kurz

LITERATURE: Johnston, FTEA 1972: 21, fig. 7. ITU: 326.

Shrub, climber or rarely a tree 2–5 m high. Forest margins, bushland, thicket, wooded grassland; 500–2700 m.

Karamoja, Ankole, Mengo, Masaka, Toro, Teso, Mbale, Busoga. Widespread in Africa and Asia.

Here assessed preliminarily as Least Concern (LC) because of its wide distribution and habitat/altitude range.

Ziziphus abyssinica A. Rich.

LITERATURE: Johnston, FTEA 1972: 27, fig. 9 ITU: 326, fig. 69.

Shrub or tree 3–6 m high. Scattered tree grassland; 400–2200 m.

West Nile, Teso, Mengo, Madi, Acholi, Lango, Karamoja, Mbale, Busoga, Toro. Widespread in Africa.

Here assessed preliminarily as Least Concern (LC) because of its wide distribution and altitude range.

Ziziphus mauritiana Lam.

LITERATURE: Johnston, FTEA 1972: 29. ITU: 328.

Shrub or tree 3–8(–16) m high. Ruderal sites; 500–1400 m.

Acholi, Lango, Karamoja, Teso. Widespread in the tropics.

Assessed as Least Concern (LC) by Sacande, Sanou & Beentje in *Guide de terrain des arbres du Burkina Faso* (in press).

Ziziphus mucronata Willd.

LITERATURE: Johnston, FTEA 1972: 25. ITU: 328.

Shrub or tree to 15(–30) m high. Open woodland, wooded grassland; 500–2000 m.

Karamoja, Ankole. Widespread in Africa.

Assessed as Least Concern (LC) by Sacande, Sanou & Beentje in *Guide de terrain des arbres du Burkina Faso* (in press).

Ziziphus pubescens Oliv.

Literature: Johnston, FTEA 1972: 24. ITU: 329.

Shrub or tree to 10(–20) m high. Riparian in wooded grassland; 500–1000 m.

Lango, Bunyoro, Mengo, Teso, Toro, West Nile, Acholi, Ankole. Widespread in Africa.

Here assessed preliminarily as Least Concern (LC) because of its wide distribution.

Ziziphus spina-christi (L.) Desf.

Literature: Johnston, FTEA 1972: 30.

Tree to 10 m high. Ruderal sites; 500–1300 m.

Karamoja. Widespread in Africa and Middle East.

Assessed as Least Concern (LC) by Sacande, Sanou & Beentje in *Guide de terrain des arbres du Burkina Faso* (in press).

RHIZOPHORACEAE

Cassipourea congoensis DC.

Literature: Lewis, FTEA 1956: 12. ITU: 330. UFT: 174.

Shrub or small tree to 13 m high. Evergreen forest; 1300–1950 m.

Kigezi, Ankole. Widespread in Africa.

Here assessed preliminarily as Least Concern (LC) because of its wide distribution.

Cassipourea gummiflua Tul.

Literature: Lewis, FTEA 1956: 15. UFT: 174.

Synonym: *Cassipourea ugandensis* (Stapf) Alston; ITU: 331.

Synonym: *Cassipourea sp. nov.* of ITU: 331.

Shrub or tree to 25 m high. Rain- and swamp forest; 1200–1500 m.

Toro, Kigezi, Masaka, Mengo, Bunyoro. Widespread in Africa.

Here assessed preliminarily as Least Concern (LC) because of its wide distribution.

Cassipourea malosana (Baker) Alston

Literature: Lewis, FTEA 1956: 11. UFT: 173.

Synonym: *Cassipourea abyssinica* (Engl.) Alston; ITU: 330.

Tree 12–45 m high. Dry and moist forest; 1700–2600 m.

Karamoja, Mbale, Acholi. Fairly widespread in NE, S central and East Africa.

Here assessed preliminarily as Least Concern (LC) because of its wide distribution and habitat/altitude range.

Cassipourea ruwensorensis (Engl.) Alston

Literature: Lewis, FTEA 1956: 14, fig. 4. ITU: 331. UFT: 174.

Synonym: *Cassipourea elliottii* (Engl.) Alston; ITU: 330.

Tree 12–45 m high. Evergreen and semi-swamp forest, elephant grassland; 1200–1700 m.

Bunyoro, Ankole, Busoga, Karamoja, Mbale, Toro, Kigezi. Fairly widespread in NE, central and East Africa.

Here assessed preliminarily as Least Concern (LC) because of its wide distribution and habitat range.

ROSACEAE

Hagenia abyssinica (Bruce) J. F. Gmel.

LITERATURE: Graham, FTEA 1960: 43, fig. 5. UFT: 207.

SYNONYM: *Hagenia anthelmintica* J. F. Gmel.; ITU: 332, photo 53.

Tree to 20 m high. Rainforest, bamboo zone, forming its own zone near the tree-line; 2400–3600 m.

Acholi, Kigezi, Mbale, Toro, Karamoja. Fairly widespread in NE, S central and East Africa.

Here assessed preliminarily as Least Concern (LC) because of its wide distribution and habitat.

Prunus africana (Hook. f.) Kalkman

LITERATURE: UFT: 141.

SYNONYM: *Pygeum africanum* Hook. f.; Graham, FTEA 1960: 45, fig. 6. ITU: 335.

Tree to 36 m high. Rainforest, riverine forest, woodland; 900–3000 m.

Acholi, Kigezi, Mbale, Mengo, Masaka, Ankole, Toro, Bunyoro, Madi, Karamoja. Widespread in Africa and Madagascar.

The stem bark is being harvested through unregulated extraction for the pharmaceutical industry in Europe. Assessed by IUCN as Vulnerable (VU A1cd), but we agree with the unofficial assessment published in Cheek (1998). In this he argues that the species is not remotely in danger of extinction, so long as some montane forest survives somewhere within the species' enormous range. He suggests a rating of LR/nt at best. We would assess this as Near Threatened (NT A1) due to widespread intense harvesting of the bark for medicinal purposes, causing severe population reduction overall.

RUBIACEAE

Afrocanthium lactescens (Hiern) Lantz

SYNONYM: *Canthium lactescens* Hiern.; Verdcourt & Bridson, FTEA 1991: 871, fig. 155. ITU: 339.

Tree 2.5–9 m high. Riverine woodland, thicket, rocky hillsides; 1000–2300 m.

Karamoja, Ankole, Mengo. Widespread in Africa.

Here assessed preliminarily as Least Concern (LC) because of its wide distribution and habitat/altitude range.

Aidia micrantha (K. Schum.) F. White

LITERATURE: Bridson & Verdcourt, FTEA 1988: 487. UFT: 187.

SYNONYM: *Randia lucidula* in the sense of ITU: 355.

SYNONYM: *Randia msonju* K. Krause; ITU: 355.

Shrub or tree 2–9 m high. Evergreen forest, semi-swamp forest, riverine forest; 1100–1800 m.

Ankole, Kigezi, Masaka. Widespread in S central Africa.

Here assessed preliminarily as Least Concern (LC) because of its wide distribution and habitat range.

Belonophora hypoglauca (Hiern) A. Chev.
LITERATURE: Bridson & Verdcourt, FTEA 1988: 728, fig. 127. ITU: 338. UFT: 182.

Shrub or tree 3–12 m high. Forest, woodland; 1050–1600 m.

West Nile, Bunyoro, Mengo. Widespread in Africa.

Here assessed preliminarily as Least Concern (LC) because of its wide distribution and habitat range.

Bertiera naucleoides (S. Moore) Bridson
LITERATURE: Bridson & Verdcourt, FTEA 1988: 482.

Shrub or tree 2–5 m high. Evergreen forest; 1100–1200 m.

Masaka, Mengo. Widespread in central Africa.

Here assessed preliminarily as Least Concern (LC) because of its wide distribution.

Bertiera racemosa (G. Don) K. Schum.
LITERATURE: Bridson & Verdcourt, FTEA 1988: 480. ITU: 338. UFT: 182.

Tree 2–5 m high. Swamp forest, lakeshore forest, forest edges; 1100–1200 m.

Masaka, Mengo. Widespread in Africa.

Here assessed preliminarily as Least Concern (LC) because of its wide distribution and habitat.

Calycosiphonia spathicalyx (K. Schum.) Robbr.
LITERATURE: Bridson & Verdcourt, FTEA 1988: 727, fig. 126.
SYNONYM: *Coffea spathicalyx* K. Schum; UFT: 188.

Shrub or tree 2–9 m high. Forest; 250–950 m.

West Nile, Toro. Widespread in Africa.

Here assessed preliminarily as Least Concern (LC) because of its wide distribution and altitude range.

Canthium oligocarpum Hiern
LITERATURE: Verdcourt & Bridson, FTEA 1991: 877, fig. 156.

Shrub or tree 1.5–20 m high. Forest; 1800–2600 m.

Karamoja, Toro, Mbale. Fairly widespread in Africa.

Here assessed preliminarily as Least Concern (LC) because of its wide distribution.

Catunaregam nilotica (Stapf) Tirveng.
LITERATURE: Bridson & Verdcourt, FTEA 1988: 497.
SYNONYM: *Randia nilotica* Stapf; ITU: 356.

Shrub or tree 2–6 m high. Riverine bushland, thicket edges, woodland, scattered tree grassland; 500–1200 m.

West Nile, Acholi. Widespread in Africa.

Here assessed preliminarily as Least Concern (LC) because of its wide distribution and habitat range.

Chassalia subochreata (De Wild.) Robyns
LITERATURE: Verdcourt, FTEA 1976: 131.

Shrub or tree 1.8–9 m high. Evergreen and riverine forest; 1650–2550 m.

Kigezi, Ankole and Mengo; Known from Bwindi, Kalinzu and Nabugabo. Also in W Kenya, W Tanzania, E DRC, Burundi.

Here assessed preliminarily as Least Concern (LC); distributed as far as 8° S and 38° E (EOO 749,222 km^2; AOO 271,140 km^2; 34 locations), with a wide altitude range.

Coffea canephora A. Froehner

LITERATURE: Bridson & Verdcourt, FTEA 1988: 710. ITU: 340. UFT: 182.

Shrub or tree 3–9 m high. Forest; 700–1400 m.

West Nile, Toro, Mbale, Mengo, Masaka, Bunyoro, Madi, Busoga. Widespread in Africa.

Here assessed preliminarily as Least Concern (LC) because of its wide distribution.

Coffea eugenioides S. Moore

LITERATURE: Bridson & Verdcourt, FTEA 1988: 713. UFT: 188.

Shrub or tree 1–5 m high. Forest; 1050–2100 m.

West Nile, Ankole, Mengo, Toro, Kigezi, Ankole. Widespread in central Africa.

Here assessed preliminarily as Least Concern (LC) because of its wide distribution.

Coffea liberica Hiern

LITERATURE: Bridson & Verdcourt, FTEA 1988: 706. UFT: 181.

SYNONYM: *Coffea excelsa* A. Chev.; ITU: 341.

Shrub or tree 3–20 m high. Forest; 750–1250 m.

West Nile, Toro, Madi, Acholi. Widespread in Africa.

Here assessed preliminarily as Least Concern (LC) because of its wide distribution.

Coptosperma graveolens (S. Moore) Degreef

SYNONYM: *Tarenna graveolens* (S. Moore) Bremek.; Bridson & Verdcourt, FTEA 1988: 598. ITU: 357.

Shrub or tree 2–7 m high. Thicket, bushland; 500–2150 m.

Karamoja, Bunyoro, Busoga, Ankole, Toro, West Nile, Madi, Acholi. Widespread in central Africa.

Here assessed preliminarily as Least Concern (LC) because of its wide distribution and habitat/altitude.

Craterispermum schweinfurthii Hiern

LITERATURE: Verdcourt, FTEA 1976: 162, fig. 16.

SYNONYM: *Craterispermum laurinum* in the sense of ITU: 341. UFT: 181.

Shrub or tree 2–15 m high. Evergreen forest along lakes or streams, swamp forest, drier forest, thickets; 1050–1500 m.

West Nile, Bunyoro, Mengo, Masaka, Ankole, Toro. Widespread in Africa.

Here assessed preliminarily as Least Concern (LC) because of its wide distribution and habitat range.

Cremaspora triflora (Thonn.) K. Schum.

LITERATURE: Bridson & Verdcourt, FTEA 1988: 733.

Shrub or tree 2–9 m high. Evergreen forest, thicket; 850–1950 m.

West Nile, Bunyoro, Mengo. Widespread in Africa.

Here assessed preliminarily as Least Concern (LC) because of its wide distribution and habitat/altitude range.

Crossopteryx febrifuga (G. Don) Benth.

Literature: Bridson & Verdcourt, FTEA 1988: 457, fig. 67. ITU: 342.

Tree 2–15 m high. Woodland, wooded grassland; 500–1350 m.

West Nile, Teso, Mengo, Madi, Acholi, Lango. Widespread in Africa.

Assessed as Least Concern (LC) by Sacande, Sanou & Beentje in *Guide de terrain des arbres du Burkina Faso* (in press).

Dictyandra arborescens Hook. f.

Literature: Bridson & Verdcourt, FTEA 1988: 686, fig. 117. ITU: 342. UFT: 184.

Shrub or tree 2–11 m high. Evergreen forest; 1050–1500 m.

Bunyoro, Kigezi, Mengo, Masaka, Ankole. Widespread in Africa.

Here assessed preliminarily as Least Concern (LC) because of its wide distribution.

Euclinia longiflora Salisb.

Literature: Bridson & Verdcourt, FTEA 1988: 495, fig. 79.

Shrub or tree 2–6 m high. Forest; ± 1200 m.

Toro. Widespread in Africa.

Here assessed preliminarily as Least Concern (LC) because of its wide distribution.

Galiniera saxifraga (Hochst.) Bridson

Literature: Bridson & Verdcourt, FTEA 1988: 696, fig. 119.

Synonym: *Galiniera coffeoides* Delile; ITU: 342. UFT: 185.

Shrub or tree 2–14 m high. Evergreen forest; 1700–3000 m.

West Nile, Kigezi, Mbale, Ankole. Widespread in central Africa.

Here assessed preliminarily as Least Concern (LC) because of its wide distribution and altitude range.

Gardenia erubescens Stapf & Hutch.

Literature: Bridson & Verdcourt, FTEA 1988: 505. ITU: 343.

Shrub or tree 1.5–3 m high. Wooded grassland; 1300 m.

West Nile, Madi, Karamoja. Widespread in Africa.

Assessed as Least Concern (LC) by Sacande, Sanou & Beentje in *Guide de terrain des arbres du Burkina Faso* (in press).

Gardenia imperialis K. Schum.

Literature: Bridson & Verdcourt, FTEA 1988: 511. ITU: 343.

Shrub or tree 3–12 m high. Forest edges, grassland by lakes and rivers and swamps; 1100–1300 m.

West Nile, Masaka. Widespread in Africa.

Here assessed preliminarily as Least Concern (LC) because of its wide distribution and habitat.

Gardenia ternifolia Schumach. & Thonn.

Literature: Bridson & Verdcourt, FTEA 1988: 508.

SYNONYM: *Gardenia jovis-tonantis* (Welw.) Hiern; ITU: 344, fig. 71.

Shrub or tree 1–6 m high. Scattered tree grassland, bushland, woodland; 500–2100 m.

West Nile, Acholi, Ankole, Teso, Mengo, Masaka, Madi, Lango, Karamoja, Mbale, Busoga, Bunyoro, Toro. Widespread in Africa.

Here assessed preliminarily as Least Concern (LC) because of its wide distribution and habitat/ altitude range.

Gardenia volkensii K. Schum.

LITERATURE: Bridson & Verdcourt, FTEA 1988: 507, fig. 81.

Shrub or tree 1–10 m high. Scattered tree grassland, thicket, dry bushland, woodland; 500–1950 m.

Karamoja. Widespread in Africa.

Here assessed preliminarily as Least Concern (LC) because of its wide distribution and habitat/ altitude range.

Grumilea buchananii in the sense of ITU (*Thomas* 845, Sese Is.) is a misapplied name — it is unclear to us which species of *Psychotria* this would be.

Hallea rubrostipulata (K. Schum.) J.-F. Leroy

LITERATURE: Bridson & Verdcourt, FTEA 1988: 449, fig. 64.

SYNONYM: *Mitragyna rubrostipulata* (K. Schum.) Havil.; ITU: 348. UFT: 180.

Tree 12–18 m high. Swamp forest, evergreen forest; 900–2200 m.

Ankole, Kigezi, Masaka, Toro, West Nile. Fairly widespread in Africa.

Here assessed preliminarily as Least Concern (LC) because of its wide distribution.

Hallea stipulosa (DC.) J.-F. Leroy

LITERATURE: Bridson & Verdcourt, FTEA 1988: 447.

SYNONYM: *Mitragyna stipulosa* (DC.) Kuntze; ITU: 349, fig. 73. UFT: 179.

Tree 6–42 m high. Swamp forest, evergreen forest; 1050–1200 m.

Bunyoro, Mengo, Toro, West Nile, Madi, Acholi. Widespread in Africa.

Here assessed preliminarily as Least Concern (LC) because of its wide distribution.

Heinsenia diervilleoides K. Schum.

LITERATURE: Bridson & Verdcourt, FTEA 1988: 730, fig. 128. ITU: 346. UFT: 187.

Shrub or tree 3–12 m high. Moist forest; 350–2400 m.

Ankole, Mbale, Mengo, Toro. Widespread in Africa.

Here assessed preliminarily as Least Concern (LC) because of its wide distribution and altitude range.

Hymenodictyon floribundum (Hochst. & Steud.) B. L. Rob.

LITERATURE: Bridson & Verdcourt, FTEA 1988: 452, fig. 66. ITU: 346, fig. 72.

Shrub or tree 4–9 m high. Rocky areas, woodland, wooded grassland; 500–2250 m.

Acholi, Ankole, Teso, Masaka, Kigezi, Mubende, Toro, West Nile, Madi, Karamoja, Mbale, Busoga. Widespread in Africa.

Here assessed preliminarily as Least Concern (LC) because of its wide distribution and habitat/ altitude range.

Hymenodictyon parvifolium Oliv.
LITERATURE: Bridson & Verdcourt, FTEA 1988: 454.
SYNONYM: *Hymenodictyon scabrum* Stapf; ITU: 348.

Shrub or tree 1–4.5(–10) m high. Open woodland, thicket, rocky hills; 600–1350 m.

West Nile, Acholi, Madi, Bunyoro, Toro. Widespread in Africa.

Here assessed preliminarily as Least Concern (LC) because of its wide distribution and habitat range.

Ixora mildbraedii K. Krause
LITERATURE: Bridson & Verdcourt, FTEA 1988: 615.

Shrub or tree 2–4.5 m high. Stream-sides; ± 900 m.

West Nile. Also in DRC, CAR, South Sudan.

Here assessed preliminarily as Least Concern; widespread in DRC and in a range of habitat from swamp or riverine forest to wooded grassland. EOO 967,735 km^2; AOO 312,022 km^2; > 19 locations.

Ixora seretii De Wild.
See restricted list, page 195.

Keetia zanzibarica (Klotzsch) Bridson is described in FTEA as occasionally being a small tree, but it is never so in Uganda, as far as we know.

Lasianthus kilimandscharicus K. Schum.
LITERATURE: Verdcourt, FTEA 1976: 142, fig. 11.
SYNONYM: *Lasianthus sp.* of ITU: 348.

Shrub or tree 1–7.5 m high. Rainforest; 1700–2400 m.

Ankole, Kigezi. Widespread in Africa.

Here assessed preliminarily as Least Concern (LC) because of its wide distribution.

Leptactina platyphylla (Hiern) Wernham
LITERATURE: Bridson & Verdcourt, FTEA 1988: 689.

Shrub or tree 2–8 m high. Evergreen forest, woodland, secondary bushland; 50–1650 m.

Kigezi, Mengo. Widespread in Africa.

Here assessed preliminarily as Least Concern (LC) because of its wide distribution and habitat/altitude range.

Macrosphyra brachystylis Hiern
LITERATURE: Bridson & Verdcourt, FTEA 1988: 493.

Shrub or tree 1.5–2.5 m high. Evergreen forest, forest edges; 1150–1350 m.

Toro, Mengo, Masaka. Also in DRC, Angola.

Here assessed preliminarily as Least Concern (LC) because of its wide distribution and habitat.

Meyna tetraphylla (Hiern) Robyns subsp. *tetraphylla*
LITERATURE: Verdcourt & Bridson, FTEA 1991: 859.

Shrub or tree 2–9 m high. Woodland, thicket; 1100–1650 m.

West Nile, Karamoja, Acholi. Also in South Sudan, Ethiopia, Kenya, S Somalia.

Here assessed preliminarily as Least Concern (LC) because of its wide distribution (EOO 657,291 km^2; AOO 175,492 km^2) and habitat range.

Morinda lucida Benth.

LITERATURE: Verdcourt, FTEA 1976: 146. ITU: 351. UFT: 182.

Shrub or tree 2–18 m high. Grassland, thicket, bushland, forest; 750–1300 m.

Kigezi, Busoga, Masaka, Mengo, Toro, Bunyoro, Mbale. Widespread in Africa.

Here assessed preliminarily as Least Concern (LC) because of its wide distribution and habitat range.

Morinda titanophylla E. M. A. Petit

LITERATURE: Verdcourt, FTEA 1976: 148. UFT: 182.

Shrub or tree 1–8 m high. Rainforest, semi-deciduous forest; 1200–1800 m.

Ankole, Kigezi. Widespread in Africa.

Here assessed preliminarily as Least Concern (LC) because of its wide distribution and habitat range.

Multidentia crassa (Hiern) Bridson & Verdc.

LITERATURE: Verdcourt & Bridson, FTEA 1991: 845, fig. 149.

SYNONYM: *Canthium crassum* Hiern; ITU: 339.

Shrub or tree 0.6–6 m high. Woodland, thicket, grassland; 900–2100 m.

West Nile, Acholi, Teso, Karamoja, Ankole. Widespread in Africa.

Here assessed preliminarily as Least Concern (LC) because of its wide distribution and habitat/altitude range.

Multidentia dichrophylla (Mildbr.) Bridson

LITERATURE: Verdcourt & Bridson, FTEA 1991: 842.

Scandent shrub or tree 4–12 m high. Evergreen forest; 1220 m.

Mengo. Widespread in central Africa.

Here assessed preliminarily as Least Concern (LC) because of its wide distribution.

Mussaenda microdonta subsp. *odorata* (Hutch.) Bridson

LITERATURE: Bridson & Verdcourt, FTEA 1988: 464.

Shrub or small tree 3–9 m high. Evergreen forest; ± 2000 m.

Mbale. Also in Kenya and Tanzania.

Assessed here as Data Deficient (DD) as the single Ugandan specimen is doubtfully placed under this name.

Nauclea diderrichii (De Wild. & T. Durand) Merr.

LITERATURE: Bridson & Verdcourt, FTEA 1988: 441, fig. 61. ITU: 351. UFT: 180.

Tree 30–40 m high. Evergreen forest; 800–1650 m.

Toro. Widespread in Africa from West Africa south to Angola. In Uganda, restricted to Semliki valley.

Here assessed preliminarily as Least Concern (LC) because of its wide distribution.

Oxyanthus formosus Planch.

Literature: Bridson & Verdcourt, FTEA 1988: 537.

Shrub or tree 1–5 m high. Forest; ± 1375 m.

Bunyoro, Toro, Kigezi. Widespread in Africa.

Here assessed preliminarily as Least Concern (LC) because of its wide distribution.

Oxyanthus lepidus S. Moore

Literature: Bridson & Verdcourt, FTEA 1988: 531.

Shrub or tree 2–7 m high. Forest; 650–1850 m.

Karamoja, Toro, Mengo, Bunyoro. Fairly widespread in central Africa.

Here assessed preliminarily as Least Concern (LC) because of its wide distribution and altitude range.

Oxyanthus speciosus DC.

Literature: Bridson & Verdcourt, FTEA 1988: 527. UFT: 184.

Synonym: *Oxyanthus speciosus* in the sense of ITU: 353, partly.

Shrub or tree 2–12 m high. Forest; 750–2300 m.

Karamoja, Ankole, Mbale. Widespread in Africa.

Here assessed preliminarily as Least Concern (LC) because of its wide distribution and altitude range.

Oxyanthus troupinii Bridson

Literature: Bridson & Verdcourt, FTEA 1988: 534, fig. 86.

Shrub or tree 1–12 m high. Forest; ± 2300 m.

Kigezi. Widespread in Africa.

Here assessed preliminarily as Least Concern (LC) because of its wide distribution.

Oxyanthus unilocularis Hiern

Literature: Bridson & Verdcourt, FTEA 1988: 537. ITU: 353. UFT: 180.

Shrub or tree 1.5–6 m high. Forest; altitude not given.

Toro, Masaka, Mengo, Madi. Widespread in Africa.

Here assessed preliminarily as Least Concern (LC) because of its wide distribution.

Pauridiantha callicarpoides (Hiern) Bremek.

Literature: Verdcourt, FTEA 1976: 157. ITU: 354. UFT: 180.

Shrub or tree 4–15 m high. Evergreen forest; 1050–1650 m.

Kigezi, Ankole, Toro. Widespread in central Africa.

Here assessed preliminarily as Least Concern (LC) because of its wide distribution.

Pauridiantha dewevrei (De Wild. & T. Durand) Bremek.

Literature: Verdcourt, FTEA 1976: 157.

Shrub or tree 2–8 m high. Evergreen forest; 1800–2400 m.

Kigezi. Widespread in Africa.

Here assessed preliminarily as Least Concern (LC) because of its wide distribution.

Pauridiantha paucinervis (Hiern) Bremek.

Literature: Verdcourt, FTEA 1976: 153, fig. 14.

Synonym: *Pauridiantha holstii* (K. Schum.) Bremek.; ITU: 354.

Shrub or tree 1.5–9(–12) m high. Evergreen forest; 1350–2300 m.

Toro, Ankole, Kigezi, Masaka. Widespread in Africa.

Here assessed preliminarily as Least Concern (LC) because of its wide distribution and altitude range.

Pauridiantha viridiflora (Hiern) Hepper

Literature: Verdcourt, FTEA 1976: 158. UFT: 180.

Synonym: *Pamplethantha viridiflora* (Hiern) Bremek.; ITU: 353.

Shrub or tree 3–7.5(–12) m high. Evergreen forest, forest edges, secondary bushland; 1100–1400 m.

Masaka. Widespread in Africa.

Here assessed preliminarily as Least Concern (LC) because of its wide distribution and habitat range.

Pavetta abyssinica Fresen.

Literature: Bridson & Verdcourt, FTEA 1988: 660.

Shrub or tree 1–7 m high. Evergreen forest; 1050–2500 m.

Karamoja, Mbale. Widespread in South Sudan, Ethiopia, Kenya, Tanzania.

Here assessed preliminarily as Least Concern (LC) because of its wide distribution and altitude range.

Pavetta bagshawei S. Moore

Literature: Bridson & Verdcourt, FTEA 1988: 658.

Shrub or tree 0.6–4.5 m high. Evergreen forest; 950–1700 m.

Toro, Ankole, Kigezi, Masaka. Widespread in central Africa.

Here assessed preliminarily as Least Concern (LC) because of its wide distribution and altitude range.

Pavetta crassipes K. Schum.

Literature: Bridson & Verdcourt, FTEA 1988: 670. ITU: 354.

Shrub or tree 1–8 m high. Wooded grassland, woodland, forest edges; 350–2000 m.

Acholi, Teso, Mengo, Mbale, Bunyoro, West Nile, Madi, Lango, Karamoja, Toro. Widespread in Africa.

Here assessed preliminarily as Least Concern (LC) because of its wide distribution and habitat/altitude range.

Pavetta gardeniifolia A. Rich.

Literature: Bridson & Verdcourt, FTEA 1988: 676.

Shrub or tree 1–7 m high. Bushland, grassland, forest, rocky hillsides; 600–2100 m.

Karamoja, Ankole, Kigezi. Widespread in central Africa.

Here assessed preliminarily as Least Concern (LC) because of its wide distribution and habitat/altitude range.

[*Pavetta insignis* Bremek. of ITU, now a synonym of *Pavetta molundensis*, is really a shrub]

Pavetta molundensis K. Krause
LITERATURE: Bridson & Verdcourt, FTEA 1988: 631.

Shrub or tree 2–7 m high. Evergreen forest; 750–1200 m.

Bunyoro, Kigezi, Mengo. Widespread in central Africa.

Here assessed preliminarily as Least Concern (LC) because of its wide distribution.

Pavetta oliveriana Hiern
LITERATURE: Bridson & Verdcourt, FTEA 1988: 644.

Shrub or tree 1.5–7 m high. Grassland, thicket, evergreen forest, rock crevices; 900–2300 m.

Karamoja, Ankole, Mengo. Widespread in Africa.

Here assessed preliminarily as Least Concern (LC) because of its wide distribution and habitat/altitude range.

Pavetta ternifolia (Oliv.) Hiern
LITERATURE: Bridson & Verdcourt, FTEA 1988: 633, fig. 108.

Shrub or tree 1–7 m high. Evergreen forest, thicket, forest edges, grassland; 1150–1950 m.

Toro, Ankole, Masaka. Fairly widespread in central Africa.

Here assessed preliminarily as Least Concern (LC) because of its wide distribution and habitat range.

Pavetta urundensis Bremek.
See restricted list, page 196.

Psychotria bagshawei E. M. A. Petit
See restricted list, page 197.

Psychotria capensis Schönland subsp. *riparia* (K. Schum. & K. Krause) Verdc.
LITERATURE: Verdcourt in *Fl. Zambesiaca* 5(1): 13 (1989).

SYNONYM: *Psychotria riparia* (K. Schum. & K. Krause) E. M. A. Petit; Verdcourt, FTEA 1976: 38.

Shrub or tree 2.5–6(–20) m high. Riverine forest, forest edges, evergreen bushland, wooded grassland; 500–1800 m.

Acholi, Mbale. Fairly widespread in S central and East Africa.

Here assessed as least concern (LC) because of its wide distribution and habitat/altitude range.

Psychotria fractinervata E. M. A. Petit
LITERATURE: Verdcourt, FTEA 1976: 50.

Shrub or tree 2–6 m high. Rainforest to *Hypericum* zone; 1800–2600 m.

Mbale. Also in Kenya, N Tanzania, where not uncommon.

Here assessed preliminarily as Least Concern (LC) because of its distribution and habitat/altitude range.

Psychotria lauracea (K. Schum.) E. M. A. Petit
LITERATURE: Verdcourt, FTEA 1976: 61.

Shrub or tree 1–6 m high. Evergreen and riverine forest, thicket; 500–1800 m.

Kigezi, Mbale. Fairly widespread in S central and East Africa.

Here assessed preliminarily as Least Concern (LC) because of its wide distribution and habitat/altitude range.

Psychotria mahonii C. H. Wright

Literature: Verdcourt, FTEA 1976: 58, fig. 3.

Synonym: *Grumilea megistosticta* S. Moore of ITU: 346.
Synonym: *Grumilea sp.* of ITU: 346.
Synonym: *Psychotria megistosticta* of UFT: 185.

Shrub or tree 1.5–15(–24) m high. Evergreen forest, swamp forest, thicket; 1100–2700 m.

Toro, Kigezi, Ankole, Masaka, Mengo. Fairly widespread in S central and East Africa.

Here assessed preliminarily as Least Concern (LC) because of its wide distribution and habitat/altitude range.

Psychotria orophila E. M. A. Petit

Literature: Verdcourt, FTEA 1976: 49.

Shrub or tree 1.5–10 m high. Evergreen forest; 1650–2700 m.

Mbale. Fairly widespread in Kenya and Ethiopia, N Tanzania.

Here assessed preliminarily as Least Concern (LC) because of its wide distribution and altitude range.

Psychotria succulenta (Hiern) E. M. A. Petit

Literature: Verdcourt, FTEA 1976: 57.

Shrub or tree 1.5–10 m high. Evergreen forest, woodland and thicket, especially along rivers and lakes; 1100–1650 m.

West Nile, Mengo. Fairly widespread in S central and East Africa.

Here assessed preliminarily as Least Concern (LC) because of its wide distribution and habitat range.

Psydrax parviflora (Afzel.) Bridson

Literature: Verdcourt & Bridson, FTEA 1991: 896.

Synonym: *Canthium rubrocostatum* Robyns; ITU: 340.
Synonym: *Canthium vulgare* (K. Schum.) Bullock; ITU: 340. UFT: 187.

Shrub or tree 2–27 m high. Forest, thicket, bushed grassland; 1050–2750 m.

Kigezi, Teso, Masaka, West Nile, Mbale, Mengo, Masaka, Ankole, Toro, Bunyoro, Acholi, Busoga. Widespread in Africa.

Here assessed preliminarily as Least Concern (LC) because of its wide distribution and habitat/altitude range.

Psydrax schimperiana (A. Rich.) Bridson

Literature: Verdcourt & Bridson, FTEA 1991: 901.

Synonym: *Canthium euryoides* in the sense of ITU: 339.
Synonym: *Canthium schimperianum* A. Rich. of ITU: 340.

Shrub or tree 2–10 m high. Forest, thicket, bushland; 500–2500 m.

Karamoja, Bunyoro, Masaka, Mubende, Toro, Ankole, Busoga. Widespread in Africa.

Here assessed preliminarily as Least Concern (LC) because of its wide distribution and habitat/altitude range.

Psydrax subcordata (DC.) Bridson

LITERATURE: Verdcourt & Bridson, FTEA 1991: 894, fig. 159.

Tree 5–15 m high. Forest; 1500 m.

Ankole, Toro, Kigezi. Widespread in Africa.

Here assessed preliminarily as Least Concern (LC) because of its wide distribution.

Pyrostria affinis (Robyns) Bridson

LITERATURE: Verdcourt & Bridson, FTEA 1991: 890.

Shrub or tree 3–5 m high. Forest; 1200 m.

Kigezi. Widespread in Africa.

Here assessed preliminarily as Least Concern (LC) because of its wide distribution.

Rothmannia longiflora Salisb.

LITERATURE: Bridson & Verdcourt, FTEA 1988: 515.

Shrub or tree 3–9 m high. Forest, woodland; 900–1200 m.

Bunyoro, Masaka, Mengo. Widespread in Africa.

Here assessed preliminarily as Least Concern (LC) because of its wide distribution and habitat range.

Rothmannia urcelliformis (Hiern) Robyns

LITERATURE: Bridson & Verdcourt, FTEA 1988: 514.

SYNONYM: *Randia urcelliformis* (Hiern) Eggeling; ITU: 356, plate 18. UFT: 185.

Shrub or tree 4–9(–15) m high. Forest; 850–1700 m.

West Nile, Kigezi, Mengo, Toro, Bunyoro, Karamoja. Widespread in Africa.

Here assessed preliminarily as Least Concern (LC) because of its wide distribution and altitude range.

Rothmannia whitfieldii (Lindl.) Dandy

LITERATURE: Bridson & Verdcourt, FTEA 1988: 518. UFT: 187.

SYNONYM: *Randia malleifera* (Hook. f.) Hook. f.; ITU: 355.

Shrub or tree 2–8 m high. Forest; 1050–1700 m.

Acholi, Bunyoro, Mengo, Toro, West Nile, Madi. Widespread in Africa.

Here assessed preliminarily as Least Concern (LC) because of its wide distribution.

Rytigynia acuminatissima (K. Schum.) Robyns

See restricted list, page 198.

Rytigynia bridsoniae Verdc.

LITERATURE: Verdcourt & Bridson, FTEA 1991: 815.

Shrub or tree 3–6 m high. Bamboo forest; 2400 m.

Kigezi, only known from two records from Bwindi; also in DRC, Rwanda and Burundi

Here assessed preliminarily as Least Concern; large numbers of collections from other countries, where the range is 1600–3000 m in montane forest, swamp forest and heath zone. EOO > 20,879 km^2; AOO > 5,261 km^2.

Rytigynia bugoyensis (K. Krause) Verdc.
Literature: Verdcourt & Bridson, FTEA 1991: 833.

Shrub or tree 1–6 m high. Evergreen forest; 1200–2400 m.

Toro, Kigezi, Mbale, Ankole. Widespread in central Africa.

Here assessed preliminarily as Least Concern (LC) because of its wide distribution and altitude range.

Rytigynia dubiosa (De Wild.) Robyns
Literature: Verdcourt & Bridson, FTEA 1991: 810.

Shrub or tree to 2.5 m high. Evergreen forest edges; 1100–1200 m.

Masaka, Mengo. Widespread in central Africa.

Here assessed preliminarily as Least Concern (LC) because of its wide distribution and habitat.

Rytigynia kigeziensis Verdc.
See restricted list, page 199.

Rytigynia neglecta (Hiern) Robyns
Literature: Verdcourt & Bridson, FTEA 1991: 813.

Shrub or tree 1–9 m high. Dry forest and forest edges; 1850–2400 m.

Karamoja, Masaka, Kigezi, Acholi. Also in N Kenya, South Sudan, Ethiopia, and possibly in Senegal and Cameroon.

Here assessed preliminarily as Least Concern, due to the wide distribution (from 1° S to 9° N and 29° to 41° E; EOO 523,355 km^2; AOO 326,537 km^2; > 20 locations) coupled to habitat type and altitude range. Quite a large number of specimens collected over this distribution range; no specific threats known.

Rytigynia ruwenzoriensis (De Wild.) Robyns
See restricted list, page 200.

Rytigynia umbellulata (Hiern) Robyns
Literature: Verdcourt & Bridson, FTEA 1991: 808.

Shrub or tree 1–5(–9) m high. Forest, woodland, thicket, often by lakes or on rock; 750–1200 m.

Masaka, Mengo, Ankole. Widespread in Africa.

Here assessed preliminarily as Least Concern (LC) because of its wide distribution and habitat range.

Rytigynia sp. *B* of FTEA
See restricted list, page 201.

Sarcocephalus latifolius (Smith) E. A. Bruce
Literature: Bridson & Verdcourt, FTEA 1988: 439, fig. 60.

SYNONYM: *Nauclea latifolia* Sm.; ITU: 352.

Shrub or tree 2–9 m high. Scattered tree grassland, scrub; 900–1500 m.

Bunyoro, Teso, Mbale, West Nile, Madi, Acholi, Lango, Busoga. Widespread in Africa.

Here assessed preliminarily as Least Concern (LC) because of its wide distribution and habitat range.

Tarenna pavettoides (Harv.) Sim

LITERATURE: Bridson & Verdcourt, FTEA 1988: 588. ITU: 357. UFT: 184.

Shrub or tree 2–10 m high. Forest edges, thicket; 1100–1600 m.

Kigezi, Masaka, Mengo, Toro, Mbale. Widespread in central Africa.

Here assessed preliminarily as Least Concern (LC) because of its wide distribution and habitat.

Tricalysia bagshawei S. Moore

See restricted list, page 202.

Tricalysia congesta (Oliv.) Hiern

SYNONYM: *Tricalysia ruandensis* Bremek.; Bridson & Verdcourt, FTEA 1988: 564.

Shrub or tree 1–8 m high. Dry thicket, woodland, riverine forest; 1000–1500 m.

Ankole, Masaka. Widespread in Africa.

Here assessed preliminarily as Least Concern (LC) because of its wide distribution and habitat range.

Tricalysia coriacea (Benth.) Hiern

LITERATURE: Bridson & Verdcourt, FTEA 1988: 546, fig. 90.

Shrub or tree to 8 m high. Swamp forest, river-banks; 500–1700 m.

Masaka. Widespread in Africa.

Here assessed preliminarily as Least Concern (LC) because of its wide distribution and altitude range.

Tricalysia elliotii (K. Schum.) Hutch. & Dalziel

LITERATURE: Bridson & Verdcourt, FTEA 1988: 544.

Tree 6–20 m high. Forest; 1050–1300 m.

Mengo. Widespread in Africa.

Here assessed preliminarily as Least Concern (LC) because of its wide distribution.

Tricalysia niamniamensis Hiern

LITERATURE: Bridson & Verdcourt, FTEA 1988: 552. ITU: 357.

Shrub or tree to 8 m high. Woodland, thicket, bushland, riverine forest; 900–1550 m.

West Nile, Busoga, Mengo, Masaka, Mubende, Bunyoro, Madi, Acholi, Lango, Teso, Karamoja, Toro. Widespread in central Africa.

Here assessed preliminarily as Least concern (LC) because of its wide distribution and habitat range.

Vangueria apiculata K. Schum.

LITERATURE: Verdcourt & Bridson, FTEA 1991: 853. ITU: 358. UFT: 181.

Shrub, climber or tree 2–12 m high. Evergreen forest, riverine or lakeside forest, bushland, scattered tree grassland, thicket; 900–2200 m.

Acholi, Busoga, Mengo, Masaka, Ankole, Kigezi, Madi, Karamoja, Mbale, Toro, Bunyoro. Widespread in Africa.

Here assessed preliminarily as Least Concern (LC) because of its wide distribution and habitat/altitude range.

Vangueria infausta Burch.

LITERATURE: Verdcourt & Bridson, FTEA 1991: 851.

SYNONYM: *Vangueria tomentosa* in the sense of ITU: 358, partly.

Shrub or tree 1–8 m high. Evergreen forest, riverine forest, woodland, bushland, scattered tree grassland, thicket; 500–2100 m.

West Nile, Karamoja. Widespread in Africa.

Here assessed preliminarily as Least Concern (LC) because of its wide distribution and habitat/altitude range.

Vangueria madagascariensis J. F. Gmel.

LITERATURE: Verdcourt & Bridson, FTEA 1991: 849, fig. 150.

SYNONYM: *Vangueria acutiloba* Robyns; ITU: 358.

Shrub or tree 1–15 m high. Evergreen forest, riverine forest and woodland, bushland, scattered tree grassland; 500–2150 m.

Karamoja, Teso, Masaka, Mengo, Bunyoro, Madi, Acholi, Mbale. Widespread in Africa.

Here assessed preliminarily as Least Concern (LC) because of its wide distribution and habitat/altitude range.

Vangueria volkensii K. Schum.

LITERATURE: Verdcourt & Bridson, FTEA 1991: 854.

SYNONYM: *Vangueria linearisepala* K. Schum.; ITU: 358.

Shrub, climber or tree 2–9(–15) m high. Evergreen forest and forest edges, thicket; 1100–2300 m.

Karamoja, Mbale, Masaka, Mengo. Widespread in Africa.

Here assessed preliminarily as Least Concern (LC) because of its wide distribution and habitat/altitude range.

RUTACEAE

Aeglopsis eggelingii M. Taylor

See restricted list, page 203.

Balsamocitrus dawei Stapf

See restricted list, page 204.

Calodendrum capense (L. f.) Thunb.

LITERATURE: Kokwaro, FTEA 1982: 11, fig. 4. ITU: 361.

Tree to 20 m high. Evergreen forest, riverine forest; 1200–2200 m.

Ankole, Mengo. Widespread in Africa.

Here assessed preliminarily as Least Concern (LC) because of its wide distribution and altitude range.

Citropsis articulata (Spreng.) Swingle & Kellerm.
Literature: Kokwaro, FTEA 1982: 33, fig. 10. UFT: 206.

Shrub or tree 2.5–5 m high. Rainforest; 650–1550 m.

Toro, Ankole, Mengo. Widespread in Africa.

Here assessed preliminarily as Least Concern (LC) because of its wide distribution and altitude range.

Clausena anisata (Willd.) Benth.
Literature: Kokwaro, FTEA 1982: 49, fig. 13. ITU: 361. UFT: 206.

Shrub or tree to 4(–10) m high. Forest, forest margins, secondary busland and woodland; 1–2450 m.

Bunyoro, Busoga, Mengo, Masaka, Ankole, Kigezi, Toro, Mubende, West Nile, Madi, Acholi, Teso, Karamoja, Mbale. Widespread in Africa.

Here assessed preliminarily as Least Concern (LC) because of its wide distribution and habitat/altitude range.

Fagaropsis angolensis (Engl.) Dale
Literature: Kokwaro, FTEA 1982: 45, fig. 12. ITU: 365. UFT: 203.

Tree 7–15 m high. Rainforest, forest edges, drier evergreen forest; 1000–2250 m.

Toro, Ankole, Mengo, Mubende, Bunyoro, Mbale, Busoga. Widespread in Africa.

Here assessed preliminarily as Least Concern (LC) because of its wide distribution and habitat/altitude range.

Teclea sp. (349) of UFT: 195, unclear what this is.

Vepris eggelingii (Kokwaro) Mziray
See restricted list, page 205.

Vepris glomerata (F. Hoffm.) Engl.
Literature: Kokwaro, FTEA 1982: 22.

Synonym: *Teclea pilosa* (Engl.) I. Verd. of ITU: 369.
Synonym: *Teclea sp. nov. 2* of ITU: 370.

Shrub or tree 2–8 m high. Wooded grassland, bushland; 1–1850 m.

Karamoja, Acholi, Mbale. Fairly widespread in East Africa.

Here assessed preliminarily as Least Concern (LC) because of its wide distribution and habitat/altitude range.

Vepris grandifolia (Engl.) Mziray
Literature: Mziray in *Symb. Bot. Upsal*. 30: 1–95 (1992).

Synonym: *Teclea grandifolia* Engl.; Kokwaro, FTEA 1982: 27. ITU: 367. UFT: 195.

Shrub or tree 4–11 m high. Rainforest, riverine forest; 900–1650 m.

Bunyoro, Mengo, Ankole, Busoga. Widespread in Africa.

Here assessed preliminarily as Least Concern (LC) because of its wide distribution.

Vepris nobilis (Delile) Mziray

Literature: Mziray in *Symb. Bot. Upsal.* 30: 1–95 (1992).

Synonym: *Teclea nobilis* Delile; Kokwaro, FTEA 1982: 26. ITU: 367, fig. 75. UFT: 195.
Synonym: *Teclea simplicifolia* in the sense of ITU: 369.

Shrub or tree 5–12 m high. Evergreen forest, wooded grassland, riverine forest; 900–2600 m.

Karamoja, Kigezi, Mbale. Widespread in Africa.

Here assessed preliminarily as Least Concern (LC) because of its wide distribution and habitat/altitude range.

Vepris stolzii I. Verd.

Literature: Kokwaro, FTEA 1982: 19.

Tree 6–15 m high. Evergreen forest; 900–2150 m.

Kigezi. Widespread in Africa.

Here assessed preliminarily as Least Concern (LC) because of its wide distribution and altitude range.

Vepris trichocarpa (Engl.) Mziray

Literature: Mziray in *Symb. Bot.Upsal.* 30: 1–95 (1992).

Synonym: *Teclea trichocarpa* (Engl.) Engl.; Kokwaro, FTEA 1982: 25. ITU: 369.
Synonym: *Teclea angustialata* Engl.; ITU: 367.

Shrub or tree 2–9 m high. Evergreen forest, wooded grassland, riverine; 500–2300 m.

Ankole, Busoga, Mengo, Toro. Widespread in Africa.

Here assessed preliminarily as Least Concern (LC) because of its wide distribution and habitat/altitude range.

Zanthoxylum chalybeum Engl.

Literature: Kokwaro, FTEA 1982: 37, fig. 11.

Synonym: *Fagara chalybea* (Engl.) Engl.; ITU: 363.

Shrub or tree to 10 m high. Dry bushland, wooded grassland; 500–1550 m.

Acholi, Lango, Karamoja, Mengo, Bunyoro, Teso, Mbale, Busoga, Masaka, Ankole, West Nile. Widespread in Africa.

Here assessed preliminarily as Least Concern (LC) because of its wide distribution and habitat/altitude range.

Zanthoxylum gilletii (De Wild.) P. G. Waterman

Literature: Kokwaro, FTEA 1982: 38.

Synonym: *Fagara macrophylla* (Oliv.) Engl.; ITU: 364. UFT: 205.

Tree 10–35 m high. Rainforest; 900–2400 m.

West Nile, Ankole, Mengo, Kigezi, Toro, Bunyoro, Mbale. Widespread in Africa.

Here assessed preliminarily as Least Concern (LC) because of its wide distribution and altitude range.

Zanthoxylum lemairei (De Wild.) P. G. Waterman
LITERATURE: Kokwaro, FTEA 1982: 39.

Tree to 30 m high. Rainforest; ± 1150 m.

Mengo. Widespread in Africa.

Here assessed preliminarily as Least Concern (LC) because of its wide distribution.

Zanthoxylum leprieurii Guill. & Perr.
LITERATURE: Kokwaro, FTEA 1982: 39.

SYNONYM: *Fagara angolensis* Engl.; ITU: 363.
SYNONYM: *Fagara leprieurii* of UFT: 206.
SYNONYM: *Fagara stuhlmannii* (Engl.) Engl.; ITU: 364.

Tree 6–15 m high. Rainforest; 900–2000 m.

Toro, Ankole, Mengo, Masaka, Toro, Bunyoro, Madi. Widespread in Africa.

Assessed as Least Concern (LC) by Sacande, Sanou & Beentje in *Guide de terrain des arbres du Burkina Faso* (in press).

Zanthoxylum mildbraedii (Engl.) P. G. Waterman
See restricted list, page 206.

Zanthoxylum rubescens Hook. f.
LITERATURE: Kokwaro, FTEA 1982: 44.

SYNONYM: *Fagara melanacantha* (Oliv.) Engl.; ITU: 364.
SYNONYM: *Fagara rubescens* (Hook. f.) Engl.; UFT: 205.

Shrub or tree to 8 m high. Rainforest or riverine; 900–1800 m.

Bunyoro, Ankole, Mengo, Masaka, Mubende, Toro, Madi, Karamoja. Widespread in Africa.

Here assessed preliminarily as Least Concern (LC) because of its wide distribution and altitude range.

SALICACEAE

Casearia battiscombei R. E. Fr.
LITERATURE: Sleumer, FTEA 1975: 49, fig. 16. ITU: 372. UFT: 144.

Tree to 40 m high. Rainforest; 1100–2450 m.

Ankole, Mbale. Also in Kenya, Tanzania, Malawi, Zimbabwe.

Here assessed preliminarily as Least Concern (LC) because of its wide distribution.

Casearia runssorica Gilg
LITERATURE: Sleumer, FTEA 1975: 48.

SYNONYM: *Casaeria engleri* in the sense of ITU: 372. UFT: 144.

Shrub or tree to 40 m high. Rainforest and semi-swamp forest; 850–2100 m.

Bunyoro, Kigezi, Masaka, Toro. Also in South Sudan, N Tanzania, DRC, Rwanda, Burundi.

Here assessed preliminarily as Least Concern (LC) because of its wide distribution and altitude range.

Dovyalis abyssinica (A. Rich.) Warb.

Literature: Sleumer, FTEA 1975: 61, fig. 21. UFT: 129.

Synonym: *Dovyalis engleri* Gilg; ITU: 147.

Shrub or tree to 8 m high. Rainforest, riverine or dry forest, wooded grassland; 1500–3000 m.

Karamoja, Mbale, Mengo, Kigezi, Bunyoro, West Nile. Widespread in E and NE Africa.

Here assessed preliminarily as Least Concern (LC) because of its wide distribution and habitat range.

Dovyalis macrocalyx (Oliv.) Warb.

Literature: Sleumer, FTEA 64. ITU: 147. UFT: 129.

Synonym: *Dovyalis glandulossissima* Gilg; ITU: 147.

Shrub or tree to 8 m high. Rainforest, riverine or dry forest, bushland, wooded grassland; 1200–2600 m.

Acholi, Busoga, Mengo, Ankole, Toro, Bunyoro. Widespread in Africa.

Here assessed preliminarily as Least Concern (LC) because of its wide distribution and habitat range.

Dovyalis spinossissima Gilg

Literature: Sleumer, FTEA 1975: 61.

Synonym: *Dovyalis abyssinica* in the sense of ITU: 147.
Synonym: *Dovyalis macrocarpa* Bamps; UFT: 129.
Synonym: *Dovyalis sp. nov.(1)* of ITU: 148.
Synonym: *Dovyalis sp. nov.(2)* of ITU: 148.

Shrub or tree 2–10 m high. Rainforest and dry forest edges, riverine forest, secondary growth, bushland; 700–2000 m.

West Nile, Toro, Kigezi, Bunyoro. Widespread in Africa.

Here assessed preliminarily as Least Concern (LC) because of its wide distribution and habitat/altitude range.

Dovyalis zenkeri Gilg

Literature: Sleumer, FTEA 1975: 62.

Shrub or tree 1.5–6(–10) m high. Rainforest, riverine or secondary forest, evergreen bushland; 1500–1700 m.

Bunyoro, Toro, Kigezi. Widespread in Africa.

Here assessed preliminarily as Least Concern (LC) because of its wide distribution and habitat range.

Flacourtia indica (Burm. f.) Merr.

Literature: Sleumer, FTEA 1975: 57, fig. 20. ITU: 148. UFT: 128.

Shrub or tree to 10 m high. Woodland, wooded grassland, bushland, often riparian; 500–2400 m.

West Nile, Teso, Mengo, Ankole, Bunyoro, Acholi, Mbale, Karamoja. Widespread in Africa and Asia.

Assessed as Least Concern (LC) by Sacande, Sanou & Beentje in *Guide de terrain des arbres du Burkina Faso* (in press).

Homalium abdessammadi Asch. & Schweinf.

Literature: Sleumer, FTEA 1975: 43, fig. 15.

Shrub or tree 5–10(–20) m high. Forest and forest edges, riverine forest; 500–2000 m.

Mbale. Widespread in Africa.

Here assessed preliminarily as Least Concern (LC) because of its wide distribution, habitat and altitude range.

Oncoba routledgei Sprague

Literature: Sleumer, FTEA 1975: 16. ITU: 149. UFT: 129.

Shrub or tree to 8 m high. Rain forest, often riverine; 900–2450 m.

Ankole, Kigezi, Mbale, Toro. Widespread in East Africa.

Here assessed preliminarily as Least Concern (LC) because of its wide distribution and altitude range.

Oncoba spinosa Forssk.

Literature: Sleumer, FTEA 1975: 16, fig. 6. ITU: 149. UFT: 129.

Shrub or tree to 10 m high. Forest and forest edges, associated bushland, woodland, wooded grassland; 500–1850 m.

Bunyoro, Busoga, Masaka, Ankole, Madi, Karamoja, Mbale, Toro. Widespread in Africa.

Assessed as Least Concern (LC) by Sacande, Sanou & Beentje in *Guide de terrain des arbres du Burkina Faso* (in press).

Salix subserrata Willd.

Literature: Wilmot-Dear, FTEA 1985: 1, fig. 1. ITU: 371.

Synonym: *Salix hutchinsii* Skan; ITU: 370.

Shrub or tree 2–10 m high. Along streams in forest, bushland and grassland; 900–2600 m.

Mbale. Widespread in Africa.

Here assessed preliminarily as Least Concern (LC) because of its wide distribution and habitat/altitude range.

Scolopia rhamniphylla Gilg

Literature: Sleumer, FTEA 1975: 35, fig. 12. ITU: 150. UFT: 128.

Shrub or tree to 12 m high. Rainforest, dry forest, associated bushland or riverine forest; 1000–2000 m.

Kigezi, Masaka, Mengo, Ankole, ?Toro. Widespread in Africa.

Here assessed preliminarily as Least Concern (LC) because of its wide distribution and habitat range.

Scolopia theifolia Gilg

Literature: Sleumer, FTEA 1975: 33.

Shrub or tree to 15 m high. Dry forest, wooded grassland; 1600–2700 m.

Karamoja. Widespread in East Africa.

Here assessed preliminarily as Least Concern (LC) because of its wide distribution and habitat range.

Scolopia zeyheri (Nees) Harv.

LITERATURE: Sleumer, FTEA 1975: 37.

SYNONYM: *Scolopia rigida* R.E. Fr.; ITU: 150.

Shrub or tree to 25 m high. Dry forest, riverine forest, bushland, wooded grassland; 500–2400 m.

Ankole. Widespread in Africa.

Here assessed preliminarily as Least Concern (LC) because of its wide distribution and habitat/altitude range.

Trimeria grandifolia (Hochst.) Warb.

LITERATURE: Sleumer, FTEA 1975: 40, fig. 14.

SYNONYM: *Trimera bakeri* Gilg; ITU: 372. UFT: 129.

Shrub or tree 2–6(–20) m high. Forest and riverine forest, associated bushland, wooded grassland; 900–2450 m.

Ankole, Busoga, Mubende, Kigezi, Mbale. Widespread in Africa.

Here assessed preliminarily as Least Concern (LC) because of its wide distribution and habitat/altitude range.

SALVADORACEAE

Dobera glabra (Forssk.) Poir.

LITERATURE: Verdcourt, FTEA 1968: 4.

Shrub or tree 1.8–7.5 m high. Dry bushland, rocky hillsides, saline river-beds; 500–1100 m.

Karamoja. Widespread in Africa and Asia.

Here assessed preliminarily as Least Concern (LC) because of its wide distribution and habitat type.

Salvadora persica L.

LITERATURE: Verdcourt, FTEA 1968: 7, fig. 3. ITU: 371.

Shrub or tree to 6 m high. Dry bushland, especially along luggas; 500–1350 m.

Karamoja. Widespread in Africa and Asia.

Here assessed preliminarily as Least Concern (LC) because of its wide distribution and habitat type.

SANTALACEAE

Osyris lanceolata Hochst & Steud.

LITERATURE: Polhill, FTEA 2005: 23, fig. 5.

SYNONYM: *Osyris abyssinica* Hochst.; ITU: 373.

Shrub or tree to 9 m high. Dry forest and associated bushland and grassland; 900–2700 m.

Ankole, Kigezi, Karamoja, Mbale. Widespread in Old World.

Here assessed preliminarily as Least Concern (LC) because of its wide distribution (7 locations; EOO 86,553 km^2; AOO 29,654 km^2) and habitat/altitude range.

SAPINDACEAE

Allophylus abyssinicus (Hochst.) Radlk.

Literature: Davies & Verdcourt, FTEA 1998: 78. ITU: 375. UFT: 195.

Tree 6–21 m high. Evergreen forest, thicket, woodland; 1050–2550 m.

Toro, Mbale, Mubende, Kigezi, Acholi, Karamoja. Widespread in Africa.

Here assessed preliminarily as Least Concern (LC) because of its wide distribution and habitat/altitude range.

Allophylus africanus P. Beauv.

Literature: Davies & Verdcourt, FTEA 1998: 80. ITU: 375.

Shrub or tree 4.5–10 m high. Forest, riverine forest, wooded grassland, thicket; 200–1900 m.

West Nile, Karamoja, Toro, Mengo, Ankole, Kigezi, Mubende, Madi, Acholi, Lango, Mbale, Busoga, Teso. Widespread in Africa.

Here assessed preliminarily as Least Concern (LC) because of its wide distribution and habitat/altitude range.

Allophylus dummeri Baker f.

Literature: Davies & Verdcourt, FTEA 1998: 80. ITU: 376. UFT: 196.

Shrub or tree 2.5–9 m high. Evergreen forest and forest clearings; 1050–1300 m.

Bunyoro, Mengo, Toro. Also in DRC and Angola.

Here assessed preliminarily as Least Concern (LC) because of its wide distribution and habitat type.

Allophylus ferrugineus Taub.

Literature: Davies & Verdcourt, FTEA 1998: 85. ITU: 376.

Synonym: *Allophylus macrobotrys* Gilg; ITU: 376. UFT: 196.

Shrub, climber or tree 1–9 m high. Evergreen forest, thicket; 1050–2400 m.

Kigezi, Mbale, Mengo, Masaka, Toro, Ankole, Bunyoro, Bugisu, Busoga. Widespread in Africa.

Here assessed preliminarily as Least Concern (LC) because of its wide distribution and habitat/altitude range.

Allophylus pseudopaniculatus Baker f.

Literature: Davies & Verdcourt, FTEA 1998: 83.

Synonym: *Allophylus crebriflorus* Baker f.; ITU: 376.

Tree 3–15 m high. Swamp forest, thicket, bushland; 1200–2250 m.

Ankole, Kigezi, Mengo. Widespread in Africa.

Here assessed preliminarily as Least Concern (LC) because of its wide distribution and habitat/altitude range.

Allophylus rubifolius (A. Rich.) Engl.

Literature: Davies & Verdcourt, FTEA 1998: 88.

Shrub or tree 3–7 m high. Scattered tree grassland, forest edges, thicket, woodland; 500–2250 m.

Karamoja, Bunyoro, Mbale, West Nile, Teso, Busoga. Widespread in Africa.

Here assessed preliminarily as Least Concern (LC) because of its wide distribution and habitat/ altitude range.

Blighia unijugata Baker

LITERATURE: Davies & Verdcourt, FTEA 1998: 25, Fig. 7. UFT: 225.

SYNONYM: *Phialodiscus unijugatus* (Baker) Radlk.; ITU: 382.

Tree 6–9(–30) m high. Forest, woodland, bushland; 500–1900 m.

Kigezi, Busoga, Mengo, Masaka, Ankole, Toro, Mubende, Bunyoro. Widespread in Africa.

Here assessed preliminarily as Least Concern (LC) because of its wide distribution and habitat/ altitude range.

Blighia welwitschii (Hiern) Radlk.

LITERATURE: Davies & Verdcourt, FTEA 1998: 26. UFT: 223.

SYNONYM: *Blighia wildemaniana* Radlk.; ITU: 377.

Tree 20–35 m high. Forest; 1050–1150 m.

Bunyoro, Mengo, Toro, Ankole. Widespread in Africa.

Here assessed preliminarily as Least Concern (LC) because of its wide distribution.

Chytranthus atroviolaceus Hutch. & Dalziel

LITERATURE: Davies & Verdcourt, FTEA 1998: 66.

Tree 5–18 m high. Evergreen forest; 1100–1300 m.

Bunyoro. Widespread in Africa.

Here assessed preliminarily as Least Concern (LC) because of its wide distribution.

Deinbollia fulvotomentella Baker f.

LITERATURE: Davies & Verdcourt, FTEA 1998: 68. ITU: 378. UFT: 223.

Tree 1–9 m high. Evergreen forest, riverine and lakeshore forest; 750–1350 m.

Masaka, Mengo, Ankole. Fairly widespread in East Africa.

Here assessed preliminarily as Least Concern (LC) because of its wide distribution and habitat range.

Deinbollia kilimandscharica Taub.

LITERATURE: Davies & Verdcourt, FTEA 1998: 69, fig. 22. ITU: 378. UFT: 223.

Tree or shrub 1–12 m high. Evergreen forest; 1100–2250 m.

Mbale, Mengo, Kigezi. Widespread in Africa.

Here assessed preliminarily as Least Concern (LC) because of its wide distribution and altitude range.

Dodonaea viscosa Jacq.

LITERATURE: Davies & Verdcourt, FTEA 1998: 8, fig. 1. ITU: 378, plate 19.

Shrub or tree 0.5–9 m high. Grassland, bushland, woodland, thicket, forest margins, ruderal sites; 500–2700 m.

Ankole, Kigezi, Mbale, Masaka, Toro. Widespread in the tropics.

Here assessed preliminarily as Least Concern (LC) because of its wide distribution and habitat/ altitude range.

Eriocoelum kerstingii Engl.

LITERATURE: Davies & Verdcourt, FTEA 1998: 28, fig. 8. ITU: 379.

Shrub or tree 8–18(–30) m high. Riverine forest; ± 800 m.

West Nile, Madi. Widespread in Africa.

Here assessed preliminarily as Least Concern (LC) because of its wide distribution.

Glenniea africana (Radlk.) Leenh.

LITERATURE: Davies & Verdcourt, FTEA 1998: 52, fig. 18.

SYNONYM: *Melanodiscus sp. nov.?* of ITU: 381. UFT: 225.

Tree 7–24 m high. Evergreen forest, riverine forest, bushland; 500–2000 m.

West Nile, Bunyoro, Busoga, Mengo. Widespread in Africa.

Here assessed preliminarily as Least Concern (LC) because of its wide distribution and habitat/altitude range.

Haplocoelum foliolosum (Hiern) Bullock

LITERATURE: Davies & Verdcourt, FTEA 1998: 44, fig. 15. ITU: 379.

Shrub or tree 4–20 m high. Scattered tree grassland, thicket, bushland, especially on rock outcrops; 750–1700 m.

West Nile, Busoga, Mengo, Ankole, Bunyoro, Madi, Karamoja, Toro. Widespread in Africa.

Here assessed preliminarily as Least Concern (LC) because of its wide distribution and habitat/altitude range.

Lecaniodiscus cupanioides Benth.

LITERATURE: Davies & Verdcourt, FTEA 1998: 39. ITU: 379.

Tree to 12 m high. Rainforest, riverine forest; 900 m.

Toro. Widespread in Africa.

Here assessed preliminarily as Least Concern (LC) because of its wide distribution.

Lecaniodiscus fraxinifolius Baker

LITERATURE: Davies & Verdcourt, FTEA 1998: 40, fig. 13. UFT: 223.

SYNONYM: *Lecaniodiscus vaughanii* Dunkley; ITU: 380.

Tree to 20(–30) m high. Riverine, lakeside or swamp forest, bushland, scattered tree grassland; 500–1400 m.

Karamoja, Busoga, Mengo. Widespread in Africa.

Here assessed preliminarily as Least Concern (LC) because of its wide distribution and habitat/altitude range.

Lepisanthes senegalensis (Poir.) Leenh.

LITERATURE: Davies & Verdcourt, FTEA 1998: 50, fig. 17.

SYNONYM: *Aphania senegalensis* (Poir.) Radlk.; ITU: 377. UFT: 225.

Tree 6–21 m high. Evergreen forest; 500–1800 m.

West Nile, Acholi, Ankole, Mengo, Toro, Bunyoro, Madi, Karamoja, Mbale, Masaka. Widespread in Africa.

Here assessed preliminarily as Least Concern (LC) because of its wide distribution and altitude range.

Lychnodiscus cerospermus Radlk.

Literature: Davies & Verdcourt, FTEA 1998: 22, fig. 6. ITU: 380. UFT: 223, fig. 412.

Tree 3–20 m high. Moist or dry forest, either primary or secondary, swamp forest, riverine forest; 450–1200 m.

Bunyoro, Toro, Masaka, Mengo, Madi. Also in DRC (including the SW, where it may be locally dominant) and South Sudan.

Though the species has a wide distribution (32 locations; EOO 1,477,920 km^2; AOO 692,859 km^2) and altitude range, some populations have declined tremendously owing to habitat loss, e.g. from Kajansi Forest. Reported by Eilu *et al.* (2004) as rare in Uganda with restricted range and low density; we assess it here preliminarily as Near Threatened (NT A1).

Majidea fosteri (Sprague) Radlk.

Literature: Davies & Verdcourt, FTEA 1998: 18. ITU: 381. UFT: 222.

Tree 15–35 m high. Rainforest; 1000–1200 m.

Bunyoro, Mengo, Masaka, Madi. Widespread in Africa.

Here assessed preliminarily as Least Concern (LC) because of its wide distribution.

Pancovia turbinata Radlk.

Literature: Davies & Verdcourt, FTEA 1998: 60, fig. 20.

Synonym: *Pancovia sp. near turbinata* in the sense of ITU: 382. UFT: 225.

Tree 4–15 m high. Rainforest, riverine forest; 1150–1500 m.

Kigezi, Masaka, Mengo, Toro, Ankole. Widespread in Africa.

Here assessed preliminarily as Least Concern (LC) because of its wide distribution.

Pappea capensis Eckl. & Zeyh.

Literature: Davies & Verdcourt, FTEA 1998: 35, fig. 11.

Synonym: *Pappea ugandensis* Baker; ITU: 382, fig. 76.

Shrub or tree to 15 m high. Scattered tree grassland, woodland, often in rocky sites; 900–2000 m.

Acholi, Ankole, Masaka, Karamoja. Widespread in Africa.

Here assessed preliminarily as Least Concern (LC) because of its wide distribution and altitude range.

Zanha golungensis Hiern

Literature: Davies & Verdcourt, FTEA 1998: 14. ITU: 384. UFT: 223.

Tree 6–24 m high. Woodland, forest; 300–1700 m.

Acholi, Bunyoro, Teso, Mengo, Madi. Widespread in Africa.

Here assessed preliminarily as Least Concern (LC) because of its wide distribution and habitat/altitude range.

SAPOTACEAE

Chrysophyllum albidum G. Don

Literature: Hemsley, FTEA 1968: 9. ITU: 391. UFT: 86.

Tree to 45 m high. Rainforest, riverine forest; 900–1700 m.

Bunyoro, Busoga, Mengo, Masaka, Ankole, Toro, Madi, Mbale. Widespread in Africa.

Here assessed preliminarily as Least Concern (LC) because of its wide distribution and habitat range.

Chrysophyllum beguei Aubrév. & Pellegr.
LITERATURE: Hemsley, FTEA 1968: 14. ITU: 392. UFT: 87.

Tree to 30 m high. Rainforest; 750–900 m.

Toro. Widespread in Africa.

Here assessed preliminarily as Least Concern (LC) because of its wide distribution.

Chrysophyllum gorungosanum Engl.
LITERATURE: Hemsley, FTEA 1968: 10, fig. 1. UFT: 86.

SYNONYM: *Chrysophyllum fulvum* S. Moore; ITU: 392.
SYNONYM: *Chrysophyllum delevoyi* De Wild.; Hemsley, FTEA 1968: 13. UFT: 87.

Tree to 40 m high. Rainforest; 1200–2250 m.

West Nile, Kigezi, Mengo, Toro, Ankole. Widespread in Africa.

Here assessed preliminarily as Least Concern (LC) because of its wide distribution and altitude range.

Chrysophyllum muerense Engl.
See restricted list, page 207.

Chrysophyllum pentagonocarpum Engl. & K. Krause
LITERATURE: Hemsley, FTEA 1968: 15. UFT: 87.

SYNONYM: *Chrysophyllum pruniforme* in the sense of ITU: 395.

Tree to 25 m high. Rainforest; ± 1500 m.

Toro. Widespread in Africa.

Here assessed preliminarily as Least Concern (LC) because of its wide distribution.

Chrysophyllum perpulchrum Hutch. & Dalziel
LITERATURE: Hemsley, FTEA 1968: 12. ITU: 395, fig. 80b. UFT: 86.

Tree to 40 m high. Rainforest; 800–1200 m.

Bunyoro, Mengo. Widespread in Africa.

Here assessed preliminarily as Least Concern (LC) because of its wide distribution.

Chrysophyllum pruniforme Engl.
LITERATURE: Hemsley, FTEA 1968: 17. ITU: 395. UFT: 87.

Tree to 30 m high. Rainforest; ± 1500 m.

Kigezi, possibly Toro. Widespread in Africa.

Here assessed preliminarily as Least Concern (LC) because of its wide distribution.

Englerophytum natalense (Sond.) T. D. Penn.
LITERATURE: Pennington 1991: 243.

SYNONYM: *Bequaertiodendron natalense* (Sond.) Heine & J. H. Hemsl.; Hemsley, FTEA 1968: 19, fig. 2.

Tree to 25 m high. Rainforest, riverine and groundwater forest; 500–1800 m.

Ankole. Widespread in Africa.

Here assessed preliminarily as Least Concern (LC) because of its wide distribution and altitude range.

Englerophytum oblanceolatum (S. Moore) T. D. Penn.

LITERATURE: Pennington 1991: 252.

SYNONYM: *Bequaertiodendron oblanceolatum* (S. Moore) Heine & J. H. Hemsl.; Hemsley, FTEA 1968: 23. UFT: 84.

SYNONYM: *Chrysophyllum natalense* of ITU: 392.

Shrub or tree to 10 m high. Rainforest; 900–1700 m.

Toro, Bunyoro, Mengo. Widespread in Africa.

Here assessed preliminarily as Least Concern (LC) because of its wide distribution.

Manilkara butugi Chiov.

LITERATURE: Hemsley, FTEA 1968: 67. UFT: 88.

SYNONYM: *Manilkara multinervis* in the sense of ITU: 399.

SYNONYM: *Manilkara sp.* of ITU: 399.

SYNONYM: *Mimusops sp.* of ITU: 402.

Tree to 35 m high. Rainforest, riverine forest; 1500–2300 m.

West Nile, Karamoja, Mbale, Ankole, West Nile. Also in South Sudan, Ethiopia, Kenya.

Here assessed preliminarily as Least Concern (LC) because of its distribution range.

Manilkara dawei (Stapf) Chiov.

LITERATURE: Hemsley, FTEA 1968: 65. ITU: 399, fig. 81a. UFT: 88.

Tree to 25 m high. Rainforest, riverine forest; 1100–1600 m.

West Nile, Busoga, Mengo, Ankole, Kigezi, Toro. Also in CAR, western DRC, NW Tanzania.

Here assessed preliminarily as Least Concern (LC) because of its distribution range.

Manilkara obovata (Sabine & G. Don) J. H. Hemsl.

LITERATURE: Hemsley, FTEA 1968: 68. UFT: 88.

SYNONYM: *Manilkara cuneifolia* (Baker) Dubard; ITU: 396, fig. 82bc.

SYNONYM: *Manilkara multinervis* (Baker) Dubard; Hemsley, FTEA 1968: 69. UFT: 87.

SYNONYM: *Manilkara schweinfurthii* (Engl.) Dubard; ITU: 399, fig. 82a.

Tree 6–35 m high. Rainforest, riverine and swamp forest, woodland; 900–1300 m.

Masaka, West Nile, Acholi, Mengo, Madi, Lango. Widespread in Africa.

Here assessed preliminarily as Least Concern (LC) because of its wide distribution and habitat range.

Mimusops bagshawei S. Moore

LITERATURE: Hemsley, FTEA 1968: 57. ITU: 400. UFT: 87.

Tree to 40 m high. Rainforest; 1100–2400 m.

Bunyoro, Ankole, Mengo, Masaka, Busoga, Toro. Also in South Sudan, W Kenya, NW Tanzania.

EOO 440,650 km^2; AOO 239,016 km^2; 52 collections from 33 locations.Here assessed preliminarily as Least Concern (LC) because of its distribution and altitude range.

Mimusops kummel A. DC.

LITERATURE: Hemsley, FTEA 1968: 54. ITU: 400, fig. 83. UFT: 87.

SYNONYM: *Mimusops sp. near warneckei* of ITU: 402.

Shrub or tree to 25 m high. Dry evergreen forest, riverine forest, wooded grassland; 500–2100 m.

Mimosops kummel in fruit in Kabwoya Wildlife Reserve (photo: James Kalema)

West Nile, Bunyoro, Teso, Mengo, Mbale, Acholi, Karamoja. Widespread in Africa.

Here assessed preliminarily as Least Concern (LC) because of its wide distribution and habitat/altitude range.

Pouteria adolfi-friedericii (Engl.) A. Meeuse

LITERATURE: fide African Plant Database

SYNONYM: *Aningeria adolfi-friedericii* (Engl.) Robyns & Gilbert; Hemsley, FTEA 1968: 28, fig. 4. ITU: 385. UFT: 84.

Tree to 50 m high. Rainforest, rarely riverine forest; 1400–2500 m.

Acholi, Toro, Mbale, Kigezi. Also in E DRC, Rwanda, W Kenya, Tanzania, Zambia, Malawi.

Here assessed preliminarily as Least Concern (LC) because of its wide distribution and altitude range.

Pouteria altissima (A. Chev.) Baehni

LITERATURE: Pennington 1991: 203.

SYNONYM: *Aningeria altissima* (A. Chev.) Aubrév. & Pellegr.; Hemsley, FTEA 1968: 27. ITU: 386, fig. 77. UFT: 83.

Tree to 50 m high. Rainforest, riverine forest; 1000–1700 m.

West Nile, Bunyoro, Mengo, Ankole, Kigezi, Toro, Madi, Acholi, Mbale, Busoga. Widespread in Africa.

Here assessed preliminarily as Least Concern (LC) because of its wide distribution.

Synsepalum brevipes (Baker) T. D. Penn.
LITERATURE: Pennington 1991: 248.
SYNONYM: *Pachystela brevipes* (Baker) Engl.; Hemsley, FTEA 1968: 36, fig. 6. ITU: 402. UFT: 88.

Tree to 35 m high. Rainforest, lakeside and riverine forest; 500–1500 m.

Ankole, Masaka, Mengo, Toro, Bunyoro. Widespread in Africa.

Here assessed preliminarily as Least Concern (LC) because of its wide distribution and habitat/altitude range.

Synsepalum cerasiferum (Welw.) T. D. Penn.
LITERATURE: Pennington 1991: 248.
SYNONYM: *Afrosersalisia cerasifera* (Welw.) Aubrév.; Hemsley, FTEA 1968: 42, fig. 8. UFT: 90.
SYNONYM: *Sersalisia edulis* S. Moore; ITU: 403.

Tree to 40 m high. Rainforest, groundwater and riverine forest; 300–1500 m.

West Nile, Bunyoro, Masaka, Mengo, Mbale, Ankole. Widespread in Africa.

Here assessed preliminarily as Least Concern (LC) because of its wide distribution and habitat/altitude range.

Synsepalum msolo (Engl.) T. D. Penn.
LITERATURE: Pennington 1991: 249.
SYNONYM: *Pachystela msolo* (Engl.) Engl.; Hemsley, FTEA 1968: 38. ITU: 403. UFT: 88.

Tree to 50 m high. Rainforest, riverine forest; 50–1400 m.

Toro, Mengo, Bunyoro, Busoga. Widespread in Africa.

Here assessed preliminarily as Least Concern (LC) because of its wide distribution and altitude range.

Vitellaria paradoxa C. F. Gaertn.
LITERATURE: Pennington 1991: 128.
SYNONYM: *Butyrospermum paradoxum* (C. F. Gaertn.) Hepper; Hemsley, FTEA 1968: 49, fig. 10.
SYNONYM: *Butyrospermum parkii* (G. Don) Kotschy of ITU: 388, fig. 78.

Tree to 20 m high. Tree grassland, also planted; 950–1500 m.

West Nile, Teso, Mengo, Madi, Acholi, Lango, Karamoja, Mbale. Widespread in Africa.

Assessed in 1998 as Vulnerable (VU A1cd); IUCN Red List, consulted 1 June 2010, due to 'overexploitation'.

While this species has a wide distribution, here we assess it preliminarily as Near Threatened (NT A4) owing to the widespread heavy commercial exploitation for charcoal burning and other uses as mentioned in many countries of its range (e.g. Boffa *et al.* 1996 for Burkina Faso, under increasing pressure from agriculture, drought and parasitism; Djossa *et al.* 2008 for Benin, who report severe reduction of saplings in farmed lands; Odebiyi *et al.* 2004 for Nigeria; Fondoun & Onana 2001 for Cameroun, used for wood/fuelwood, charcoal, fruit/food; across its range there is heavy use for fuelwood and charcoal, and low regeneration rates are giving cause for concern. In Uganda, charcoal burning is a big threat to this species. It is slow growing and farmers do not really seem to like planting it, although they usually spare the seedlings on-farm.

Vitellaria paradoxa: tree felling for charcoal burning is the biggest threat to the species in Uganda (photo: James Kalema, Apr. 2010).

SCROPHULARIACEAE

Buddleja polystachya Fresen.

Literature: Bruce & Lewis, FTEA 1960; ITU: 165; UFT: 331 (all as *Buddleia*, in Loganiaceae).

Shrub or tree to 5 m. Upland grassland and bushland, forest clearings; 1000–2700 m.

Karamoja, Mbale. From Eritrea to Tanzania.

Here assessed preliminarily as Least Concern (LC) because of its wide distribution and habitat range.

SIMAROUBACEAE

Brucea antidysenterica J. F. Mill.

Literature: Stannard, FTEA 2000: 7, fig. 3. ITU: 408.

Shrub or tree to 10(–15) m high. Evergreen forest and forest edges, secondary growth, grassland; 1400–2800 m.

Toro, Kigezi, Mengo, Ankole, Karamoja. Widespread in Africa.

Here assessed preliminarily as Least Concern (LC) because of its wide distribution and habitat/altitude range.

Harrisonia abyssinica Oliv.

Literature: Stannard, FTEA 2000: 2, fig. 1. ITU: 409. UFT: 207.

Synonym: *Harrisonia occidentalis* Engl.; ITU: 409.

Shrub or tree 1–13 m high. Dry evergreen forest and forest edges, wooded grassland, thickets, riverine; 500–1550 m.

Karamoja, Kigezi, Teso, probably all districts. Widespread in Africa.

Here assessed preliminarily as Least Concern (LC) because of its wide distribution and habitat/altitude range.

Quassia undulata (Guill. & Perr.) D. Dietr.
LITERATURE: Stannard, FTEA 2000: 11, fig. 4.
SYNONYM: *Hannoa longipes* (Sprague) G. C. C. Gilbert; UFT: 207.
SYNONYM: *Hannoa undulata* (Guill. & Perr.) Planch.; Lebrun & Stork.
SYNONYM: *Odyendea longipes* Sprague; ITU: 411.

Shrub or 'large' tree. Thickets, scattered tree grassland, wooded grassland, forest, riverine and semi-swamp forest; 150–2500 m.

Toro, Kigezi, Ankole. Widespread in Africa.

Assessed as Least Concern (LC) by Sacande, Sanou & Beentje in *Guide de terrain des arbres du Burkina Faso* (in press).

SOLANACEAE

Brugmansia suaveolens (Willd.) Bercht. & J. Presl
LITERATURE: Edmonds, FTEA 2012: 47.

Woody herb or tree to 6 m. Grown as ornamental, but a common escape in forest; 800–1800 m.

Masaka. Originally from Central America.

Not assessed here.

Discopodium eremanthum Chiov.
LITERATURE: Edmonds, FTEA 2012: 64.

Shrub or tree to 5 m high. Moorlands, forest and woodland; 3000–3500 m.

Mbale. Also in Ethiopia, Kenya, Tanzania, and possibly in DRC.

Here assessed preliminarily as Least Concern (LC) because of its distribution and habitat/altitude range.

Discopodium penninervium Hochst.
LITERATURE: Edmonds, FTEA 2012: 64; ITU: 413.

Shrub or tree to 6 m high. Forest; 1400–2500 m.

Mengo, Ankole, Kigezi, Toro, Bunyoro, Acholi, Karamoja, Mbale. Also from Nigeria to Ethiopia and south to Malawi.

Here assessed preliminarily as Least Concern (LC) because of its wide distribution and habitat/altitude range.

Lycium shawii Roem. & Schult.
LITERATURE: Edmonds, FTEA 2012: 50; ITU: 413.
SYNONYM: *Lycium europaeum* of many authors.

Shrub or tree to 4.5 m. Wooded grassland, grass/bushland and ruderal sites; 0–2000 m.

Karamoja. Also in Egypt, Sudan, South Sudan, Somalia, Eritrea and Ethiopia southwards to Botswana, Zambia, Zimbabwe, Malawi, Namibia and NE South Africa; widespread from Mediterranean Europe, through the Middle East and Arabia to western India.

Here assessed preliminarily as Least Concern (LC) because of its wide distribution and habitat/altitude range.

Solanum aculeastrum Dunal

Literature: Vorontsova & Knapp in Edmonds, FTEA 2012: 198; ITU: 413.

Shrub or tree to 6 m high. Wooded grassland, secondary bushland, cultivated; 1060–2600 m.

Kigezi, Toro, Mbale, Busoga. Widely distributed in the tropics, and frequently cultivated as a hedge plant.

Here assessed preliminarily as Least Concern (LC) because of its wide distribution and habitat/altitude range.

Solanum betaceum Cav.

Literature: Edmonds, FTEA 2012: 104; ITU: 413.

Synonym: *Cyphomandra betacea* (Cav.) Sendtn.

Shrub or tree to 7 m. Grown as an ornamental, and sometimes an escape in forest; 1100–2050 m.

Toro, Mengo. Originally from South America.

Not assessed here.

Solanum giganteum Jacq.

Literature: Edmonds, FTEA 2012: 148; ITU: 413.

Shrub or tree to 8 m high. Forest edge, secondary bushland; 800–2450 m.

Mengo, Toro, Bunyoro, Mbale, Busoga. Probably native to the Cape area of South Africa, but now common from Cameroon to Ethiopia, and south to South Africa.

Here assessed preliminarily as Least Concern (LC) because of its wide distribution and habitat/altitude range.

Solanum mauritianum Scop.

Literature: Edmonds, FTEA 2012: 146; ITU: 413.

Shrub or tree to 7 m. Grown as an ornamental, often naturalised in forest and roadsides; 1150–2800 m.

Mengo, Jinja. Originally from South America.

Not assessed here.

Solanum tettense Klotzsch

Literature: Edmonds, FTEA 2012: 158; ITU: 413.

Shrub or tree to 4 m. Grassland and riverine habitats; 800–1400 m.

Teso, Ankole. Also in DRC, Rwanda, Tanzania, Zambia, Malawi and Mozambique.

Here assessed preliminarily as Least Concern (LC) because of its wide distribution and habitat/altitude range.

STILBACEAE

Halleria lucida L.

Literature: Ghazanfar *et al.*, FTEA 2008.

Shrub or tree 3–10 m. Montane forest, especially common along forest margins where it may be dominant; 900–2700 m.

Acholi, Karamoja, Mbale. Extending from Ethiopia to South Africa and to Angola; SW Arabia.

Here assessed preliminarily as Least Concern (LC) because of its wide distribution and habitat type and altitude range.

Part of Matiri Forest Reserve being cleared for cultivation (photo: James Kalema, Apr. 2010).

THYMELAEACEAE

Craterosiphon beniense Domke

LITERATURE: Peterson, FTEA 1978: 8, fig. 3.

Liana or shrubby tree (?size). Evergreen forest and forest edges; 1200 m.

Bunyoro, Mengo, Masaka. Widespread in central Africa.

Here assessed preliminarily as Least Concern (LC) because of its wide distribution and habitat type.

Dicranolepis buchholzii Engl. & Gilg

LITERATURE: Peterson, FTEA 1978: 4.

Shrub or tree to 4(–9) m high. Evergreen forest; 1400–1800 m.

Kigezi. Widespread in central Africa.

Here assessed preliminarily as Least Concern (LC) because of its wide distribution.

Dicranolepis incisa A. Robyns

See restricted list, page 208.

Gnidia glauca (Fresen.) Gilg

LITERATURE: Peterson, FTEA 1978: 32.

SYNONYM: *Lasiosiphon glaucus* Fresen.; ITU: 424, plate 20.

Shrub or tree to 15(–24) m high. Forest edges and associated wooded grassland and bushland; 1500–3300 m.

Acholi, Mbale. Widespread in Africa.

Here assessed preliminarily as Least Concern (LC) because of its wide distribution and habitat/altitude range.

Gnidia lamprantha Gilg

Literature: Peterson, FTEA 1978: 31.

Synonym: *Lasiosiphon lampranthus* (Gilg) H. Pearson; ITU: 425.

Shrub or tree to 3(–5) m high. Wooded grassland, bushland; 1050–2100 m.

Acholi, Ankole, Mbale, Toro. Fairly widespread in East Africa.

Here assessed preliminarily as Least Concern (LC) because of its wide distribution and habitat/altitude range.

Peddiea fischeri Engl.

Literature: Peterson, FTEA 1978: 12, fig. 4. ITU: 425. UFT: 158.

Shrub or tree to 10 m high. Rainforest, secondary forest, riverine or lakeshore forest, swamp forest, bamboo zone, rarely in thicket; 1050–2400 m.

Kigezi, Mengo, Masaka, Toro. Also in E DRC, Rwanda, Burundi, Angola, Zambia.

EOO 1,952,370 km^2; AOO 1,109,230 km^2; 107 collections from 75 locations, here assessed preliminarily as Least Concern (LC) because of its wide distribution and habitat/altitude range.

Peddiea rapaneoides Engl.

Literature: Peterson, FTEA 1978: 14.

Shrub or tree to 10 m high. Forest, associated bushland and thickets; 950–2400 m.

Kigezi, Busoga, Mengo. Widespread in Africa.

Here assessed preliminarily as Least Concern (LC) because of its wide distribution and habitat/altitude range.

ULMACEAE

Celtis adolfi-fridericii Engl.

Literature: Polhill, FTEA 1966: 9. ITU: 430, fig. 89a. UFT: 108.

Tree to 36 m high. Rainforest; 600–900 m.

Bunyoro, Toro. Widespread in Africa.

Here assessed preliminarily as Least Concern (LC) because of its wide distribution.

Celtis africana Burm. f.

Literature: Polhill, FTEA 1966: 4. UFT: 106.

Celtis kraussiana Bernh. of ITU: 434, fig. 89b.

Tree 5–35 m high. Moist, dry and riverine forest; 500–2400 m.

Acholi, Toro, Mengo, Ankole, Mubende, Bunyoro, Madi, Karamoja, Mbale, Busoga. Widespread in Africa.

Here assessed preliminarily as Least Concern (LC) because of its wide distribution and habitat/altitude range.

Celtis gomphophylla Baker

LITERATURE: fide African Plant Database

SYNONYM: *Celtis durandii* Engl.; Polhill, FTEA 1966: 5, fig. 2. UFT: 106.
SYNONYM: *Celtis durandii* var. *ugandensis* (Rendle) Rendle of ITU: 432.

Tree 5–25 m high. Rainforest; 300–2000 m.

Bunyoro, Mengo, Kigezi, Toro, Busoga, Mbale. Widespread in Africa.

Here assessed preliminarily as Least Concern (LC) because of its wide distribution and altitude range.

Celtis mildbraedii Engl.

LITERATURE: Polhill, FTEA 1966: 7. UFT: 105.

SYNONYM: *Celtis soyauxii* in the sense of ITU: 435, fig. 89c, photo 55.

Tree 3–45 m high. Rainforest; 300–1600 m.

Bunyoro, Busoga, Mengo, Toro. Widespread in Africa.

Here assessed preliminarily as Least Concern (LC) because of its wide distribution and altitude range.

Celtis philippensis Blanco

LITERATURE: fide African Plant Database.

SYNONYM: *Celtis wightii* Planch.; Polhill, FTEA 1966: 9. UFT: 108.
SYNONYM: *Celtis brownii* Rendle; ITU: 430, fig. 91b.

Shrub or tree 3–20 m high. Rainforest, swamp and riverine forest; 500–1200 m.

Bunyoro, Toro, Mengo. Widespread in Africa and Asia.

Here assessed preliminarily as Least Concern (LC) because of its wide distribution and habitat/altitude range.

Celtis toka (Forssk.) Hepper & J. R. I. Wood

LITERATURE: fide African Plant Database.

SYNONYM: *Celtis integrifolia* Lam.; Polhill, FTEA 1966: 7. ITU: 432.

Tree to 25 m high. Riverine forest, woodland; 600–1200 m.

West Nile, Acholi, Madi. Widespread in Africa.

Here assessed preliminarily as Least Concern (LC) because of its wide distribution and habitat range.

Celtis zenkeri Engl.

LITERATURE: Polhill, FTEA 1966: 8. ITU: 435, fig. 91a. UFT: 106.

Tree 10–30 m high. Rainforest; 250–1200 m.

Bunyoro, Mengo, Kigezi, Toro, Busoga. Widespread in Africa.

Here assessed preliminarily as Least Concern (LC) because of its wide distribution and altitude range.

Chaetacme aristata Planch.

LITERATURE: Polhill, FTEA 1966: 12, fig. 4. ITU: 436. UFT: 156.

Shrub or tree to 10 m high. Rainforest, riverine forest; 900–2100 m.

Karamoja, Ankole, Mengo, Kigezi, Toro, Mubende, Bunyoro, West Nile, Madi, Acholi, Mbale, Busoga. Widespread in Africa.

Here assessed preliminarily as Least Concern (LC) because of its wide distribution and habitat/altitude range.

Holoptelea grandis (Hutch.) Mildbr.

Literature: Polhill, FTEA 1966: 3, fig. 1. ITU: 436, fig. 92. UFT: 108.

Tree 12–40 m high. Moist, dry and riverine forest; 750–1200 m.

Bunyoro, Mbale, Mengo, Toro, West Nile, Madi, Busoga. Widespread in Africa.

Here assessed preliminarily as Least Concern (LC) because of its wide distribution and habitat range.

Trema orientalis (L.) Blume

Literature: Polhill, FTEA 1966: 10, fig. 3. UFT: 108.

Synonym: *Trema guineensis* (Schumach. & Thonn.) Ficalho; ITU: 438.

Shrub or tree to 12 m high. Rainforest and its clearings, riverine forest; 500–2100 m.

West Nile, Ankole, Mengo, Masaka, Kigezi, Toro, Bunyoro, Madi, Acholi, Karamoja, Teso, Mbale, Busoga. Widespread in Africa and Asia.

Here assessed preliminarily as Least Concern (LC) because of its wide distribution and habitat/altitude range.

URTICACEAE

Obetia radula (Baker) B. D. Jacks.

Literature: Friis, FTEA 1989: 10, fig. 3.

Synonym: *Obetia pinnatifida* of ITU: 441.

Tree 5–13 m high. Forest edges, evergreen bushland on rocky hills; 700–2000 m.

Ankole, Busoga, Mengo, Toro, Bunyoro, Mbale. Widespread in East Africa and Madagascar.

Here assessed preliminarily as Least Concern (LC) because of its wide distribution, habitat type and altitude range.

VERBENACEAE

Duranta erecta L.

Literature: Verdcourt, FTEA 1992: 48, fig. 7.

Shrub or tree 2–6 m high. Forest edges, secondary bushland, riverine thicket; 1050–1950 m.

Moroto, Teso, Mengo, Ankole. Originally from the Americas, naturalised.

Here assessed preliminarily as Least Concern (LC) because of its wide distribution and habitat/altitude range.

VIOLACEAE

Rinorea angustifolia (Thouars) Baill.
LITERATURE: Grey-Wilson, FTEA 1986: 20, fig. 6.

Shrub or tree to 10 m high. Evergreen forest; 1150–1500 m.

Ankole. Widespread in Africa.

Here assessed preliminarily as Least Concern (LC) because of its wide distribution.

Rinorea beniensis Engl.
See restricted list, page 209.

Rinorea brachypetala (Turcz.) Kuntze
LITERATURE: Grey-Wilson, FTEA 1986: 7. UFT: 130.

SYNONYM: *Rinorea poggei* Engl.; ITU: 448.

Shrub or tree to 7 m high. Evergreen forest; 850–1900 m.

Acholi, Bunyoro, Mengo, Toro. Widespread in Africa.

Here assessed preliminarily as Least Concern (LC) because of its wide distribution and altitude range.

Rinorea dentata (P. Beauv.) Kuntze
LITERATURE: Grey-Wilson, FTEA 1986: 12. ITU: 446, partly. UFT: 130.

Shrub or tree to 6 m high. Evergreen forest, semi-swamp forest; 1200–1300 m.

Masaka, Mengo, Kigezi. Widespread in Africa.

Here assessed preliminarily as Least Concern (LC) because of its wide distribution.

Rinorea ilicifolia (Oliv.) Kuntze
LITERATURE: Grey-Wilson, FTEA 1986: 4, fig. 1. ITU: 448. UFT: 130.

Shrub or tree to 5 m high. Evergreen forest; 500–1800 m.

Bunyoro, Mengo, Toro, Madi, Busoga. Widespread in Africa.

Here assessed preliminarily as Least Concern (LC) because of its wide distribution and altitude range.

Rinorea oblongifolia (C. H. Wright) Chipp
LITERATURE: Grey-Wilson, FTEA 1986: 9. ITU: 448. UFT: 130.

Shrub or tree to 15 m high. Evergreen forest; 1150–1450 m.

Ankole, Mengo, Masaka, Kigezi, Bunyoro. Widespread in Africa.

Here assessed preliminarily as Least Concern (LC) because of its wide distribution.

Rinorea tschingandaensis Taton
See restricted list, page 210.

ZAMIACEAE

Encephalartos equatorialis P. J. H. Hurter
See restricted list, page 211.

Encephalartos macrostrobilus Scott Jones & Wynants
See restricted list, page 212.

Encephalartos septentrionalis Schweinf.
See restricted list, page 213.

Encephalartos whitelockii P. J. H. Hurter
See restricted list, page 214.

Encephalartos equatorialis — female cones (photo: James Kalema.)

SPECIES WITH RESTRICTED DISTRIBUTION — FULL CONSERVATION ASSESSMENTS

Note: altitudes on this list are over the global distribution of the species; in the main list, they are of the Ugandan or East African altitude range!

Flowering and fruiting times are given in graphical format, with the months of the year along the horizontal axis, and the percentage of records along the vertical. The number of records is given as "n=".

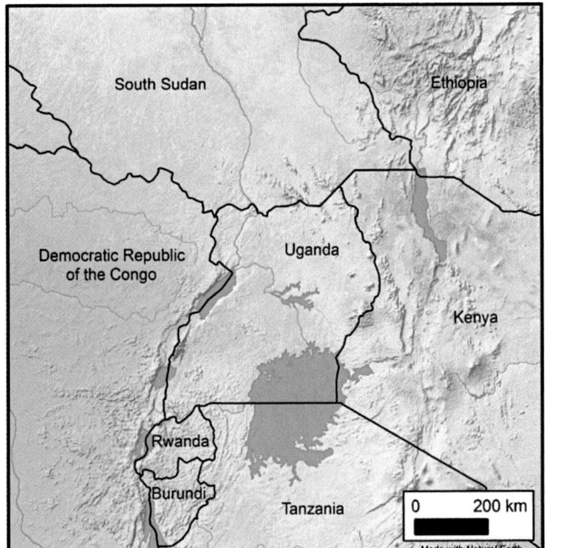

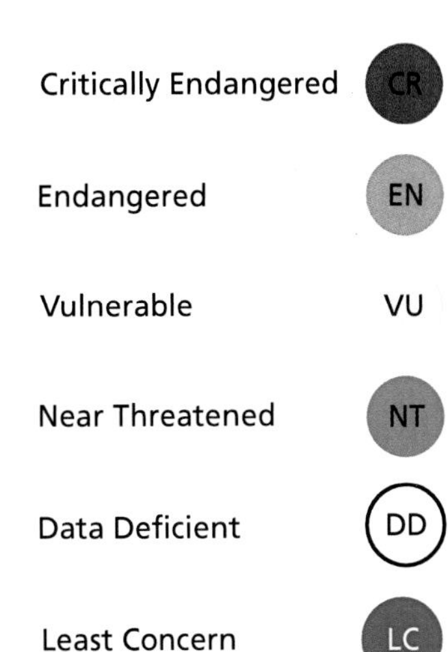

LC **_Dasylepis eggelingii_** J. B. Gillett **Achariaceae**

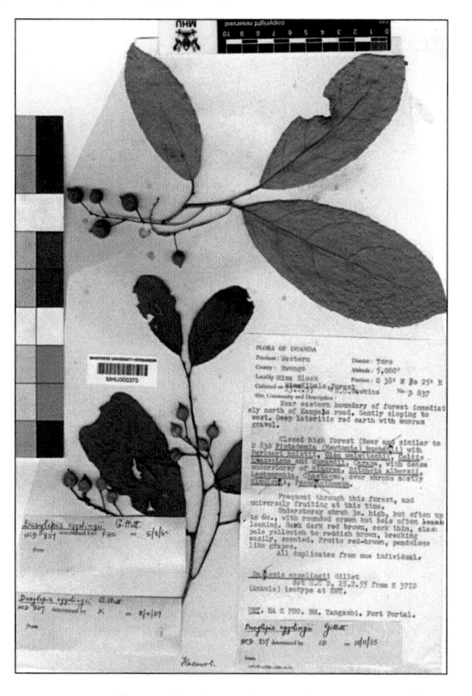

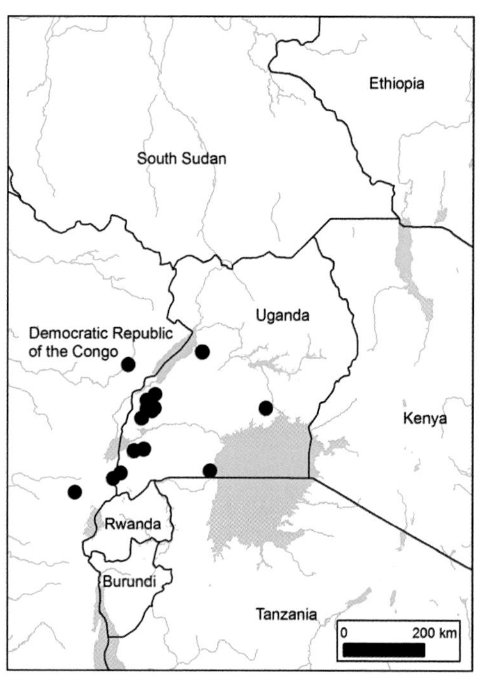

TYPE: Uganda, Ankole, Kalinzu forest, *Eggeling* 3710 (K holo., BM, BR iso.)

LITERATURE: Sleumer, FTEA 1975: 7. ITU: 146. UFT: 126.

DESCRIPTION: Shrub or tree to 10 m high; bark dark red-brown. Leaves oblong, 10–19 × 3–7 cm, finely toothed on margin. Flowers on pendulous stalks, with white or pink sepals ± 7 mm long; petals absent. Fruit purple-red, ± globose, 18 mm across.

HABITAT: Primary to secondary rainforest; 1050–1700 m.

DISTRIBUTION: Toro, Masaka, Mengo, Ankole, Bunyoro, Kigezi; also in eastern DRC.

PROTECTED AREA OCCURRENCE: Kibale & Bwindi Impenetrable National Parks; Kalinzu, Kasyoha-Kitomi, Mabira, Budongo, Sango Bay Forest Reserves.

RED LIST ASSESSMENT: Said to be common (1938) to locally abundant in Kalinzu Forest (1969), and frequent in Kibale (1955); EOO 96,147 km²; AOO 29,250 km²; 28 collections from 18 locations. Assessed here as Least Concern (LC) due to being widespread with a large EOO and AOO, the number of locations and habitat + altitude range and the reported size of populations; many of the locations occur in protected areas, and we know of no specific threats.

FLOWERING AND FRUITING TIMES:

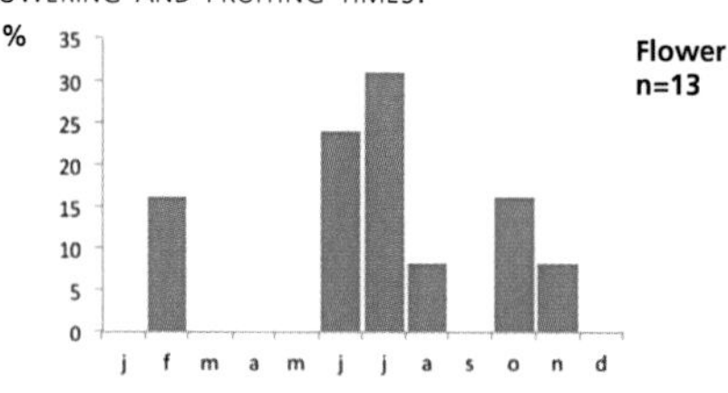

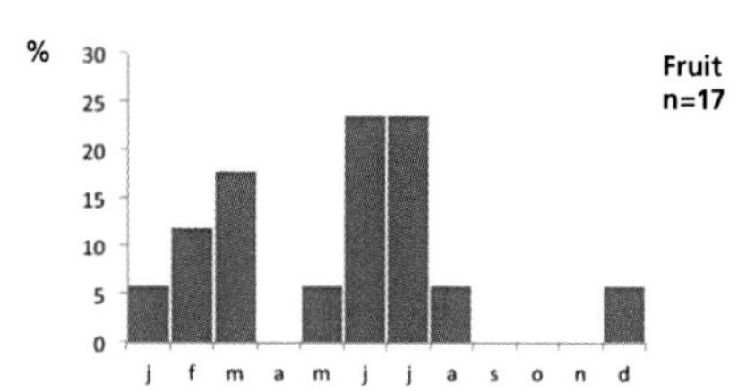

Uvariodendron magnificum Verdc. **Annonaceae** EN

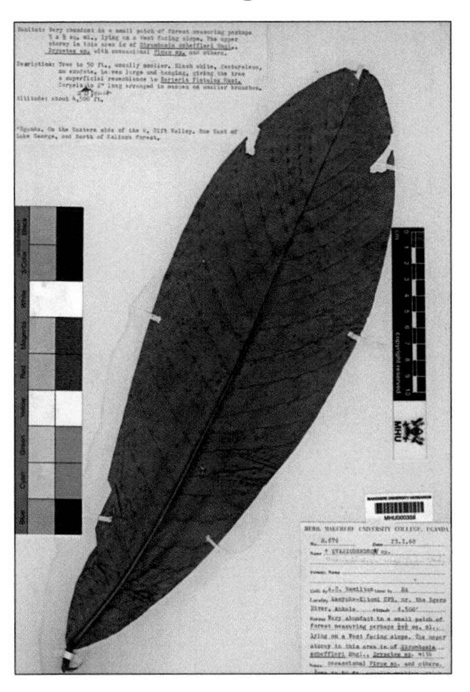

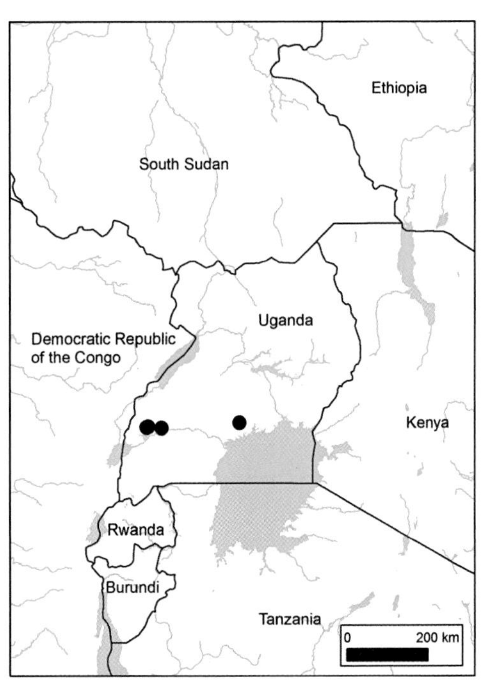

TYPE: Uganda, Kashoya-Kitomi Central Forest reserve, *Okodi in Hamilton* 696 (K, holo.)

LITERATURE: Verdcourt, FTEA 1971: 38.

DESCRIPTION: Tree 7–15 m high; bark thin and gray. Leaves 20–75 × 6–21 cm, red and drooping when young. Flowers on trunk and thicker branches, usually solitary, with sepals 30–55 mm long and white to purple petals 60–70 × 40–50 mm. Fruit of very many 'part-fruits' or monocarps, each obovoid and 3–6 × 1–3 cm, beaked at apex.

HABITAT: Evergreen forest in valleys; 1050–1350 m.

DISTRIBUTION: Ankole, Masaka. Uganda endemic.

PROTECTED AREA OCCURRENCE: Kasyoha-Kitomi Forest Reserve is the only protected area it is known from.

LOCAL NAMES AND USES: None recorded.

RED LIST ASSESSMENT: Highly localised forest obligate with disjunct distribution in two localities which form two locations; Kasyoha-Kitomi (4 collections, 1968–1969, said to be very common in small patch), threatened with continuing habitat loss and degradation through forest cutting; and Lutoboka on Ssese Island (2001), threatened with growing of oil palms by private developers in Ssese, where expanding tourism development at the beaches in Lutoboka leads to an inferred decline in population, area of occupancy and habitat quality. EOO= 419.04 km², AOO= 12.00 km²; locations 2; Here assessed as Endangered EN B1ab(ii,iii,v)+B2ab(ii,iii,v).

FLOWERING AND FRUITING TIMES:

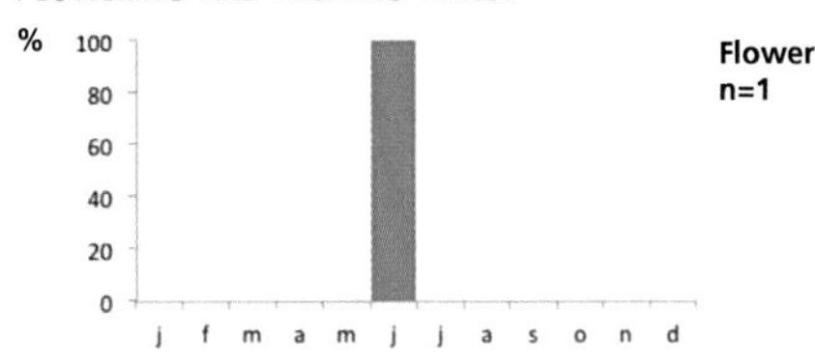

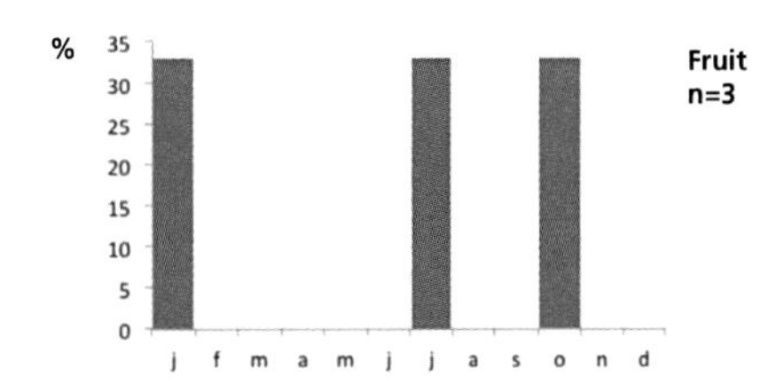

LC **_Dendrosenecio adnivalis_** (Stapf) E. B. Knox **Asteraceae /Compositae**

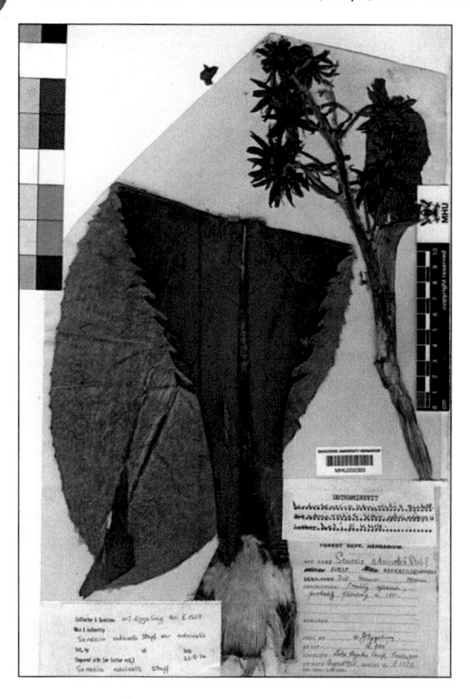

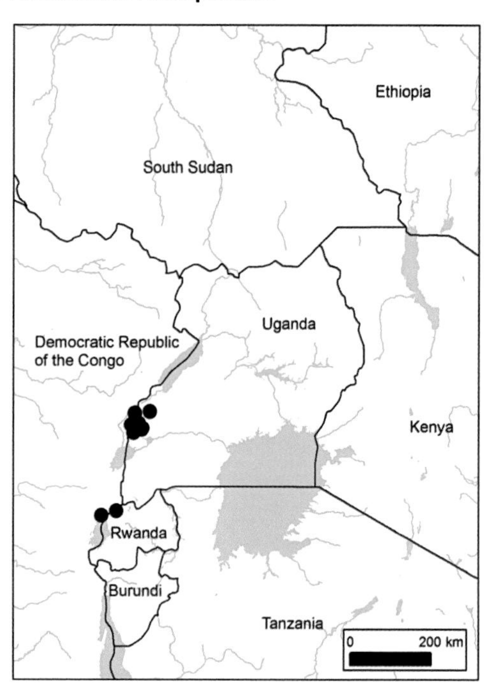

TYPE: Uganda, Ruwenzori, *Dawe* 663 (K lecto.)

LITERATURE: Knox in Beentje *et al.*, FTEA 2005: 560.

SYNONYMS: *Senecio adnivalis* Stapf; ITU: 96. UFT: 79.
Senecio erioneuron Cotton; ITU: 97.
Senecio petiolatus Hauman; ITU: 98.

DESCRIPTION: Tree to 10 m high, trunk to 40 cm in diameter; stem with densely packed leaf rosettes (each of 20–60 leaves) and old, dead leaves hanging down. Leaves to 100 × 26 cm. Flowers in panicles to 160 × 60 cm, with hanging capitula of yellow heads, with or without ray florets. Fruits minute, hidden in the involucre.

HABITAT: Afro-alpine zone, in valley bottoms and bogs, where it can form dense stands; 3000–4400 m.

DISTRIBUTION: Toro. Restricted range: Rwenzori endemic, mainly from around Bujuku.

PROTECTED AREA OCCURRENCE: Rwenzori Mountains National Park.

LOCAL NAMES AND USES: None recorded.

RED LIST ASSESSMENT: EOO 8,311 km²; AOO 6,042 km²; 63 collections from 22 locations. Very common locally, forming stands; protected in the National Park with no known threats. Conservation assessment for species: despite the restricted EOO, which would justify a vulnerable assessment, the occurrence in a protected area coupled with the large altitude range and the large numbers occurring, lead us to classify this as Least Concern (LC). This agrees with FTEA, which has all taxa within this species as Least Concern. We assess each of the subspecies and varieties accordingly:

- ***Dendrosenecio adnivalis*** (Stapf) E. B. Knox subsp. ***adnivalis*** var. ***adnivalis***: Endemic to Ruwenzori Mts (both Uganda and DRC), on small hummocks in sedge bogs, at altitudes from 3250–4500m. It occurs in the Ruwenzori Mts National Park in rather large numbers, and assessed here as Least Concern (LC).
- ***Dendrosenecio adnivalis*** (Stapf) E. B. Knox subsp. ***adnivalis*** var. ***petiolatus*** (Hedberg) E. B. Knox: endemic to Ruwenzori Mts (both Uganda and DRC), on moist slopes and along drainage lines, at altitudes from 3600–4250 m. It occurs in the National Park in rather large numbers, and assessed here as Least Concern (LC).
- ***Dendrosenecio adnivalis*** (Stapf) E. B. Knox subsp. ***friesiorum*** (Mildbr.) E. B. Knox: endemic to Ruwenzori Mts (both Uganda and DRC), on the W flank of Mt Stanley down to Lac Vert and Kiondo, and the Batoda Plateau down to Lake Batoda, at altitudes from 3900–4200 m. It occurs in the Ruwenzori Mts National Park in rather large numbers, and assessed here as Least Concern (LC).

FLOWERING AND FRUITING TIMES: Of the species as a whole.

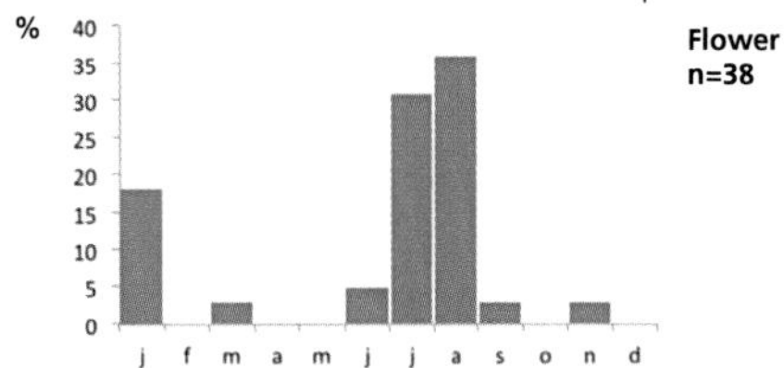

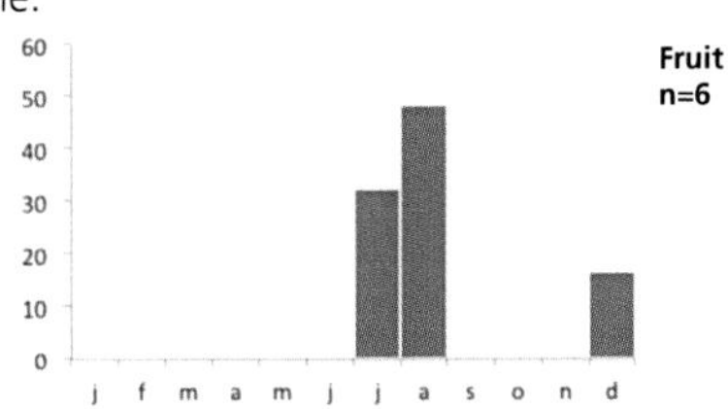

LC **_Dendrosenecio elgonensis_** (T. C. E. Fr.) E. B. Knox **Asteraceae /Compositae**

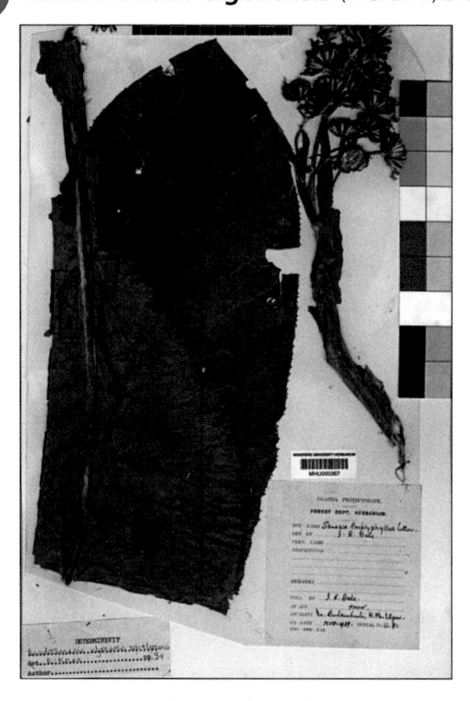

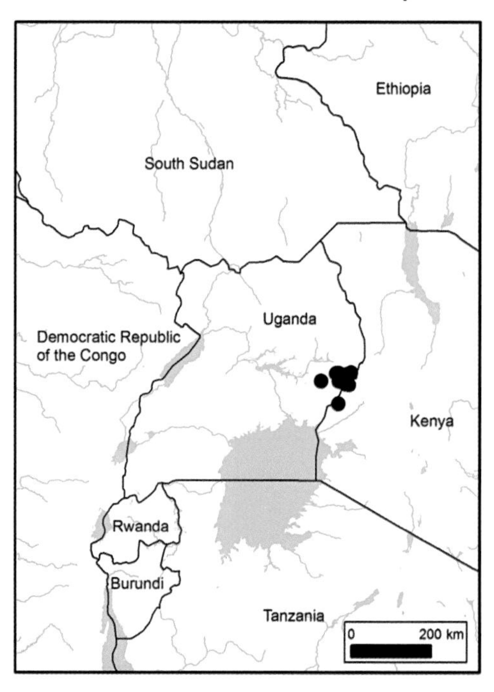

LITERATURE: Knox in Beentje *et al.*, FTEA 2005: 558.

DESCRIPTION: Tree to 8 m high, the trunk to 50 cm across; stem with densely packed leaf rosettes (each of 30–50 leaves) and old, dead leaves hanging down. Leaves to 100 × 30 cm. Flowers in panicles to 120 × 80 cm, with hanging or spreading capitula of yellow heads, with or without ray florets. Fruits minute, hidden in the involucre.

HABITAT: Upper montane forest to moorlands; 2700–4250 m

DISTRIBUTION: Uganda, Kenya, confined to upper slopes of Mt Elgon.

PROTECTED AREA OCCURRENCE Mt Elgon National Park.

LOCAL NAMES AND USES: None recorded.

RED LIST ASSESSMENT FOR SPECIES: EOO 3060 km^2; AOO 737 km^2; confined to a single mountain, Mt Elgon, but reported to be common to abundant (reports from 1960s). More recent information from a visit by James Kalema and Mary Namaganda in 2004 indicated a healthy population with no reported threats. We assess this as Least Concern (LC). Knox in FTEA conservation notes has both subspecies as Least Concern.

- subsp. ***elgonensis***:

TYPE: Uganda, Mt Elgon, *Dummer* 3382 (B† holo., K, UPS, US iso.)

LITERATURE: Knox in Beentje *et al.*, FTEA 2005: 558.

SYNONYM: *Senecio elgonensis* T. C. E. Fr.; ITU: 97, photo 13. UFT: 79.

DESCRIPTION: Ray florets present; leaf mainly hairy along midrib on lower surface, without a clear pseudo-petiole.

HABITAT: Moist sites in upper montane forest, heath zone, moorlands, with large stands along W, S and E flanks of Elgon slopes and in crater; 2750–4200 m.

DISTRIBUTION: Bugisu; also in Kenya; confined to upper slopes of Mt Elgon.

PROTECTED AREA OCCURRENCE: Mt Elgon National Park.

LOCAL NAMES AND USES: None recorded.

RED LIST ASSESSMENT: EOO 1634.5 km^2; AOO 469.24 km^2, 11 collections from 8 locations. Our assessment for the subspecies: occurs at very high elevation in protected area (National Park) with very little access by the communities and no known threats; Least Concern (LC).

- subsp. ***barbatipes*** (Hedberg) E. B. Knox

TYPE: Kenya, Mt Elgon, *Honoré* 2520 (K holo., BM, EA iso.)

SYNONYMS: *Senecio amblyphyllus* Cotton; ITU: 97. UFT: 79.
Senecio barbatipes Hedberg; UFT: 79.
Senecio gardneri Cotton; ITU: 98.

DESCRIPTION: Ray florets absent; leaf felty-hairy on lower surface, constricted towards the base, forming a pseudopetiole.

HABITAT: Moorland, afro-alpine zone, also in bamboo zone; 2700–4250 m.

DISTRIBUTION: Mbale; also in Kenya; confined to upper slopes of Mt Elgon.

PROTECTED AREA OCCURRENCE: Mt Elgon National Park.

RED LIST ASSESSMENT: EOO 39.5 km^2; AOO 28.33 km^2, 4 collections from 3 locations. Protected within the National Park. Least Concern (LC) due to protected status, but needs more recent information on status of populations.

FLOWERING AND FRUITING TIMES: For whole species.

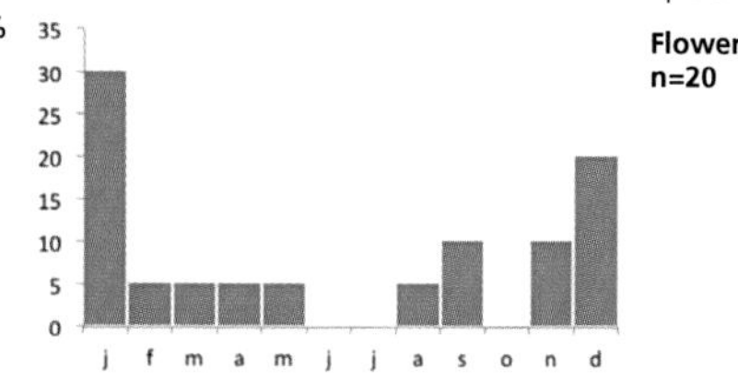

Fruit – no data

LC

Dendrosenecio erici-rosenii (R. E. Fr. & T. C. E. Fr.) E. B. Knox **Asteraceae /Compositae**

No image available

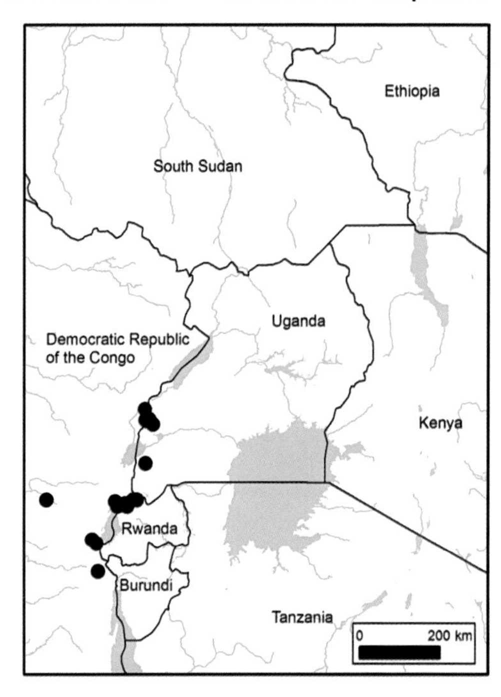

LITERATURE: Knox in Beentje *et al.*, FTEA 2005: 562.

DESCRIPTION: Tree to 9 m, the trunk to 50 cm across; stem with densely packed leaf rosettes (each of 10–30 leaves) and leaf bases of dead leaves retained. Leaves to 70 × 26 cm. Flowers in panicles to 100 × 40 cm, with hanging or spreading capitula of yellow heads, with ray florets. Fruits minute, hidden in the involucre.

HABITAT: Upper montane forest zone to moorland; 2600–4200 m.

DISTRIBUTION: Uganda, DRC, Rwanda; see subspecies for details.

LOCAL NAMES AND USES: Stem pith eaten by gorilla (fide Knox).

RED LIST ASSESSMENT FOR SPECIES: EOO 47,095 km²; AOO 18,451 km², 64 collections from 22 locations. Least Concern, based on large EOO and AOO coupled to altitude range and occurrence in protected area. Knox in FTEA conservation notes has both subspecies as Least Concern.

- ***Dendrosenecio erici-rosenii*** (R. E. Fr. & T. C. E. Fr.) E. B. Knox subsp. ***erici-rosenii***

TYPE: DRC, Mt Nyiragongo, *Fries* 1716 (UPS holo.)

SYNONYM: *Senecio erici-rosenii* R. E. Fr. & T. C. E. Fr.; ITU: 97. UFT: 79.

DESCRIPTION: Ray florets present; leaf mainly hairy along midrib on lower surface.

HABITAT: In alpine island habitats, described as occasional to locally dominant; 2750–4200 m.

DISTRIBUTION: Toro, Kigezi; also in DRC, Rwanda. Endemic to Mt Muhi, Mt Kahuzi, the Virunga Mts and the Ruwenzori Mts.

PROTECTED AREA OCCURRENCE: Rwenzori Mts, Mgahinga Gorilla National Parks and Parc National des Volcans.

RED LIST ASSESSMENT: EOO 108.15 km²; AOO 38.62 km², 9 collections from 4 locations; limited range but protected within National Parks at very high elevation; assessed here as Least Concern (LC). FTEA has both subspecies as Least Concern.

- ***Dendrosenecio erici-rosenii*** (R. E. Fr. & T. C. E. Fr.) E. B. Knox subsp. ***alticola*** (T. C. E. Fr.) E. B. Knox

TYPE: Uganda, Mt Muhavura, *Hedberg* 2086 (UPS neo., EA, K, S iso.)

SYNONYM: *Senecio alticola* T. C. E. Fr.; ITU: 97.

DESCRIPTION: Ray florets absent or inconspicuous; leaf floccose, lanate or tomentose on lower surface.

HABITAT: Moorland, afro-alpine zone, in upper forest zone confined to gullies; may be common or dominant in alpine belt; 2600–4050 m.

DISTRIBUTION: Kigezi; also in DRC and Rwanda; endemic to Virunga Mts (Muhavura, Karisimbi, Mikeno).

PROTECTED AREA OCCURRENCE: Mgahinga Gorilla National Parks, Parc National des Volcans.

RED LIST ASSESSMENT: EOO 11,276 km^2; AOO 4,666 km^2, 5 collections from 5 locations. Alpine island habitats, described as occasional to locally dominant; limited range but protected within National Parks at very high elevation; assessed here as Least Concern (LC). FTEA has both subspecies as Least Concern.

FLOWERING AND FRUITING TIMES: For species.

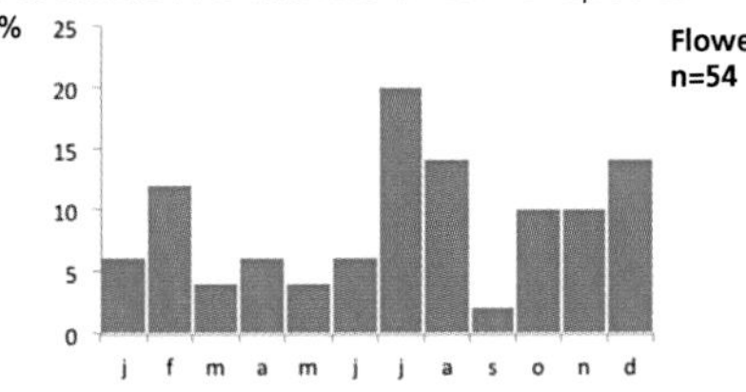

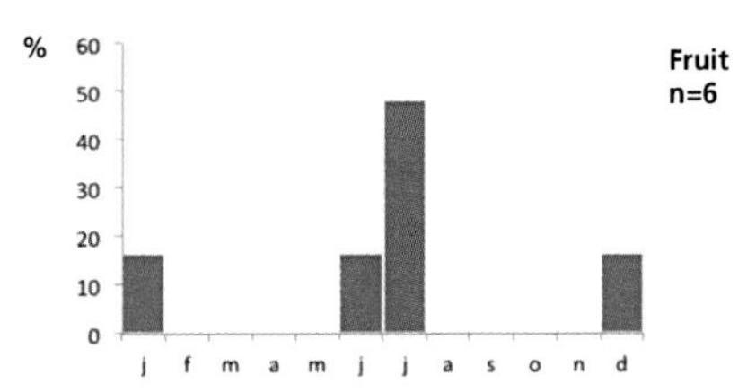

LC **_Allanblackia kimbiliensis_** Spirlet **Clusiaceae**

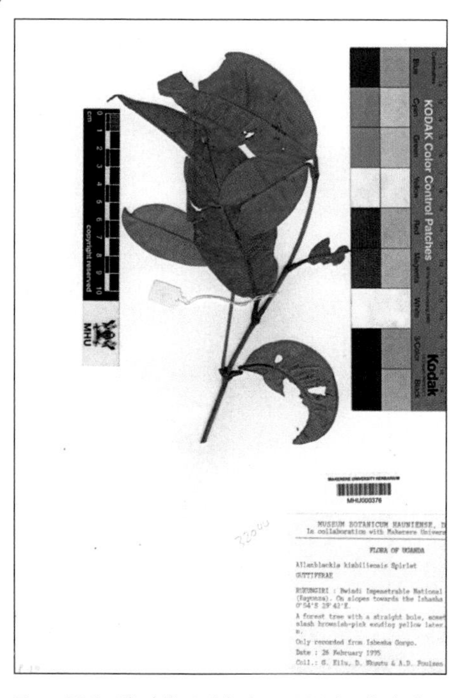

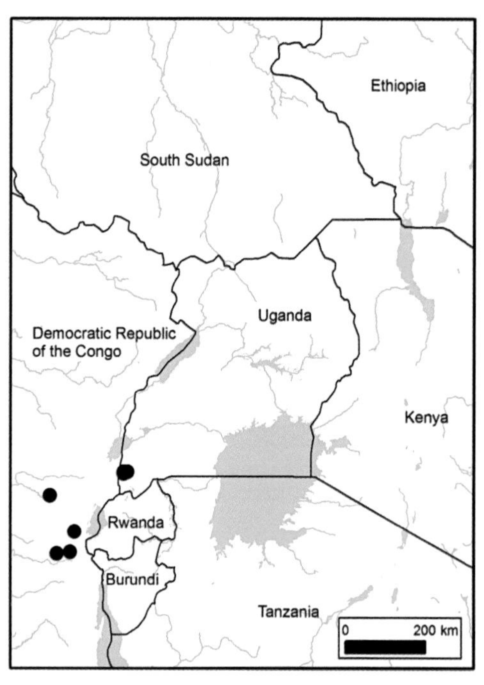

TYPE: DRC, Kimbili, *Michelson* 766 (BR holo.)

LITERATURE: Bamps, Robson & Verdcourt, FTEA 1978: 8. UFT: 169.

SYNONYM: *Allanblackia floribunda* in the sense of ITU: 152.

DESCRIPTION: Tree 18–36 m high; bark smooth; slash with yellow sap. Leaves narrowly oblong, 8–22 × 2–5 cm, with yellow sap. Flowers a few together near branch tip, male and female separate, petals white or pink and ± 2 cm across; stamens many. Fruit reddish, ovoid, ± 10 cm across, ribbed.

HABITAT: Rainforest or forest margin; 1250–1800 m.

DISTRIBUTION: Kigezi. Limited range in Uganda – only known from Ishasha Gorge. Also in eastern DRC.

PROTECTED AREA OCCURRENCE: Bwindi Impenetrable National Park.

LOCAL NAMES AND USES: Omutaka, Omuruguya (Lukiga)

RED LIST ASSESSMENT: Said to be locally common at Ishasha Gorge (in 1945); forest habitat specific. EOO 17,291 km^2; AOO 4,195 km^2; 20 collections from 7 locations. Assessed here as Least Concern due to the size of EOO and AOO; the number of locations and the altitude range; many populations in protected areas; but within Uganda this species is very rare.

FLOWERING AND FRUITING TIMES:

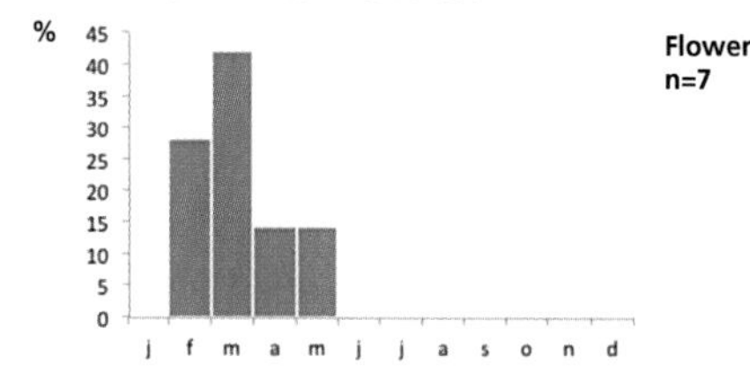

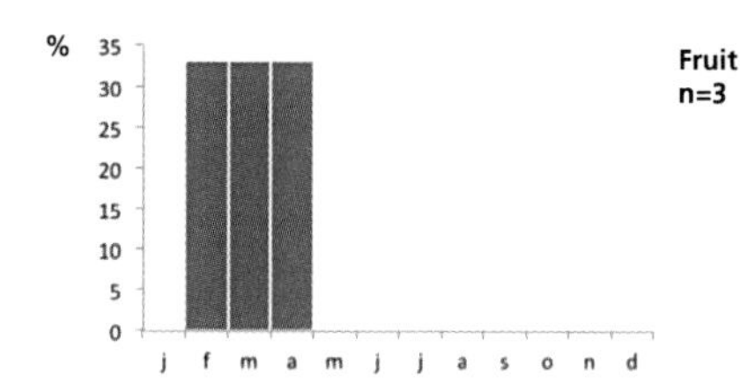

Cnestis mildbraedii Gilg

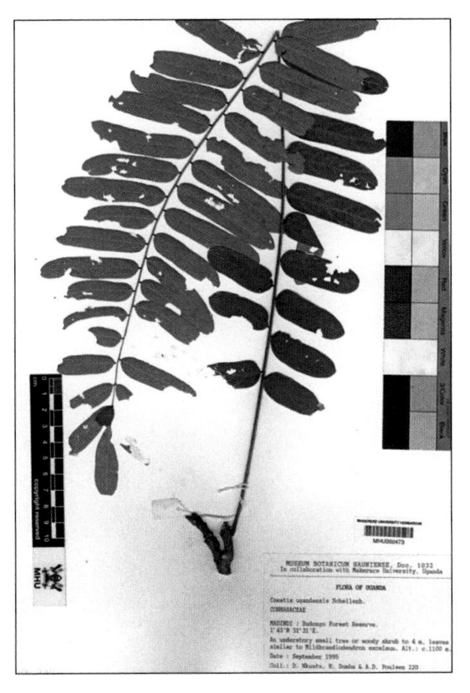

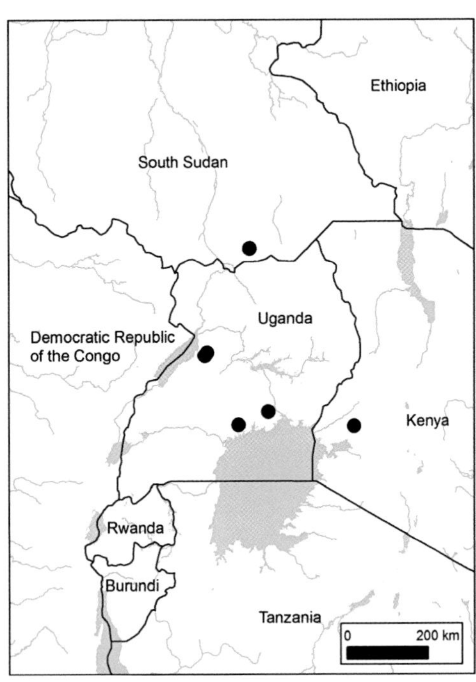

TYPE: Uganda, Imatong Mts, Lotti, *Thomas* 1759 (K, isoneo.)

LITERATURE: Breteler (ed.) 1989: 212. Lebrun & Stork 1992: 231.

SYNONYM: *Cnestis ugandensis* Schellenb.; Hemsley, FTEA 1956: 2. ITU: 100. UFT: 226.

DESCRIPTION: Tree to 7 m high, with hairy branches. Leaves with 5–15 pairs of leaflets and a terminal leaflet, hairy beneath. Flowers in dense axillary fascicles with sepals to 6.5 mm long and white or cream petals to 8 × 3 mm. Fruit red, to 2.5 × 1 cm.

HABITAT: Rainforest, also in regenerating forest; 900–1650 m.

DISTRIBUTION: Bunyoro, Mengo. Budongo, Mpanga and Mabira; forest obligate; also in South Sudan (one record from the Imatong) and Kenya (two records from Kakamega).

PROTECTED AREA OCCURRENCE: Budongo, Mpanga, Mabira, Imatongs and Kakamega Forest Reserves.

LOCAL NAMES AND USES: None recorded.

RED LIST ASSESSMENT: EOO 80,977 km^2; AOO 12,429 km^2, 9 collections from 6 locations. Most populations are in protected areas; however, Mabira is still under very heavy pressure to be partly degazetted for sugarcane growing; Budongo is also illegally felled in parts. In South Sudan, neither threat status nor protection status is known. The species appears to be well protected only in Kakamega, Kenya. We here assess the species as Near Threatened (NT) due to the low number of locations.

FLOWERING AND FRUITING TIMES:

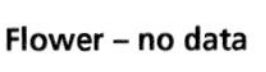

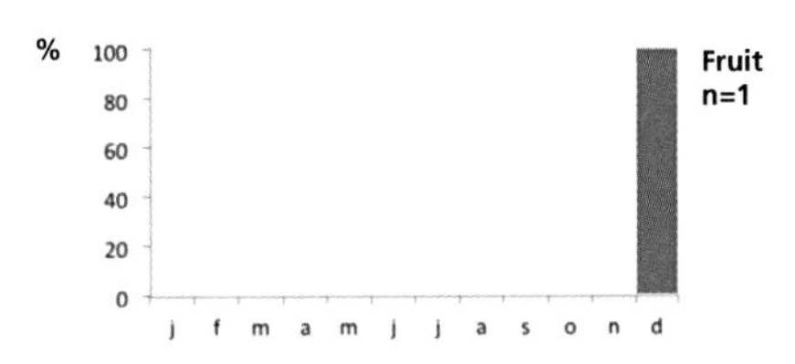

VU ***Cyathea mildbraedii*** (Brause) Domin. **Cyatheaceae**

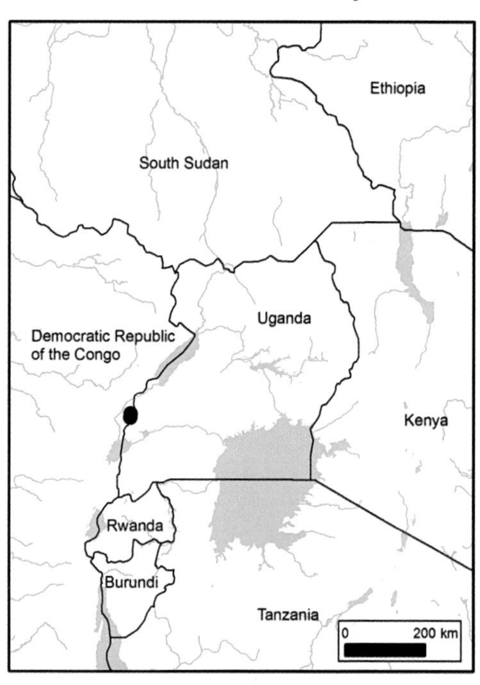

TYPE: DRC, Ruwenzori, Butahu Valley, *Mildbraed* 2545 (B holo.).

LITERATURE: Edwards, FTEA 2005: 13.

SYNONYM: *Alsophila mildbraedii* Brause

DESCRIPTION: Tree fern with trunk to 4 m high. Leaves with spiny stalk to 56 cm long and bipinnate, the pinnae to 40 cm long.

HABITAT: Giant heath zone; 2700–3300 m.

DISTRIBUTION: Toro: endemic to Ruwenzori, also in DRC. Known from only four collections, made in 1894, 1905, 1929 and 1954, two from the Butahu Valley and two without a precise locality.

PROTECTED AREA OCCURRENCE: Rwenzori Mountains National Park.

LOCAL NAMES AND USES: Previously harvested and used for building owing to its resistance to termite attack; not known if this is still continuing.

RED LIST ASSESSMENT: The heath 'forest' in the Rwenzori National Park is under no particular threat as far as we know, although isolated cases of illegal harvesting do occur; due to the few locations (at most three) and occasional illegal extraction this must be at least Vulnerable (VU D2); more data is needed to move it into the Endangered category, but this sounds like it might be a candidate if numbers of specimens are low. All *Cyathea* species are CITES Appendix II listed.

SPORULATING TIMES:

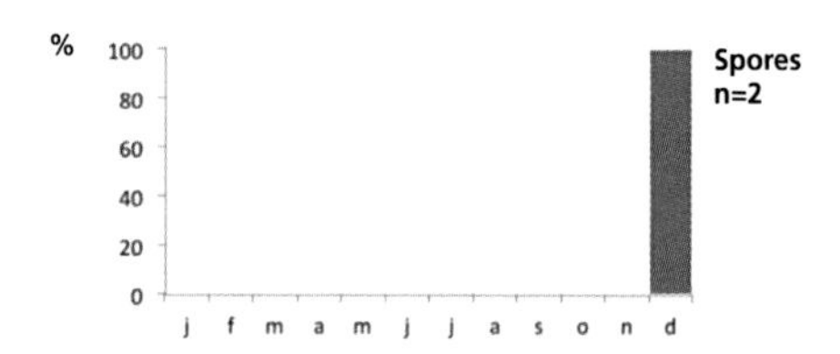

Diospyros katendei Verdc.

Ebenaceae CR PEx

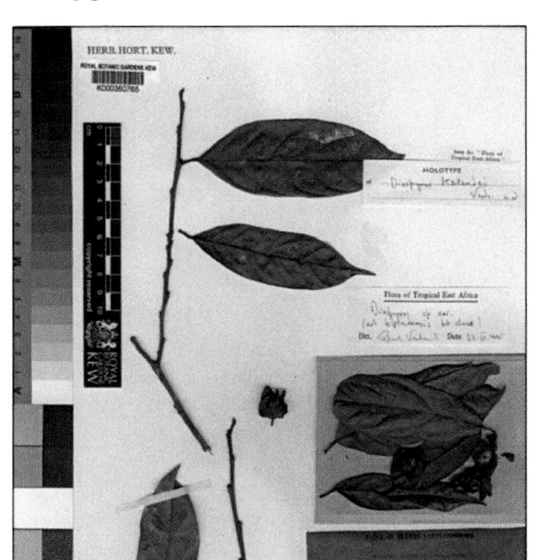

TYPE: Uganda, central Kashoya-Kitomi forest, *Katende* 3195 (K holo., MHU iso.)

LITERATURE: White & Verdcourt, FTEA 1996: 42.

DESCRIPTION: Medium-sized tree; bole irregularly fluted; slash thin and orange. Leaves narrowly oblong. Flowers unknown. Fruit possibly on older branches, ellipsoid, 10 × 8 mm, with accrescent calyx to 2 cm long.

HABITAT: Forest; ± 1800 m.

DISTRIBUTION: Ankole, only known from a single collection from Kasyoha Forest in 1987.

PROTECTED AREA OCCURRENCE: Kasyoha-Kitomi Forest Reserve.

LOCAL NAMES AND USES: None recorded.

RED LIST ASSESSMENT: Assessed as Critically Endangered by Katende in 1997. Ugandan endemic of very restricted range and presumably quite rare. The forest reserve is almost entirely surrounded by agricultural land and human population pressure appears to be high; human activity is frequent with illegal, uncontrolled and unregulated exploitation from illegal pit-sawyers. The forest is being extensively used by the surrounding local communities as a source of bushmeat, firewood, poles, medicinal compounds and natural fibres.

Very small and restricted population of just a single individual tree (Tony Katende pers. comm. to James Kalema); only known from the type despite deliberate efforts to recollect it. EOO<<100 km^2, AOO<<10 km^2, one location, number of mature individuals <50; here assessed as Critically Endangered [CR B1ab(iii)+2ab(iii), D] and Possibly Extinct.

FLOWERING AND FRUITING TIMES:

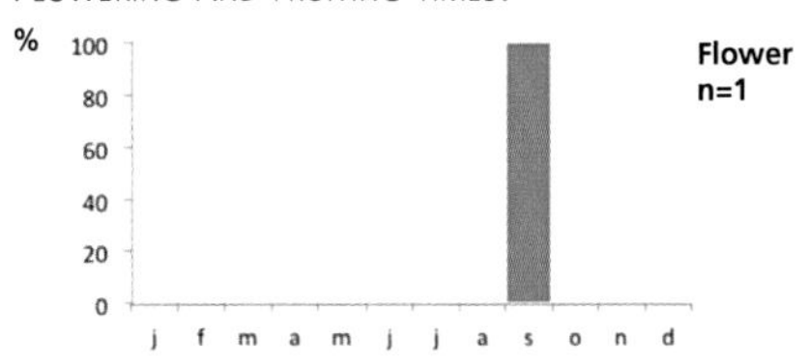

Fruit – no data

LC **_Duvigneaudia leonardii-crispi_** (J. Léonard) Kruijt & Roebers **Euphorbiaceae**

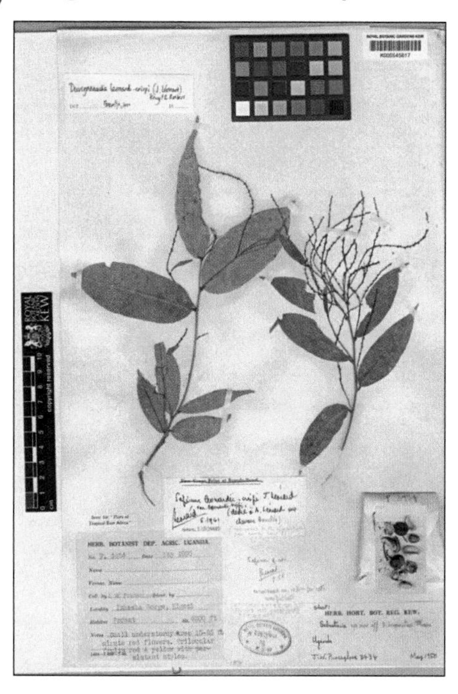

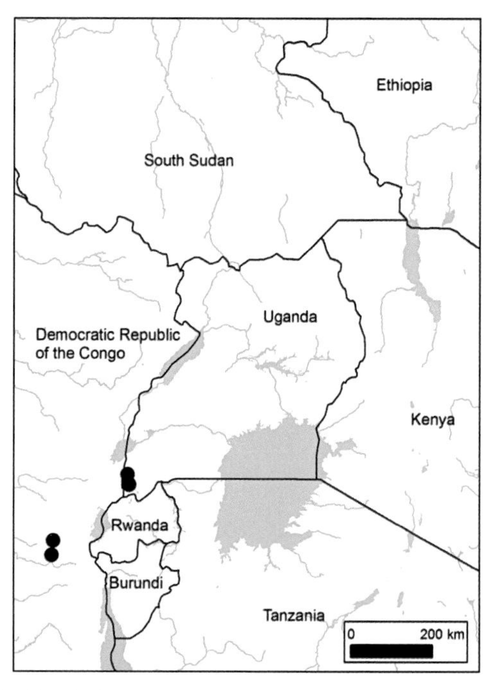

TYPE: DRC, Kivu, Shabunda, Kigulube, *Léonard* 3826 (BR holo.)

LITERATURE: Kruijt & Roebers in *Biblioth. Bot.* 146: 12–15 (1996).

SYNONYM: *Sapium leonardii-crispi* J. Léonard; Radcliffe-Smith, FTEA 1987: 391, fig. 74. UFT: 137.

DESCRIPTION: Shrub or tree 4–8 m high. Leaves narrowly ovate to elliptic, 3–14 × 1–5 cm, with minute marginal glands. Flowers in terminal spikes with 1–3 ♀ at base and others ♂, minute (± 1 mm). Fruit 3-lobed, ± 1 cm across, yellow with reddish keels.

HABITAT: Primary or secondary evergreen forest; 700–1400 m.

DISTRIBUTION: Kigezi, Ankole: only known from Bwindi and the Ishasha Gorge; also in eastern DRC (four collections).

PROTECTED AREA OCCURRENCE: Bwindi Impenetrable and Maiko National Parks.

LOCAL NAMES AND USES: None recorded.

RED LIST ASSESSMENT: EOO 60,522 km^2; AOO 9059 km^2, 7 collections from 8 locations. In DRC, covered in some protected areas e.g. Maiko National Park. Though the tree is described as 'common locally' on slopes of the Ishasha Gorge, this is obviously not a *generally* common species. Assessed here as Least Concern (LC) due to large EOO and AOO coupled with several locations and the habitat range; but in the future might move to Near Threatened (NT) due to continuing degradation of good forest habitat in the region.

FLOWERING AND FRUITING TIMES:

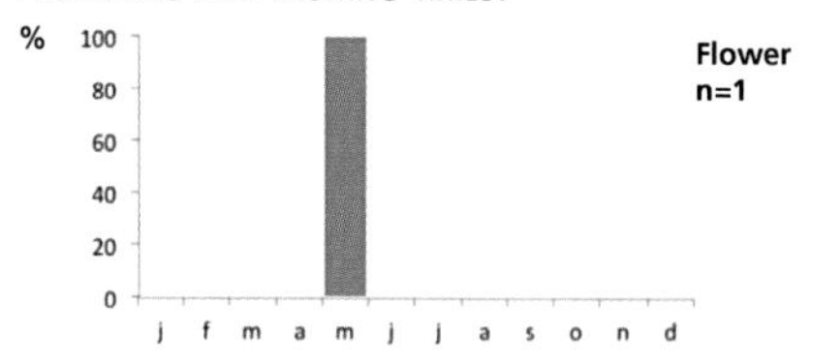

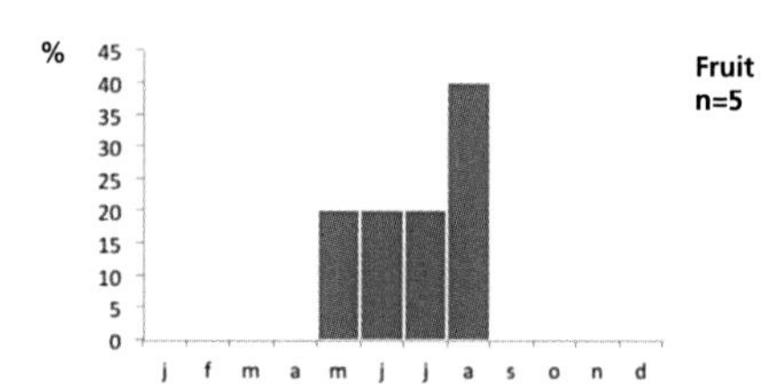

Euphorbia bwambensis S. Carter

Euphorbiaceae LC

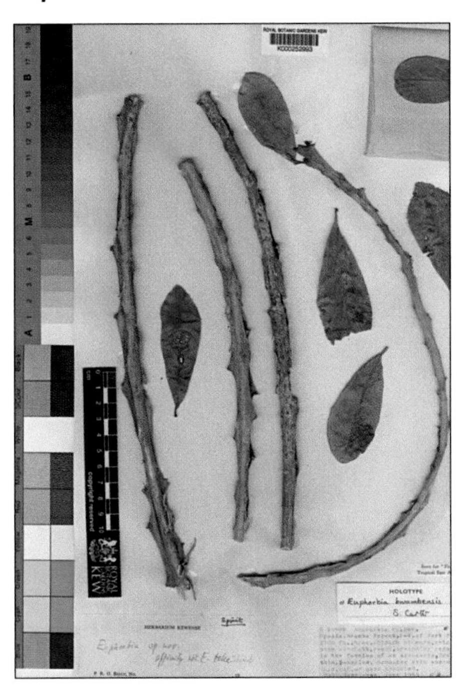

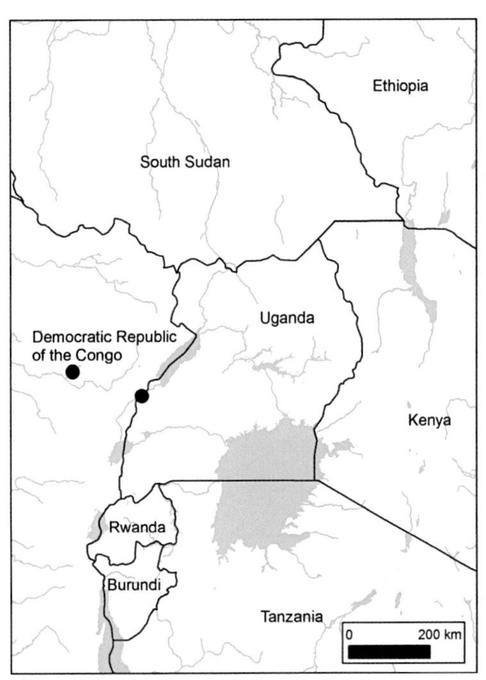

TYPE: Uganda, Bwamba Forest, *Carcasson* 31 in *Bally* 10588 (K holo, EA, G iso.)

LITERATURE: Carter & Smith, FTEA 1988: 483.

SYNONYMS: *Euphorbia neglecta* in the sense of ITU: 127.
Euphorbia sp. of UFT: 81.

DESCRIPTION: Tree to 6.5 m high with ± drooping branches; branches fleshy with spines to 3 mm long on the angles. Leaves at branch ends, obovate and to 9 × 5 cm. Flowers mixed ♀ and ♂ in small cup-shaped 'cyathia' to 6.5 mm across. Fruit on a 8 mm stalk, reddish and 9 × 14 mm.

HABITAT: Forest, forest/bare rock margin, possibly only in forest along streams or swamp forest; 730–900 m.

DISTRIBUTION: Known from Bwamba Forest in Toro (3 collections), Uganda's only truly lowland rainforest, and one collection from DRC in the Ituri Forest Reserve.

PROTECTED AREA OCCURRENCE: Semuliki National Park, Ituri Forest Reserve.

LOCAL NAMES AND USES: None recorded.

RED LIST ASSESSMENT: Assessed as Vulnerable (B1 + 2c) by Makerere University IENR in 1997. EOO= 1,560.85 km², AOO=1,414.39 km²; locations = 2. The species occurs in protected areas, such as Semuliki National Park in Uganda, and the Ituri Forest reserve in DRC. As there seem to be no current threats we assess this as Least Concern (LC); this is a downgrade from the 1997 assessment, as we believe that despite the number of locations being so small and the size of the AOO, the degree of protection in the protected areas does not allow the use of criteria B (a or c). The species could move to the Endangered category under B2 if exposed to any threat in future, on account of the very small number of locations.

FLOWERING AND FRUITING TIMES:

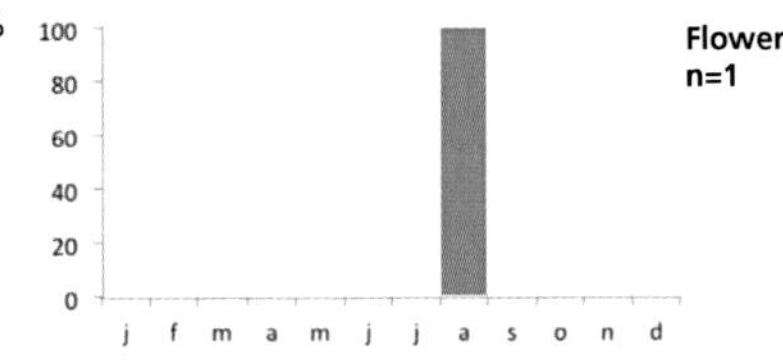

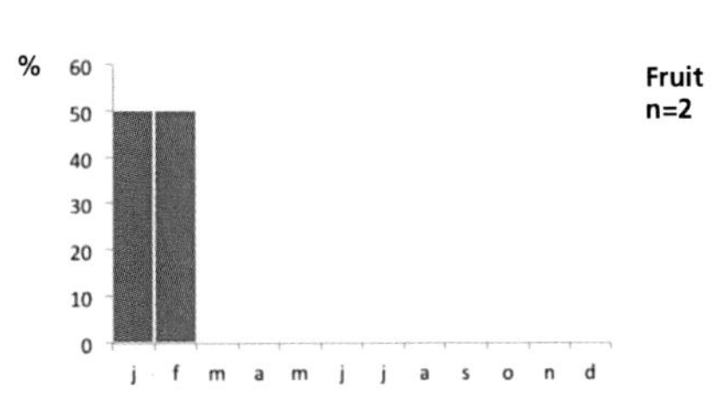

LC **_Euphorbia magnicapsula_** S. Carter **Euphorbiaceae**

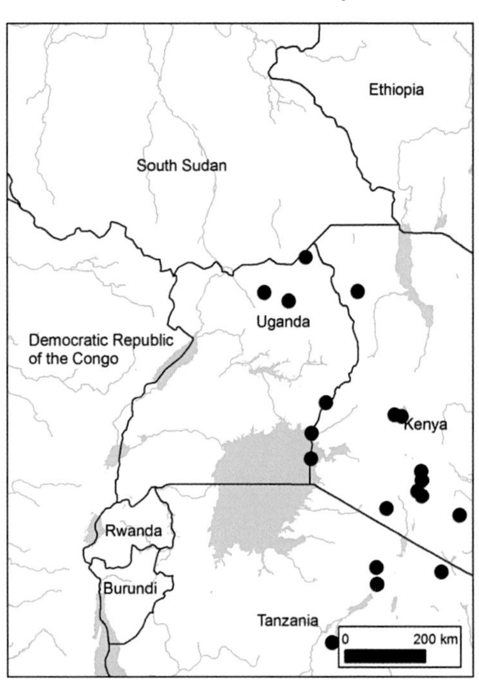

TYPE: Tanzania, Ngorongoro, foot of Windy gap road, *Greenway & Kanuri* 12532 (K holo.)

LITERATURE: Carter & Smith, FTEA 1988: 489.

SYNONYM: *Euphorbia* sp. aff. *thi* in the sense of ITU: 128, photo 21.

DESCRIPTION: Tree to 12 m high; trunk to 45 cm across; branches fleshy, winged, segmented, with spines to 15 mm long. Leaves minute and ± invisible. Flowers mixed ♀ and ♂ in small cup-shaped 'cyathia' to 6 mm across. Fruit on a 8 mm stalk, bright red and 12 × 25 mm.

HABITAT: Open bushland or woodland, where it may form pure stands on rocky hillsides; 900–1950 m.

DISTRIBUTION: Acholi, Karamoja, West Nile, Bukedi; also in Kenya, Tanzania, South Sudan.

PROTECTED AREA OCCURRENCE: Kidepo Valley National Park, Oldiang'arangar, E Serengeti.

LOCAL NAMES AND USES: 'Emuss' (Turkana); said to be poisonous.

RED LIST ASSESSMENT: Rare in Uganda, and commercial extraction of limestone in the neighbourhood of Tororo Rock threatens the species; elsewhere rather widely distributed, to Lake Manyara in Tanzania, and the Loita Hills and Kikuyu Escarpment in Kenya; EOO 247,619 km^2; AOO 127,995 km^2; 20 collections from 19 locations; a wide altitude range; no specific threats known from other countries; and therefore assessed here as Least Concern (LC). All succulent *Euphorbia* species are CITES listed on Appendix II.

FLOWERING AND FRUITING TIMES:

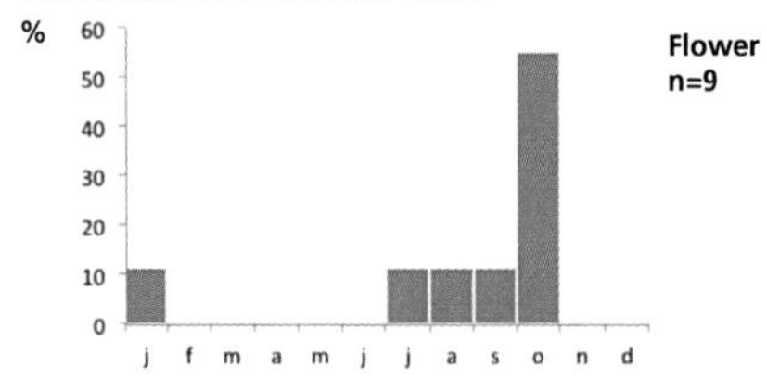

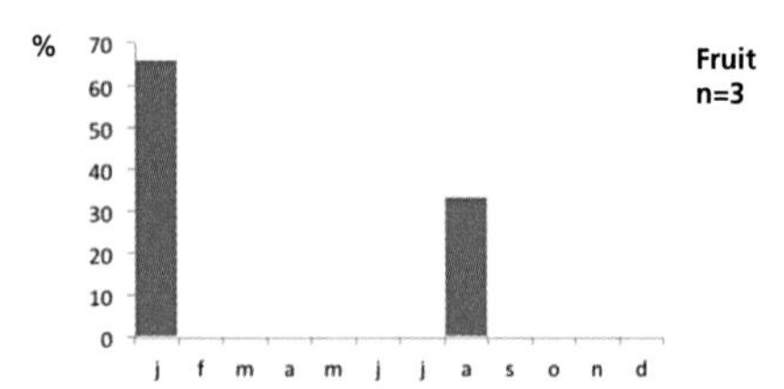

Pseudagrostistachys ugandensis (Hutch.) Pax & K. Hoffm. — **Euphorbiaceae** LC

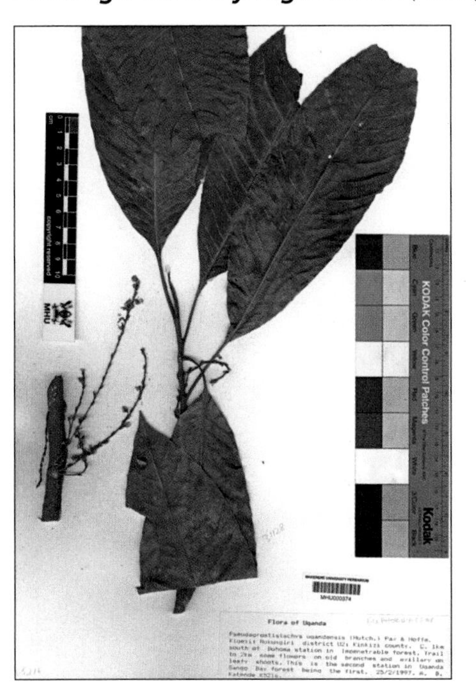

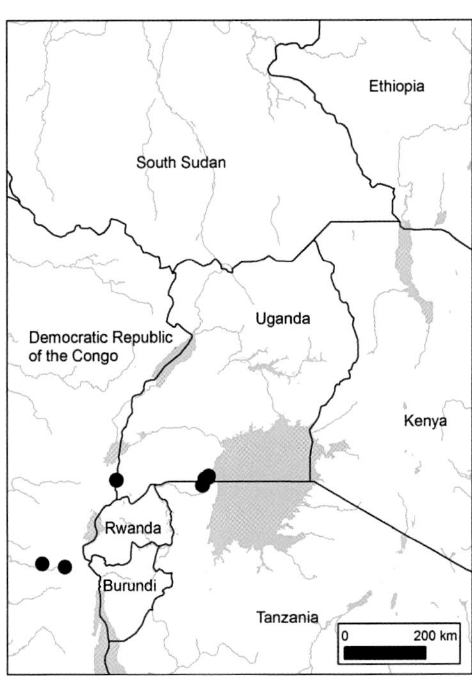

TYPE: Uganda, Masaka, Namalala forest, *Fyffe* 19 & 64 (K syn.)

LITERATURE: Radcliffe-Smith, FTEA 1987: 167, fig. 30. ITU: 138. UFT: 136.

DESCRIPTION: Tree to 6 m high. Leaves oblong to elliptic, 25–45 × 12–20 cm, margins minutely toothed. Flowers in axillary groups, with male and female flowers 3–5 mm long. Fruit dull reddish brown, 3-lobed and depressed, 2 cm across.

HABITAT: Swamp or riverine forest; 470–1200 m.

DISTRIBUTION: Masaka, Kigezi; limited distribution in Sango Bay and Bwindi; forest obligate. Also in eastern DRC and NE Tanzania (said to be 'occasional' in Minziro Forest).

PROTECTED AREA OCCURRENCE: Bwindi Impenetrable National Park; Malabigambo, Minziro and Namalala Forest Reserves (Sango Bay).

LOCAL NAMES AND USES: None recorded.

RED LIST ASSESSMENT: EOO 347,899 km²; AOO 147,412 km²; 17 collections from 11 locations. Assessed here as Least Concern (LC) due to EOO, AOO, number of locations and altitude range; the species is described as uncommon in Sango Bay, but protected in forest reserves.

FLOWERING AND FRUITING TIMES:

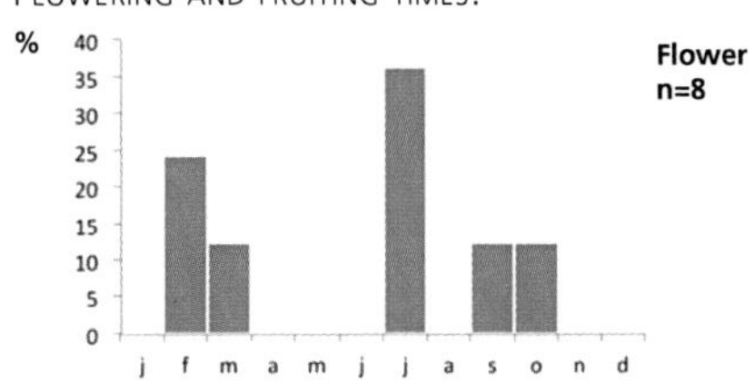

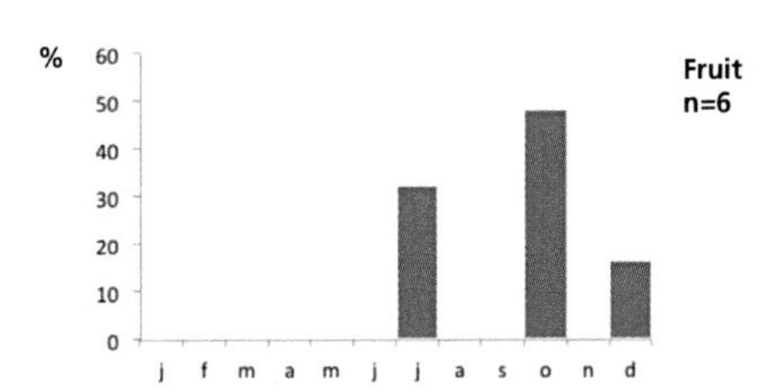

LC **_Cordyla richardii_** Milne-Redh. **Fabaceae (Leguminosae), subfam. Caesalpinioideae**

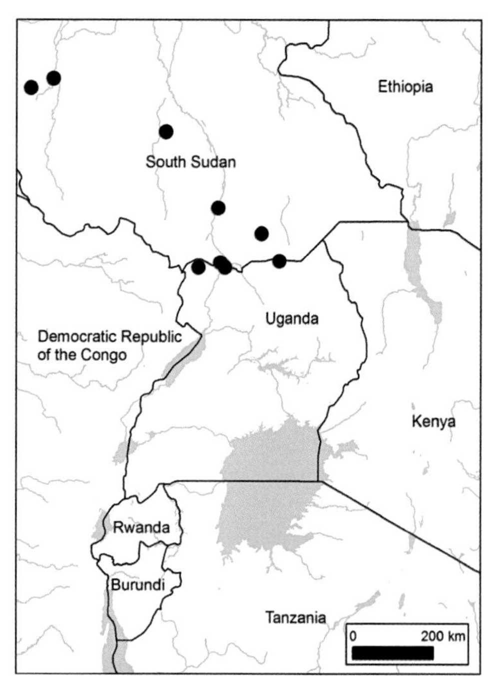

TYPE: Sudan, S of Gondokoro, *Speke & Grant* s.n. (K holo.)

LITERATURE: Brenan, FTEA 1967: 223. ITU: 298.

DESCRIPTION: Tree 3–12 m high; bark rough. Leaves with stalk 10–30 cm long and ± 20 leaflets, each 2–4 × 1–2 cm. Flowers on stalks on twigs below the leaves, petals absent, stamens many. Fruit ellipsoid, yellow, 3–5 × 2–4 cm.

HABITAT: Woodland in rocky sites; 700–1200 m.

DISTRIBUTION: West Nile, Acholi, Madi. Limited range, but locally common in northern Uganda; also in South Sudan.

PROTECTED AREA OCCURRENCE: Itie & Imatong Forest Reserves.

LOCAL NAMES AND USES: Chululoich (Dinka), used for furniture and fruit eaten; 'Homeid' (Arabic); Danjwek (Gur); choice wood for canoes, durable in water; 'Malindi' (Madi), used for canoes; Boli (Golo), used for gum.

RED LIST ASSESSMENT: Assessed as Vulnerable (VU B1 +2c) in 1998 by Makerere IENR. Re-assessed here as Least Concern (LC) due to the size of EOO, AOO, and number of locations: EOO 92,498 km^2; AOO 43,809 km^2; 17 collections from 9 locations; we believe the lack of recent collections is due to political instability, but most sub-populations occur outside protected areas in both Uganda and South Sudan, while the tree is exploited for its wood and its gum.

FLOWERING AND FRUITING TIMES:

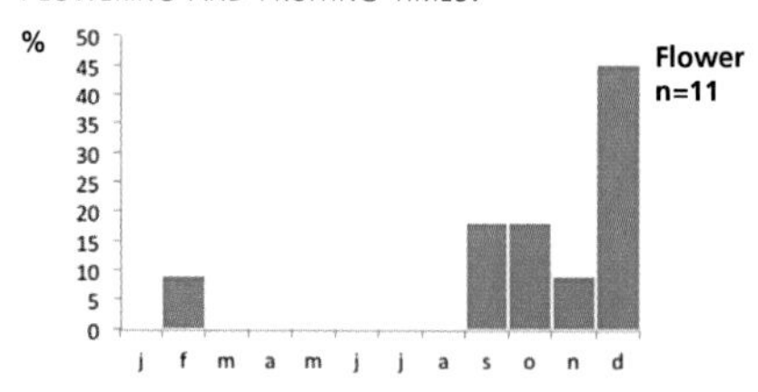

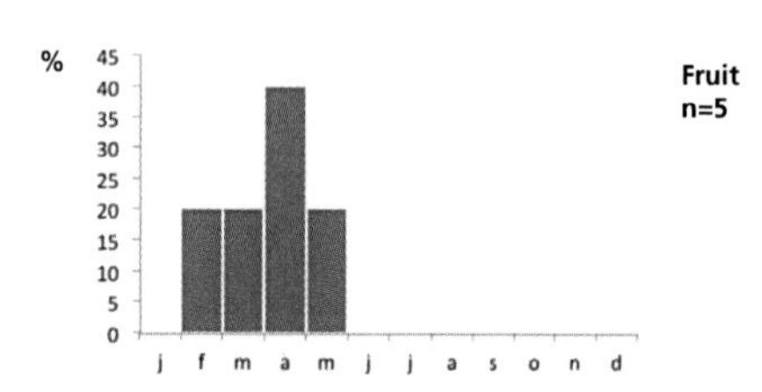

Baphia wollastonii Baker f.

Fabaceae (Leguminosae), subfam. Papilionoideae LC

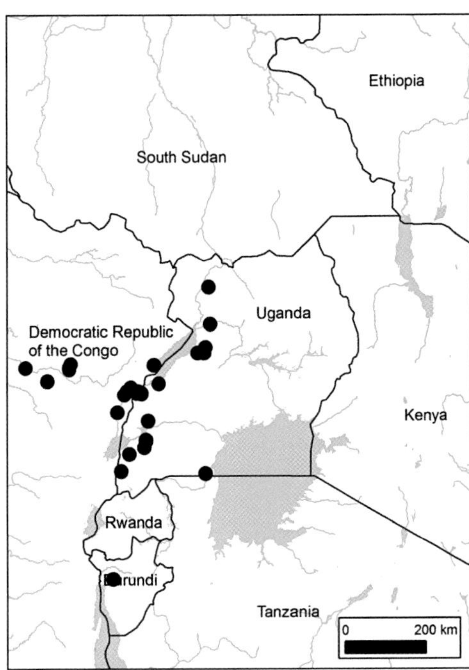

TYPE: DRC, Ruwenzori, *Wollaston* s.n. (BM holo.)

LITERATURE: Gillett, Polhill & Verdcourt, FTEA 1971: 60. ITU: 297. UFT: 152.

DESCRIPTION: Shrub or tree to 10 m high. Young stems glabrous. Leaf stalk 2–5 mm long, lamina ovate- or elliptic- or lanceolate-acuminate. Flowers solitary or in pairs, petals white with a yellow blotch near base. Pods brown, 6–9 × 1.5–2.2 cm.

HABITAT: Rainforest and its edges, swamp forest, in or near riverine forest in wooded grassland; 600–1900 m.

DISTRIBUTION: West Nile, Bunyoro, Toro, Madi, Ankole, Kigezi (Katende *et al.* 1995 also mention Mengo and Masaka). Also in East DRC, Burundi, possibly South Sudan (but no specimens found).

PROTECTED AREA OCCURRENCE: Semuliki, Queen Elizabeth and Bwindi Impenetrable National Parks; Semliki Wildlife Reserve; Budongo, Malabigambo, Kasyoha-Kitomi, Maramagambo and Zoka Forest Reserves.

LOCAL NAMES AND USES: None recorded.

RED LIST ASSESSMENT: EOO 187,420 km^2; AOO 92,511 km^2; 36 collections from 29 locations. Assessed here as Least Concern (LC) due to the large EOO and AOO, numerous locations, varied habitats and wide altitude range, with a lack of major threats.

FLOWERING AND FRUITING TIMES:

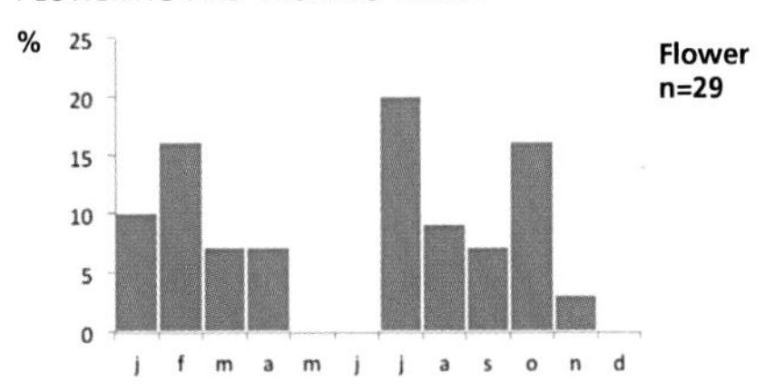

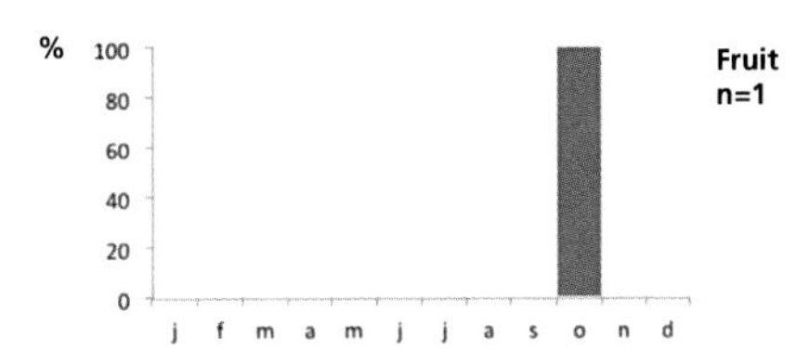

LC **_Sesbania dummeri_** E. Phillips & Hutch. **Fabaceae (Leguminosae), subfam. Papilionoideae**

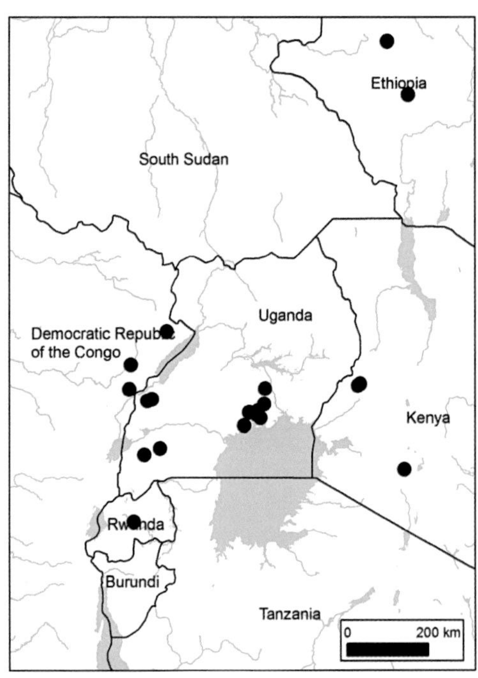

TYPE: Uganda, Mengo, Kirerema, *Dummer* 225 pro parte (BM, K syn.) & without locality, *Whyte* s.n. (K syn.)

LITERATURE: Gillett, Polhill & Verdcourt, FTEA 1971: 336. ITU: 311.

DESCRIPTION: Shrub or tree to 5 m high. Stem glabrous or sparsely pilose. Leaf rachis to 28 cm long; leaflets in 10–24 pairs, glabrous. Racemes 4–15-flowered. Pod 20–30 cm long, ± 4 mm wide, 40–55-seeded. Seeds brown, suboblong, elliptical in cross-section.

HABITAT: Swamp forest, swamps, riverine forest; 1050–2000 m. May be locally common.

DISTRIBUTION: Toro, Ankole, Mengo. Also in eastern DRC, western Kenya, SW Ethiopia (reported as dominant in a swamp in Kaffa, 1970).

PROTECTED AREA OCCURRENCE: Very low level of protection; Kifu Forest in Uganda is much degraded.

LOCAL NAMES AND USES: None recorded.

RED LIST ASSESSMENT: EOO 498,093 km²; AOO 154,839 km²; 24 collections from 20 locations. Assessed here as Least Concern (LC) due to EOO, AOO, number of locations and habitat and altitude range. Some of its main habitats (swamps) in Uganda, e.g. Namanve swamp are now nearly gone; its location, Kifu Forest is heavily degraded; Kirerema swamp area near Kampala has been converted; this rapid loss of its main habitat poses a real threat to the species within Uganda. This species might move into Near Threatened (NT) soon, if this habitat destruction continues and localities diminish.

FLOWERING AND FRUITING TIMES:

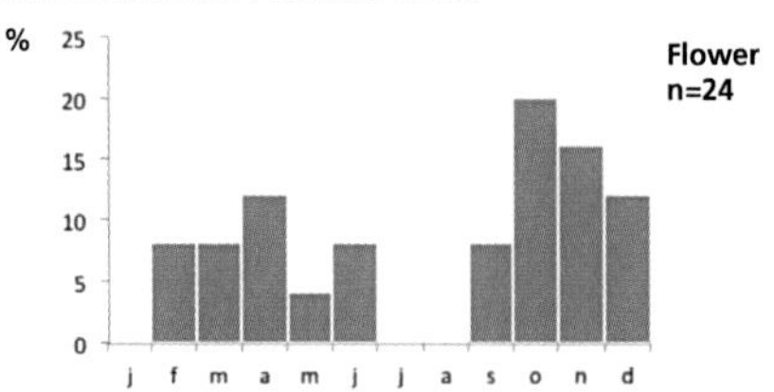

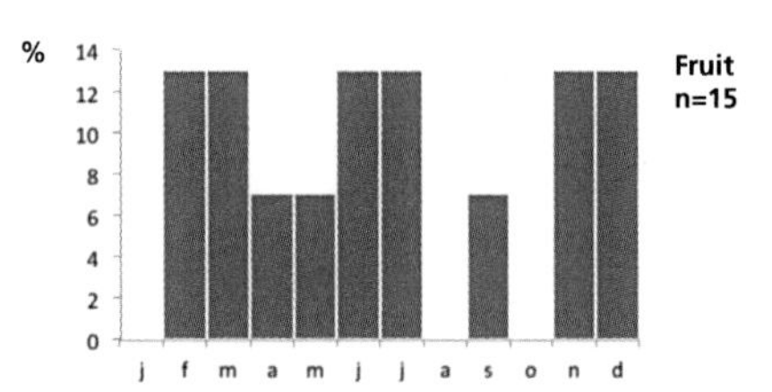

Hypericum bequaertii De Wild.

Hypericaceae LC

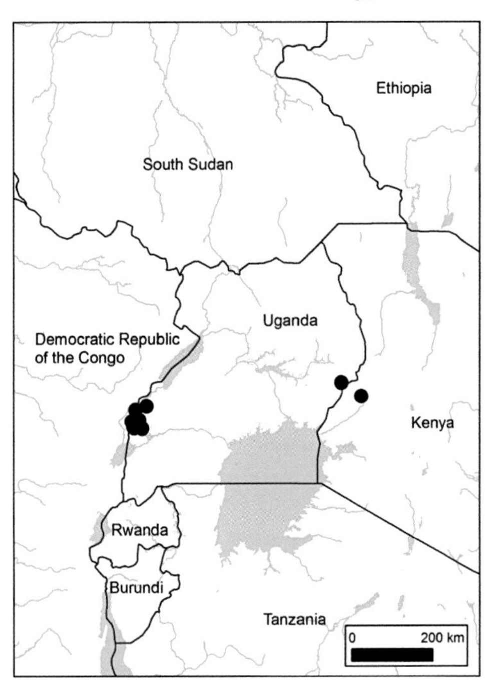

TYPE: DRC, Ruwenzori, Butahu valley, *Bequaert* 3757 (BR holo., K iso.)

LITERATURE: Milne-Redhead, FTEA Hypericaceae 1953: 5. UFT: 169.

SYNONYM: *Hypericum keniense* in the sense of ITU: 157.

DESCRIPTION: Shrub or tree to 12 m high. Leaves to 5 × 1 cm, with 3–5 veins from leaf base, with many linear glands parallel to midrib. Flowers single at branch tip, petals orange-yellow (red outside) and ± 3 cm long; stamens many. Fruit a 5-valved capsule, size unknown.

HABITAT: Giant heath zone, evergreen bushland, Senecio woodland, afroalpine zone; may be locally common; (2800–)3150–4300 m.

DISTRIBUTION: Toro, Bugisu; endemic to Rwenzori and Elgon; also in DRC.

PROTECTED AREA OCCURRENCE: Rwenzori Mts and Mt Elgon National Parks.

LOCAL NAMES AND USES: None recorded.

RED LIST ASSESSMENT: May be locally common. EOO 24,673 km²; AOO 16,404 km²; 41 collections from 19 locations. Assessed here as Least Concern (LC) due to the size of EOO and AOO; the number of locations; the fact that populations at higher elevations occur in protected areas; and the fact that we know of no specific threats.

FLOWERING AND FRUITING TIMES:

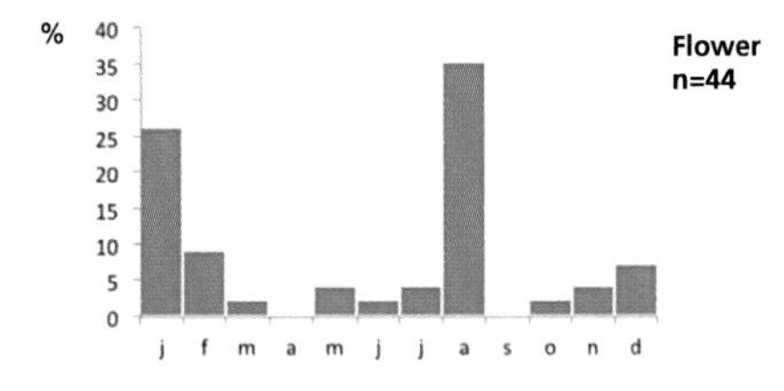

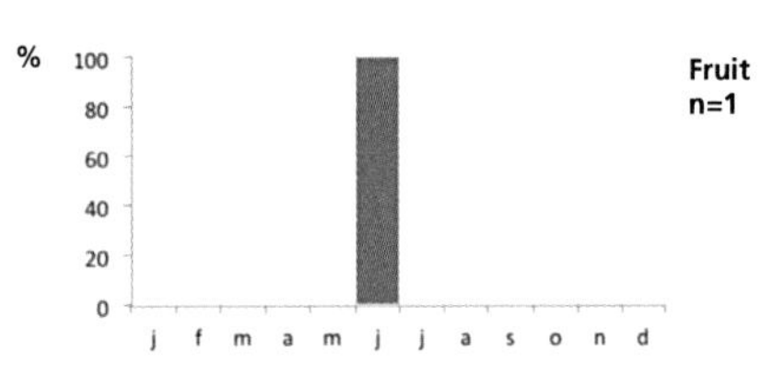

LC **_Brazzeia longipedicellata_** Verdc. **Lecythidaceae**

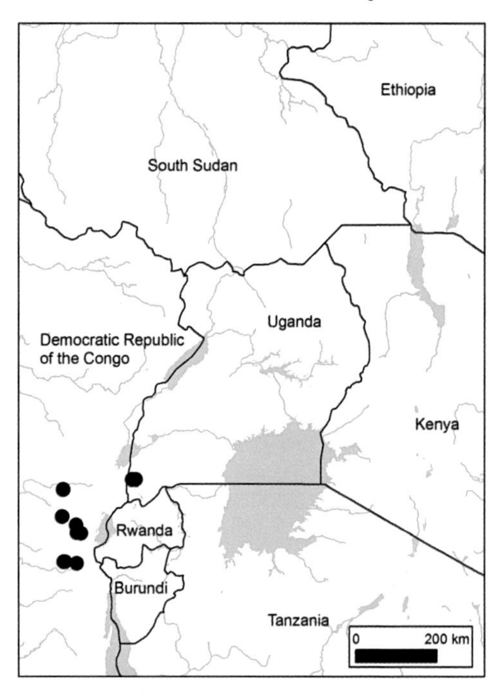

TYPE: Uganda, Kigezi, Ishasha Gorge, *Purseglove* 2002 (K holo., EA iso.)

LITERATURE: Verdcourt, FTEA 1968: 1, fig. 1. ITU: 404.

DESCRIPTION: Shrub or tree 4–12 m high. Leaf lamina elliptic to oblong-elliptic, (2.3–)4–13 cm long, (1.3–)2–5.5 cm wide; petiole short and thick, ± 3 mm long. Flowers fleshy; corolla white or rose. Ovary globose, 4 mm in diameter. Fruit woody, brown, globose, ± 2.8–3 cm in diameter, 6-ribbed. Seeds ± 8, in globose mass.

HABITAT: Moist forest, often by streambanks; 650–1500 m.

DISTRIBUTION: Kigezi, only known from the Ishasha Gorge area; also in DRC.

PROTECTED AREA OCCURRENCE: Bwindi Impenetrable National Park.

LOCAL NAMES AND USES: 'Busangango', 'Buunga', 'Kabungo', 'Kamatundu' (Kirriga/Kirega), 'Kakaki', 'Kakei', Shekakoro' (Kinyanga), 'Bushangangwa' (Kitembo).

RED LIST ASSESSMENT: Assessed as Endangered (EN B1+2c) in 1998 by Makerere IENR.

EOO 19,458 km^2; AOO 5,646 km^2; 15 collections from 4 locations; protected in Bwindi National Park. While we felt we should re-assess this as Vulnerable (VU D2) on account of locations being less than five, we could not come up with a specific threat to either the species or the habitat; the relatively high number of collections thus leads us to re-assess this as Least Concern (LC) due to its EOO and AOO, its altitude range; its occurrence in protected areas; and the lack of a specific major threat.

FLOWERING AND FRUITING TIMES:

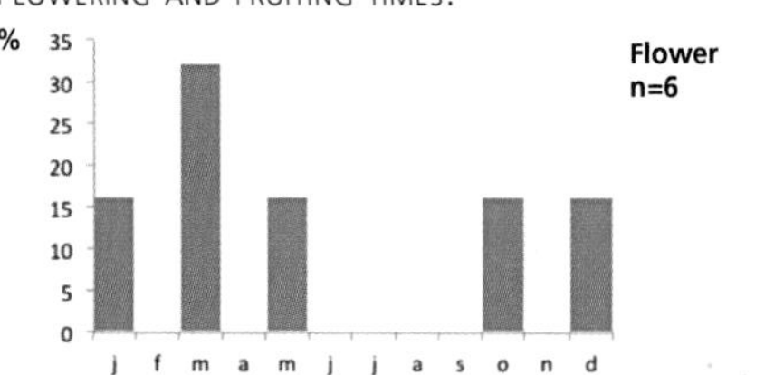

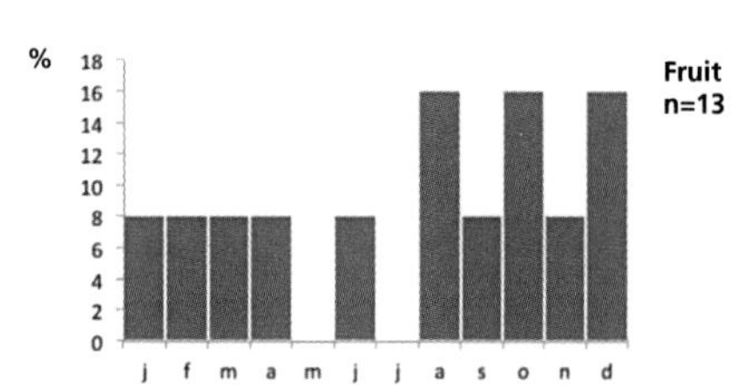

Desplatsia mildbraedii Burret **Malvaceae** VU

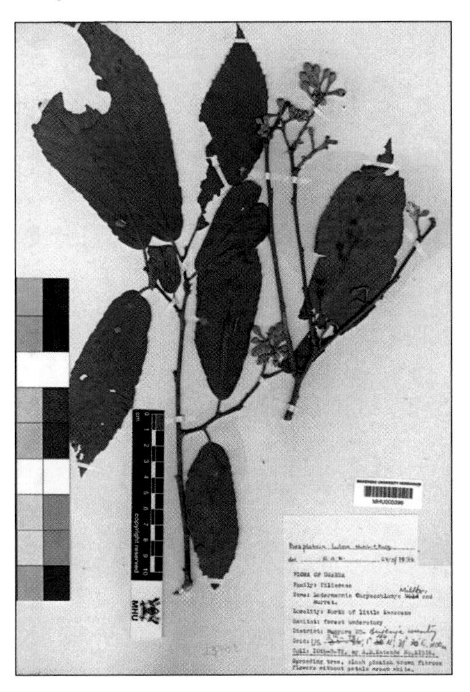

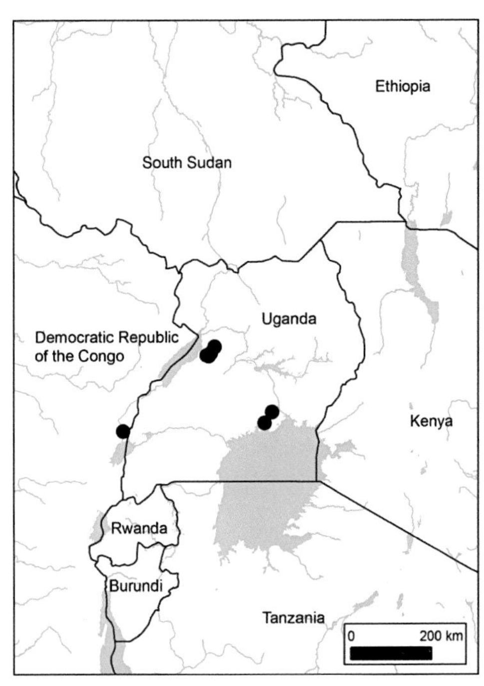

TYPE: DRC, between Beni and Irumu, near Pojo, *Mildbraed* 2836 (B† holo., BM, BR iso.)

LITERATURE: Whitehouse, Cheek, Andrews & Verdcourt, FTEA 2001: 66.

SYNONYM: *Desplatsia lutea* in the sense of ITU: 426, partly.

DESCRIPTION: Tree 9–18 m high. Branches stellate-tomentellous and with scattered simple hairs. Leaves oblong, 10–21 cm long, 2.8–8 cm wide; petiole 0.8–1.2 cm long. Cymes axillary and terminal, 4–8-flowered. Fruit oblong-globose, ± 6.5–9 cm long, 5–7.5 cm wide, turning rusty-brown within a few seconds when cut. Seeds obovate in outline, 1.7 cm long, 1 cm wide, coarsely hairy, distinctly winged at broad end.

HABITAT: Moist forest, regenerating forest; 750–1200 m.

DISTRIBUTION: Bunyoro, Mengo, Toro. Also in DRC.

PROTECTED AREA OCCURRENCE: Semuliki National Park; Budongo, Mabira Forest Reserve.

LOCAL NAMES AND USES: None recorded.

RED LIST ASSESSMENT: Mostly from Budongo, but only very few collections from elsewhere and some of these are from now developed localities with little chance of the species' presence. Said to be common in Bwamba area in 1945. EOO 36,872 km^2; AOO 6,990 km^2; 10 collections from 7 locations, but some of these are historic only. With no more than five current locations and a continuing degradation of habitat in some of these, we here assess it as Vulnerable (VU D2).

FLOWERING AND FRUITING TIMES:

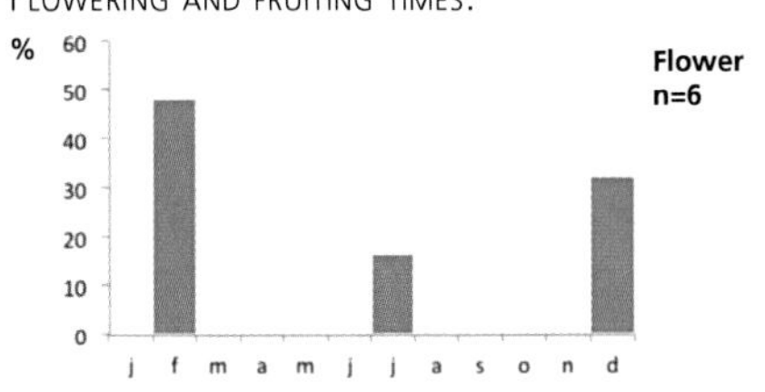

Fruit – no data

LC

Grewia rugosifolia De Wild. **Malvaceae**

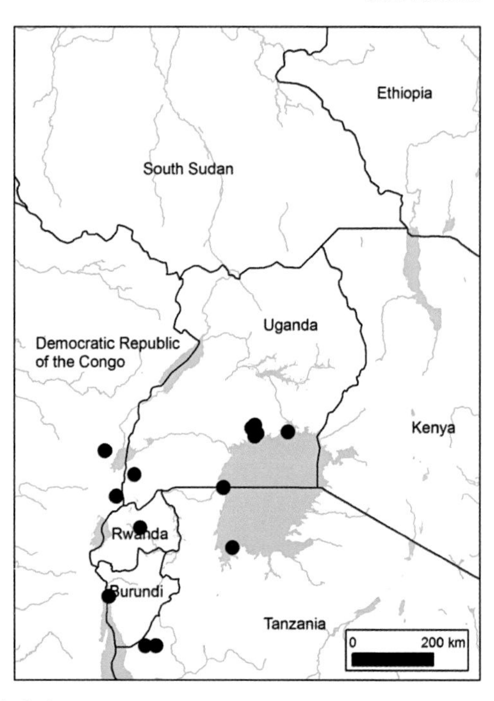

TYPE: DRC, Beni, 30 July 1914, *Bequaert* s.n. (BR holo.)

LITERATURE: Whitehouse, Cheek, Andrews & Verdcourt, FTEA 2001: 35.

DESCRIPTION: Shrub, liana or tree to 12 m high. Young stems stellate-pubescent, sometimes only sparsely so, older branches 3–4-angled. Leaves oblong to oblong-ovate, 6.5–13 cm long, 3–8.5 cm wide, margin serrate; petiole 4–10 mm long, stellate-pubescent. Cyme 1–3-flowered. Flowers yellow to orange-yellow. Fruit indistinctly 1–4-lobed, subglobose, 7–10 mm long, 9–14 mm wide.

HABITAT: Wooded grassland, forest edges, riverine forest. lakeshores, forest/thicket interface; 1100–1350 m.

DISTRIBUTION: Kigezi, Mengo. Also in W Tanzania, DRC, Burundi, ?Zambia.

PROTECTED AREA OCCURRENCE: Minziro Forest Reserve.

LOCAL NAMES AND USES: 'Kapalamakanga' (Kibemba); 'Motandaseke' (Kisimba); 'Mukoma' (Luganda). Fruit edible.

RED LIST ASSESSMENT: EOO 251,886 km^2; AOO 157,046 km^2; 18 collections from 16 locations. Assessed here as Least Concern (LC) as it has wide distribution, and occurs in a wide range of habitats, though with a narrow altitudinal range.

FLOWERING AND FRUITING TIMES:

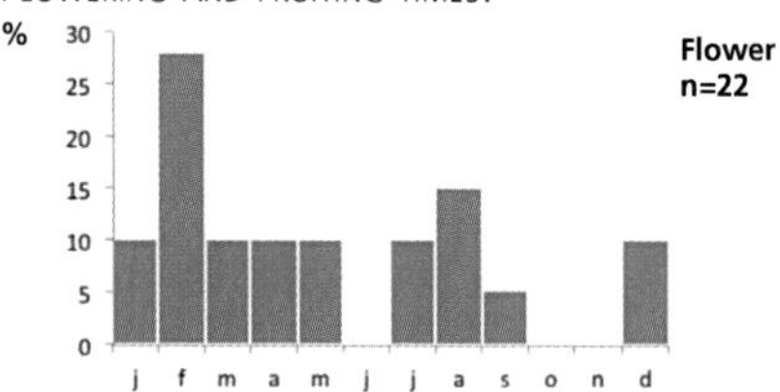

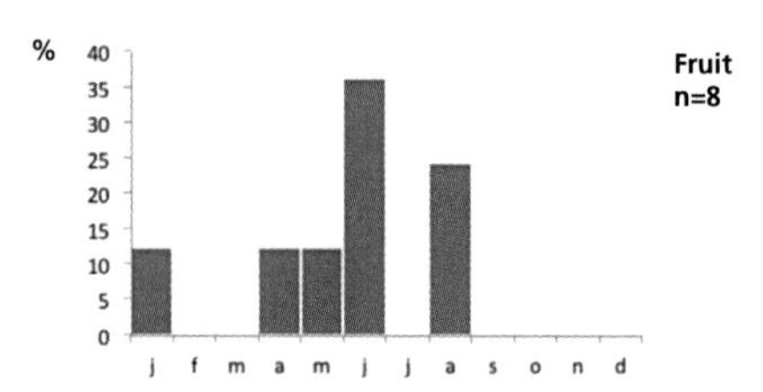

Grewia ugandensis Sprague

Malvaceae LC

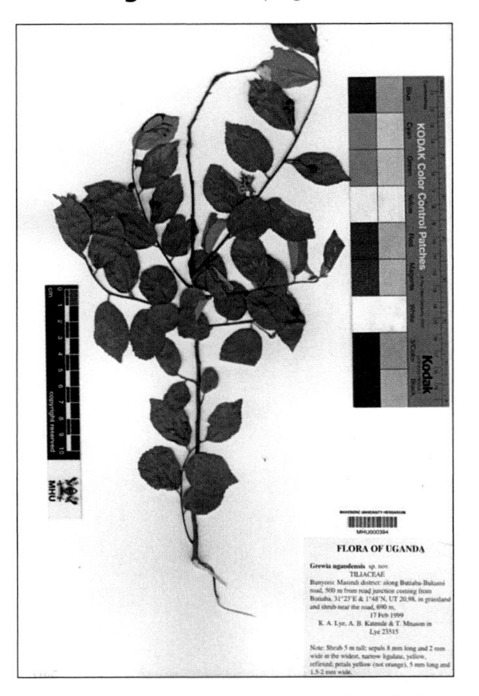

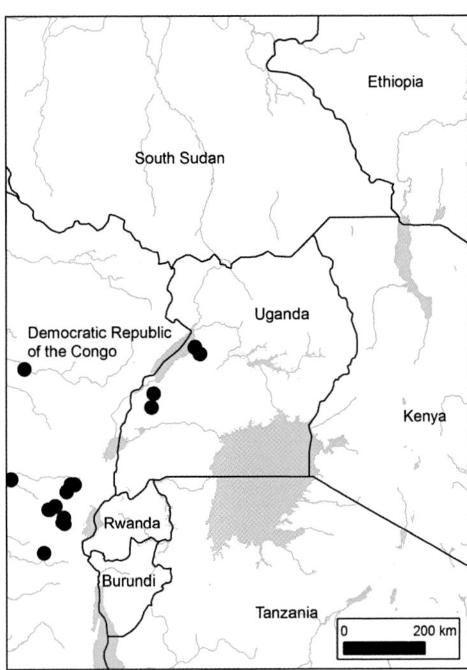

TYPE: Uganda, Bunyoro, *Dawe* 918 (K holo.)

LITERATURE: Whitehouse, Cheek, Andrews & Verdcourt, FTEA 2001: 62.

DESCRIPTION: Tree 5 m tall. Branches rusty brown-pubescent. Leaves ovate to elliptic, 5.5–14 cm long, 3–7 cm wide, margin slightly undulate, virtually glabrous, silvery-pubescent beneath; petiole 4–10 mm long, stellate-pubescent. Inflorescence a terminal or axillary panicle, with a rusty brown stellate pubescence. Fruit unlobed, ellipsoid, laterally compressed, 20–24 mm long, 8–13 mm wide, sparsely pubescent. Seed ± 9 mm long, 3.5 mm wide.

HABITAT: Forest, grassland and secondary bushland; 650–1500 m.

DISTRIBUTION: Toro. Also in DRC, South Sudan and reported from W Tanzania — but no specimens seen from there.

PROTECTED AREA OCCURRENCE: Kibale National Park; Itwara and Tongwe East Forest Reserves.

LOCAL NAMES AND USES: 'Chaunga' (Katembo); 'Duka' (Mambasa area, DRC); 'Nbafia' (Bondo area, DRC). Used for axe handles; wood is light but does not break.

RED LIST ASSESSMENT: EOO 716,783 km^2; AOO 208,789 km^2; 23 collections from 19 locations; assessed here as Least Concern (LC) due to its large EOO and AOO, its wide altitudinal range; the common habitat, and the lack of a specific major threat.

FLOWERING AND FRUITING TIMES:

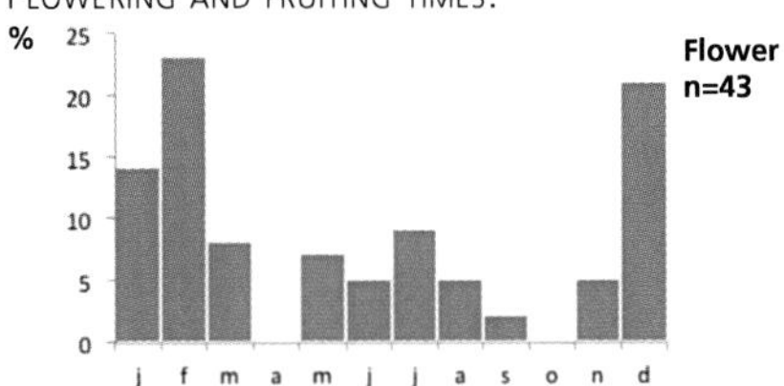

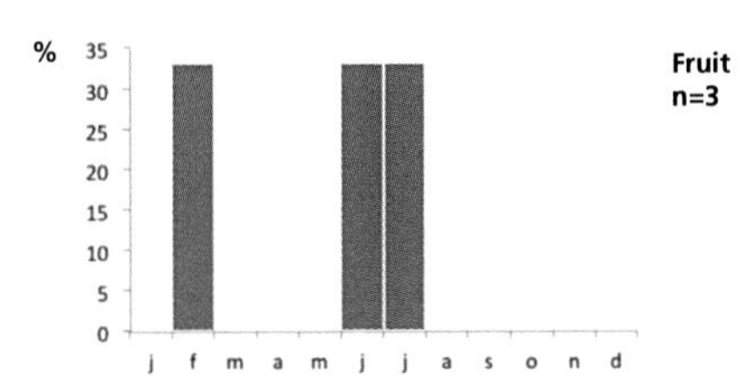

Grewia sp. A of FTEA **Malvaceae**

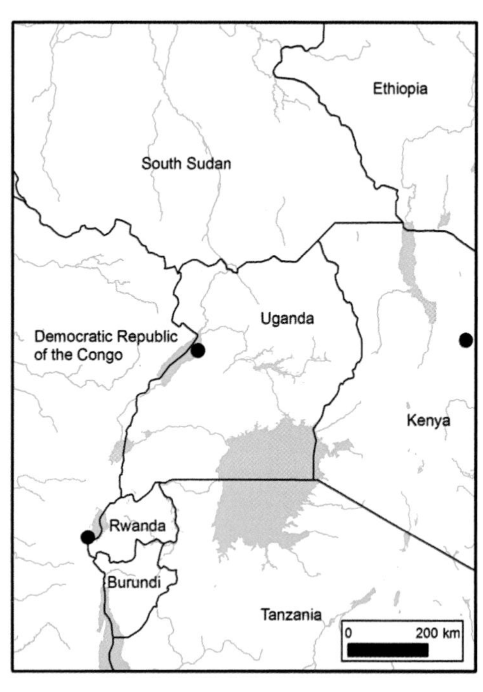

Type: none (yet) indicated; based on one Ugandan (Bunyoro: Butiaba flats) and two Kenyan specimens.

Literature: Whitehouse, Cheek, Andrews & Verdcourt, FTEA 2001: 51.

Description: Shrub or tree to 4 m high. Young branches densely pubescent. Leaves ovate to ovate-oblong, 1.7–7.7 cm long, 0.8–4.2 cm wide, margin serrate; petiole 1–4 mm long, densely pubescent. Inflorescence 1–3-flowered cyme. Flowers yellow. Fruit 2-lobed, or unlobed by abortion, sparsely pilose, orange when ripe.

Habitat: Riverine forest edge; 650–1200 m.

Distribution: Bunyoro. Also in DRC, Kenya.

Protected area occurrence: None.

Local names and uses: None recorded.

Red List assessment: Status doubtful; Data Deficient

Flowering and fruiting times:

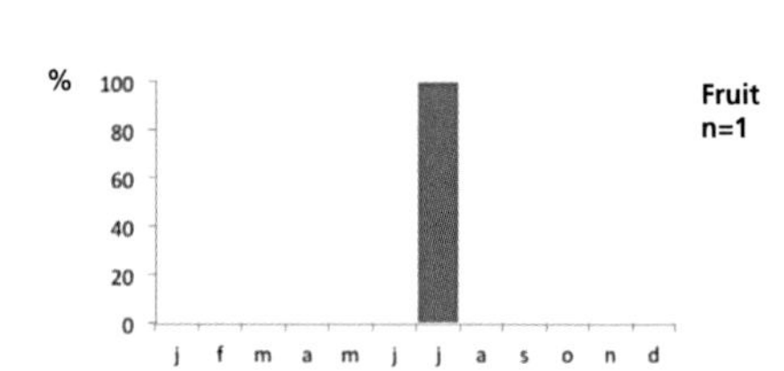

Leptonychia mildbraedii Engl.

Malvaceae LC

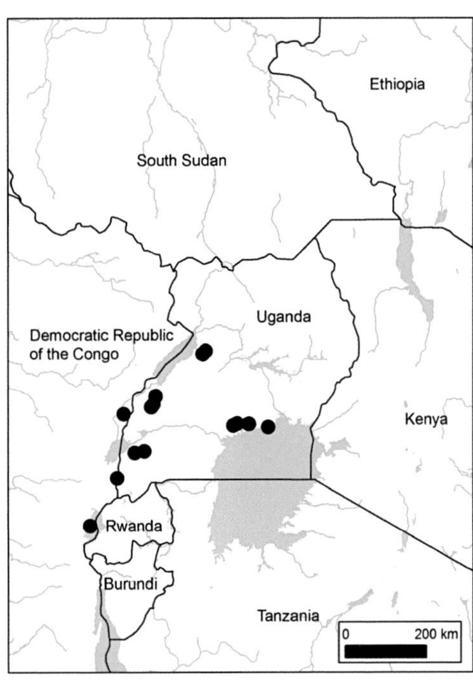

TYPE: DRC, Kwidjwi Island in Lake Kivu, *Mildbraed* 1205 (B† lecto.)

LITERATURE: ITU: 421. UFT: 120. Dorr in FTEA 2007: 99.

SYNONYM: *Leptonychia multiflora* in the sense of ITU: 421.

DESCRIPTION: Shrub or tree to 6 m high. Bark smooth, grey or red-brown; young brachlets brownish-red, glabrous. Leaves elliptic to oblong-elliptic, 7–12.5(–15) cm long, 2.7–5 cm wide, margin entire to repand; petiole 4–10 mm long, sparsely stellate-puberulous to glabrous. Cymes sessile, ± stellate-puberulous, 1–2(–3), 2–3-flowered dichasia. Petals suborbicular. Ovary 3(–4)-locular; 3 ovules per locule. Capsule ±. 2.5 cm long, 2–2.5 cm in diameter, brownish-grey, densely tomentellous.

HABITAT: Moist forest and riverine forest, also in forest remnants, sometimes associated with *Parinari*; 650–1800 m.

DISTRIBUTION: Mengo, Ankole, Bunyoro, Toro. Also in Tanzania, DRC and possibly CAR.

PROTECTED AREA OCCURRENCE: Kibale, Bwindi and Semuliki National Parks; Kalinzu, Kasyoha-Kitomi, Budongo, Itwara, Bugoma, Bukaleba, Mabira & Mpanga Forest Reserves.

LOCAL NAMES AND USES: 'Nkomakoma' (Lunynkole).

RED LIST ASSESSMENT: Said to have been abundant in Nakiza Forest (1950) but this forest has since been badly degraded. It was locally frequent in Kibale Forest (1955). EOO 200,955 km^2; AOO 72,260 km^2; 39 collections from 15 locations. Cheek's preliminary assessment in FTEA was Near Threatened (NT), due to its being 'widespread although not common ... in most of the range there are currently few threats to its habitat'. We believe the extent of EOO and AOO coupled to current threat levels do not warrant NT status, and we prefer LC (Least Concern) due to the extent of distribution coupled to its occurrence in a series of protected areas.

FLOWERING AND FRUITING TIMES:

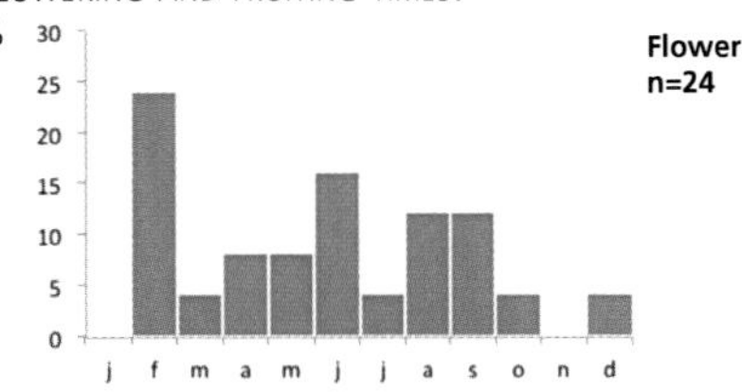

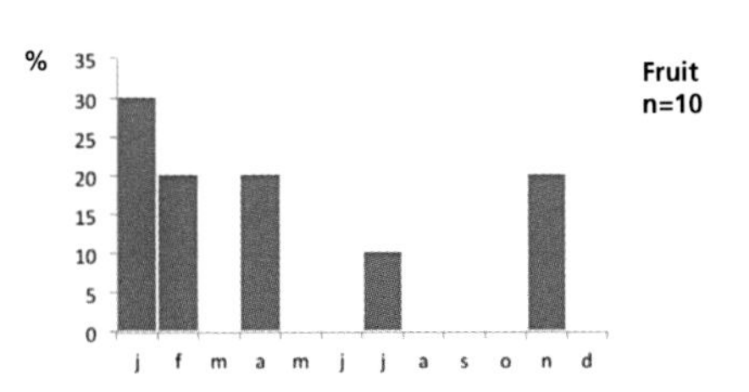

LC **_Lijndenia bequaertii_** (De Wild.) Borhidi **Melastomataceae**

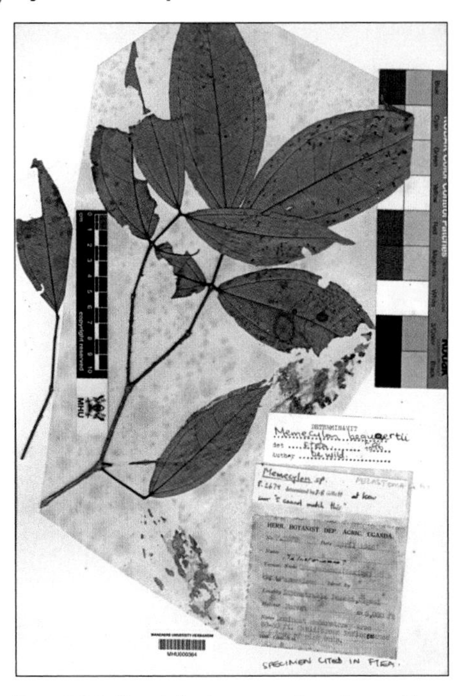

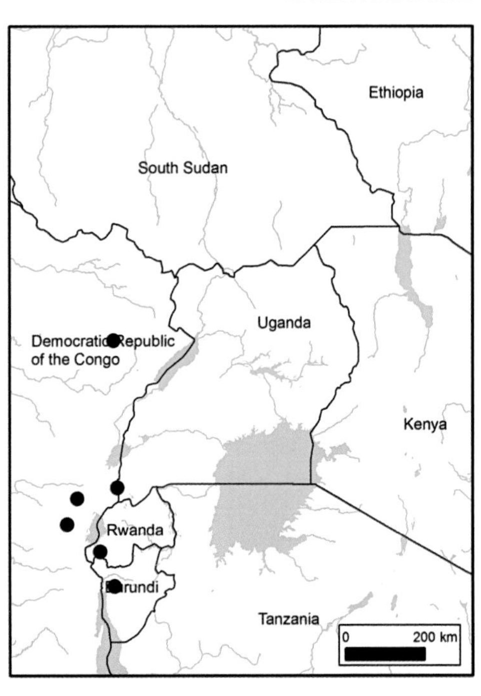

TYPE: DRC, Nandefu, between Penghe and Irumu, *Bequaert* 2666 (BR holo.)

LITERATURE: Borhidi in *Opera Bot*. 121: 151 (1993).

SYNONYMS: *Warneckea bequaertii* (De Wild.) Jacq.-Fél.; Lebrun & Stork (1991): 159.
Memecylon bequaertii De Wild.; Wickens, FTEA 1975: 83.

DESCRIPTION: Tree 9–15 m high. Branchlets obscurely 4-winged. Leaf blade elliptic, 9–15.5(–19) cm long, 3–7(–9) cm wide, petiole 2–3 mm long. Inflorescence ± 12-flowered. Petals bluish-pink. Fruit globose, 9–10 mm in diameter.

HABITAT: Moist forest; 850–2100 m.

DISTRIBUTION: Kigezi, one record from Bwindi; also in DRC.

PROTECTED AREA OCCURRENCE: Bwindi Impenetrable National Park.

LOCAL NAMES AND USES: None recorded.

RED LIST ASSESSMENT: May be locally common; said to have been dominant at 1500 m in Bwindi in 1948. EOO 37,669 km²; AOO 21,129 km²; 9 collections from 6 locations. Assessed here as Least Concern (LC) due to the size of EOO and AOO; the number of locations and the altitude range, coupled with 'locally common' notes. We do not know of any specific threats to the species.

FLOWERING AND FRUITING TIMES:

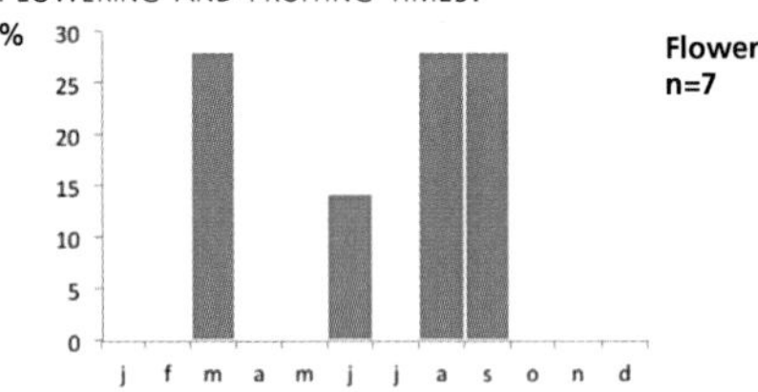

Fruit – no data

Ficus katendei Verdc.

Moraceae CR

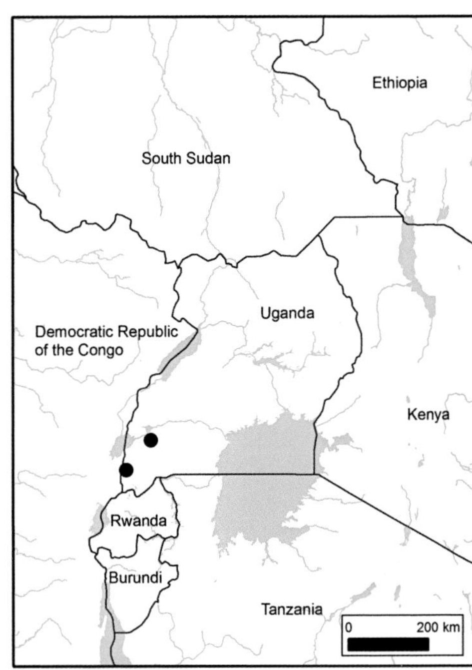

TYPE: Uganda, Kasyoha-Kitomi, Kyambura R., *Katende* 4941 (K holo., C, MHU, iso.)

LITERATURE: Verdcourt in *Kew Bull.* 53 (1): 233 (1998).

DESCRIPTION: Tree (starting as epiphyte) 5–25 m high; bark smooth, grey. Leaves obovate, 12–29 × 7–13 cm, petiole 2–7 cm. Figs in dense, head-like groups of 15–20 on bosses on the trunk; individual figs mottled red and white, obovoid, to 7 × 3 cm.

HABITAT: Moist or riverine forest; 1400–1500 m.

DISTRIBUTION: Kigezi, Bushenyi; not known elsewhere.

PROTECTED AREA OCCURRENCE: Bwindi Impenetrable National Park and Kasyoha-Kitomi Forest Reserve.

LOCAL NAMES AND USES: None recorded.

RED LIST ASSESSMENT: Known from two collections, one at the Ishasha River, collected in 1995; and two collections from a single tree on the South Kyambura river in 1991 and 1994, since cut down to make a bridge across the river (Katende, pers. comm.) This is obviously a very rare species, and we assess it here as Critically Endangered (CR D1, C2 (a1)) as the number of mature individuals is estimated at less than 50 with a decline of known individuals of 50% over one generation.

FLOWERING AND FRUITING TIMES:

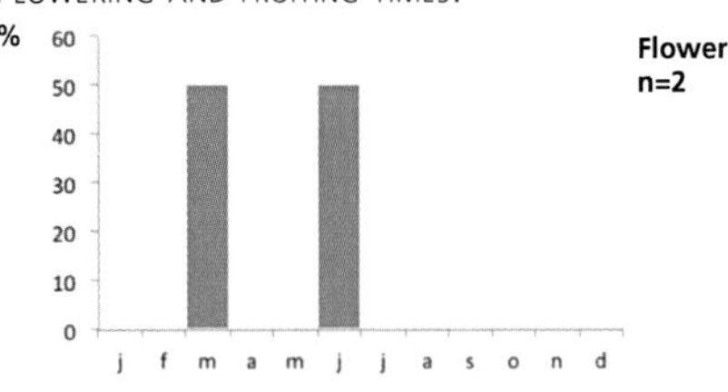

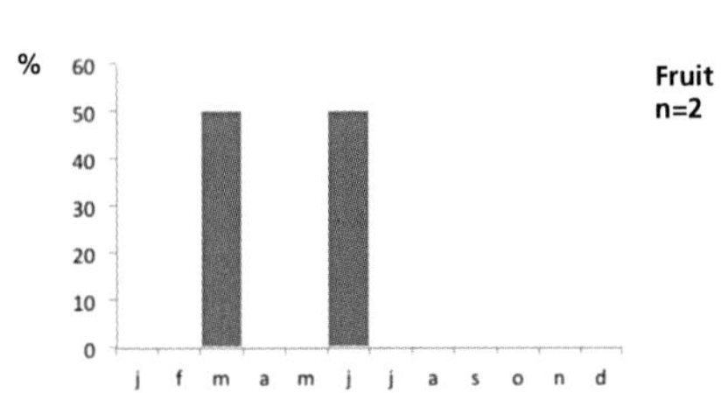

EN **Gomphia mildbraedii** (Gilg) Verdc. **Ochnaceae**

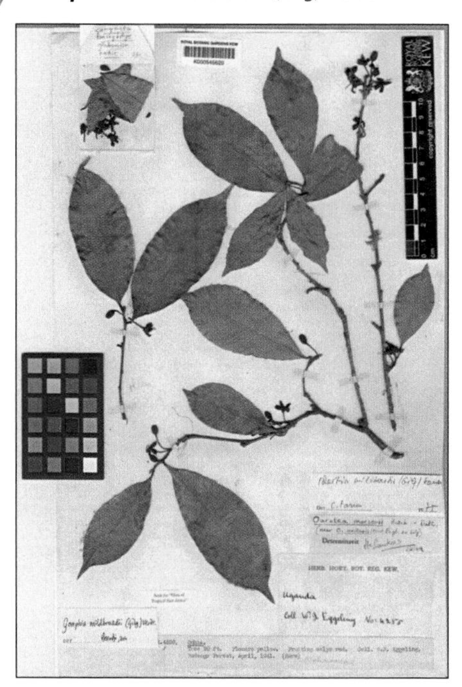

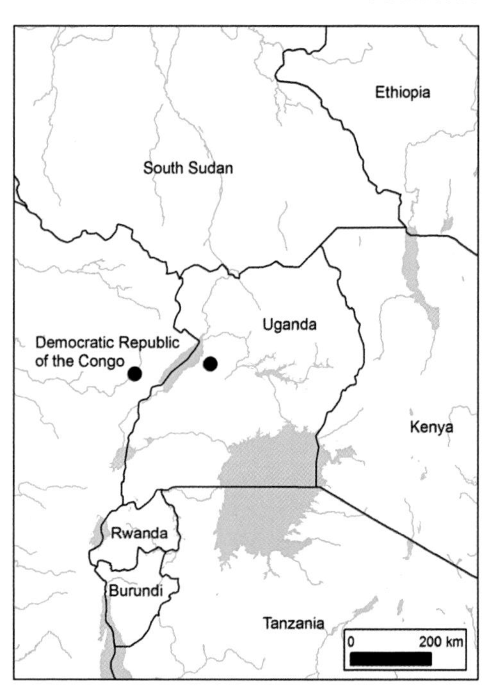

TYPE: DRC, Irumu-Mawambi, *Mildbraed* 2918 & 2931 (both B† syn.)

LITERATURE: Verdcourt, FTEA 2005: 53.

SYNONYM: *Ouratea morsoni* in the sense of ITU: 282.

DESCRIPTION: Shrub or tree to 6 m high. Leaf lamina elliptic or obovate-elliptic, 5.5–13(–15) cm long, 2.5–5.5 cm wide; petiole 4–7 mm long. Inflorescences axillary, short, sessile, 1–3-flowered; Petals yellow. Drupelets ellipsoid, 13 mm long, 6–7 mm wide.

HABITAT: Evergreen forest; 1050 m.

DISTRIBUTION: Bunyoro. Also in DRC.

PROTECTED AREA OCCURRENCE: Budongo Forest Reserve.

LOCAL NAMES AND USES: None recorded.

RED LIST ASSESSMENT: Known from only three collections; the two from DRC are from the same locality and date back to before 1913; the collection from Uganda is from 1941. Numbers must be very low as this taxon has not been recollected; at least the Ugandan site (Budongo Forest) has been visited by numerous botanists. Known from only two locations, with the Ugandan Budongo facing encroachment in the recent past. We assess this species as Endangered (EN D); the species might move to Critically Endangered (CR) if more information on the number of mature individuals becomes available; but only a search in its original locations will prove or disprove the latter.

FLOWERING AND FRUITING TIMES: No data at all.

Ochna leucophloeos A. Rich. **Ochnaceae**

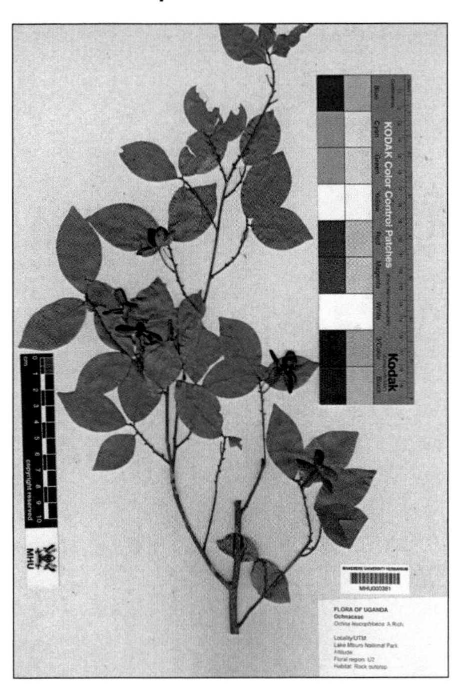

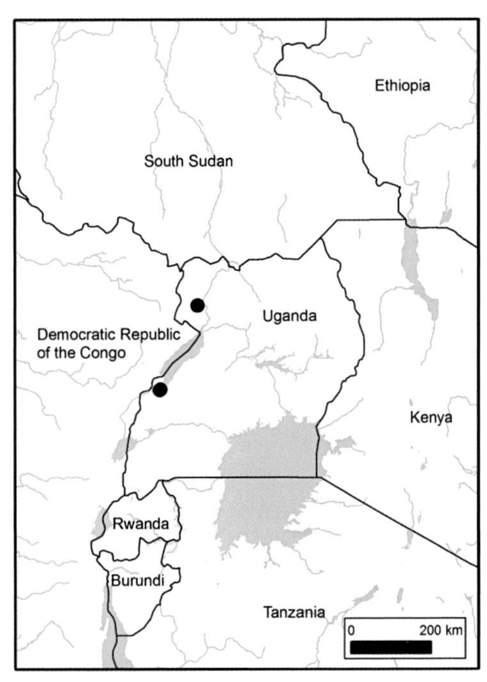

LITERATURE: Verdcourt, FTEA 2005: 21.

SYNONYM: *Ochna ovata* of ITU: 280, partly.

DESCRIPTION: Shrub or tree 3–8 m high. Bark grey or yellowish, often powdery or scaly; branches grey-brown or yellowish; youngest shoots purplish brown, lenticellate. Leaves elliptic to slightly obovate, 4–7.5(–21) cm long, 2.3–4(–8) cm wide, margin serrate; petiole 2–4 mm long, winged from decurrent leaf base. Petals bright yellow. Drupelets ellipsoid, 8–13 mm long.

DISTRIBUTION: Distributed in South Sudan and Ethiopia.

RED LIST ASSESSMENT: We assess the species as Least Concern (LC), due to its wide distribution.

- subspecies ***ugandensis*** Verdc. is endemic to Uganda.

EN

TYPE: Uganda, Adilang, *Greenway & Hummel* 7338 (K holo., EA iso.)

HABITAT: Open woodland; 900 m.

LOCAL NAMES AND USES: None recorded.

DISTRIBUTION: Acholi, West Nile, Toro; limited distribution in Semliki and Ajai.

PROTECTED AREA OCCURRENCE: Ajai, Semliki Wildlife Reserves.

RED LIST ASSESSMENT: We assess the subspecies as Endangered (EN D1) as the two collections (from different locations) are both pre-1955 and the taxon has only been collected once since (*Kalema* 2372); EOO and AOO cannot be computed on the basis of only two localities. Both Semliki and Ajai Wildlife Reserves have undergone degradation in the past, although attempts are now being made to reverse the trend. The assessment is based on the very small number of individuals, estimated as less than 250.

FLOWERING AND FRUITING TIMES: No data at all for the subspecies.

DD **_Ochna_ sp. 40** of FTEA **Ochnaceae**

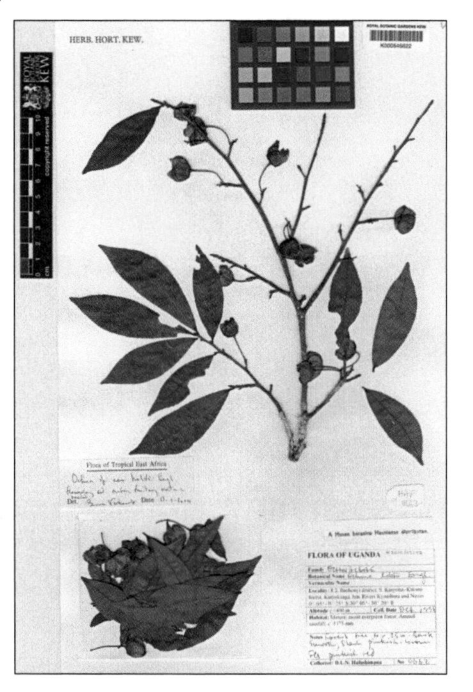

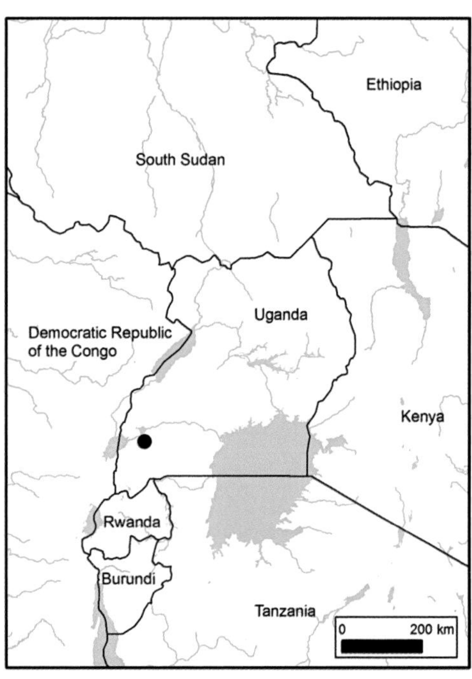

TYPE: Species not yet described officially.

LITERATURE: Verdcourt, FTEA 2005: 37.

DESCRIPTION: Tree to 15 m high. Bark smooth, slash pinkish brown. Leaves thin, narrowly elliptic to oblanceolate, 6–8 cm long, 1.6–2.5 cm wide, apex acuminate, base cuneate, margin closely serrulate. Flowers 2–5 at end of short spur shoots.

HABITAT: Evergreen forest; 1400 m.

DISTRIBUTION: Ankole.

PROTECTED AREA OCCURRENCE: Kashyoha-Kitomi Forest Reserve.

LOCAL NAMES AND USES: None recorded.

RED LIST ASSESSMENT: Only known from a single specimen, taxonomy unclear: Data Deficient (DD).

FLOWERING AND FRUITING TIMES: Young fruit in October.

Pandanus chiliocarpus Stapf

Pandanaceae NT

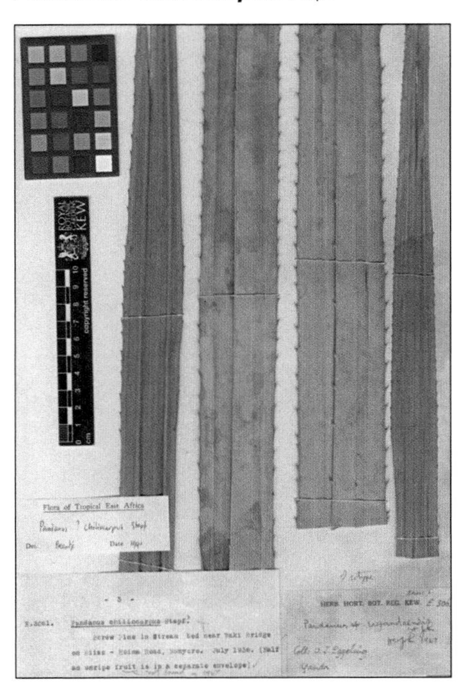

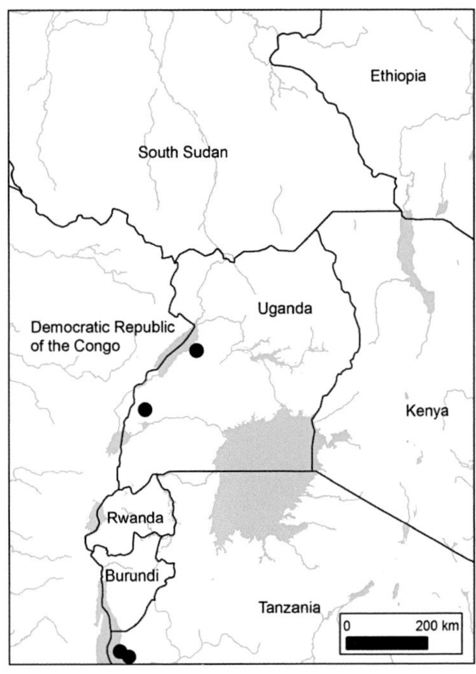

TYPE: Uganda, Toro, by Nsonge and Dura rivers, *Dawe* 523 (type missing).

LITERATURE: Beentje, FTEA 1993: 6. ITU: 294, photos 49a,b.

SYNONYM: *Pandanus ugandaensis* of UFT: 76.

DESCRIPTION: Tree 7–15 m high; bark grey or brown with conical spines; stilt roots numerous, 50–100 cm high, with sharp spines. Leaves in twisted dichotomous arrangement, 135–270 cm long, 4–8 cm wide and clasping at base, with marginal spines; midrib with spines. Male inflorescence 14–26 cm long, with 4–6 spathes. Female inflorescences unknown.

HABITAT: Swamps, streams in forest; 1150–1200 m.

DISTRIBUTION: West Nile, Toro, Ankole, Bunyoro, Madi, Busoga. Also in Tanzania.

PROTECTED AREA OCCURRENCE: Kibale, Queen Elizabeth and Semuliki National Parks.

LOCAL NAMES AND USES: None recorded.

RED LIST ASSESSMENT: This species was described as 'common' at the Malagarasi River in 1950, and as 'dominant by all rivers' in the Kigoma/Kasulu area in 1961. The last collections are from the 1960s, when the species was said to be common by riverbanks, even next to cultivation. Pandans survive reasonably well in cleared areas, but on the other hand the area of occupation is small and the habitat is declining due to increasing cultivation and conversion of its suitable habitat. It however still occurs in some protected areas with sizeable populations in Semuliki National Park and smaller populations in Queen Elizabeth National Park and Kibale National Park (Kalema's observations 1994, 2005 and 2009 respectively). EOO 54,543 km^2; AOO 31,020 km^2; 10 collections from 6 locations. Assessed here as Near Threatened (NT (D2)) due to the small number of locations and the increasing agricultural use of the specific habitat in parts of its range.

FLOWERING AND FRUITING TIMES:

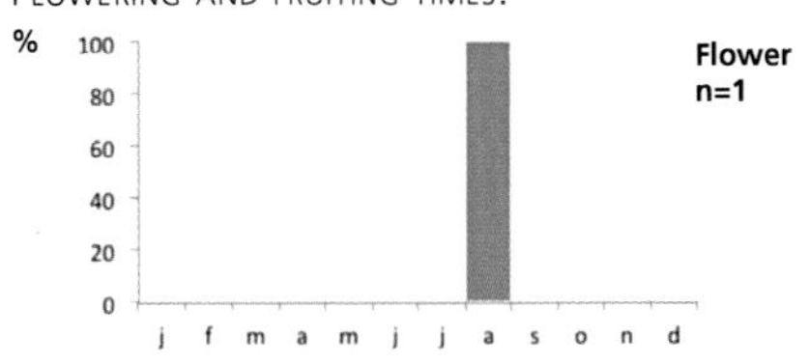

Fruit – no data

LC **_Balthasaria schliebenii_** (Melch.) Verdc. **Pentaphylacaceae**

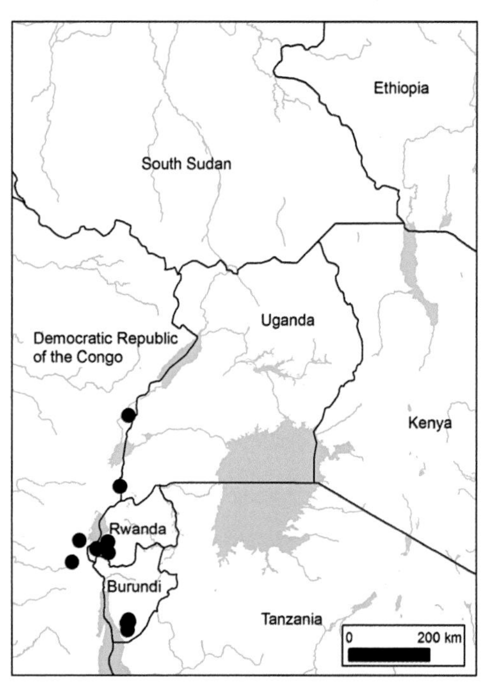

TYPE: Tanzania, NW Uluguru Mts, Lupanga, *Schlieben* 3175 (B† holo., A, BM, BR, K iso.)

LITERATURE: Verdcourt in *Kew Bull*. 23: 469 (1969).

SYNONYMS: *Melchiora schliebenii* (Melch.) Kobuski; Verdcourt, FTEA 1962: 4, fig. 2. UFT: 140.
Adinandra schliebenii in the sense of ITU: 424.

DESCRIPTION: Tree 6–40 m high. Trunk straight; bark rough, reticulate. Leaves oblanceolate or elliptic-oblanceolate to broadly elliptic, finely glandular-serrate, glabrous above; petiole 5–6(–10) mm long. Flowers with two opposite bracts; petals mostly orange or orange-red with paler tips. Fruit ovoid, 2–3 cm long and 1 cm in diameter. Seeds numerous, 1–1.5 mm in diameter.

HABITAT: Moist forest, often on streambanks but also found on ridge-tops; 1000–2450 m.

DISTRIBUTION: Kigezi, in Uganda only known from Bwindi and no collections since 1940. Also in DRC, Rwanda, Burundi, Tanzania.

PROTECTED AREA OCCURRENCE: Bwindi Impenetrable National Park, West Usambara Mts Forest Reserve.

LOCAL NAMES AND USES: Eaten by pigeons. 'Msambia nr. 2' (Tanzania); 'Msambu' (Kishambaa), 'Umujugeshi', Umushegashi' (Kinyarwanda).

RED LIST ASSESSMENT: Assessed as Lower Risk/Near Threatened [NT] by Lovett & Clarke in 1998.

Where notes on occurrence are given, these say 'occasional' or 'five trees seen'; said to be locally common in South Pare (1942), common in Shagayu (1953) and codominant in Bondwa (1953). EOO 422,240 km²; 133,106 km²; 33 collections from 22 locations. Several times described as locally common; wide altitudinal range; assessed here as Least Concern (LC) due to its EOO and AOO, its altitudinal range; its occurrence in protected areas; and the lack of a specific major threat.

FLOWERING AND FRUITING TIMES:

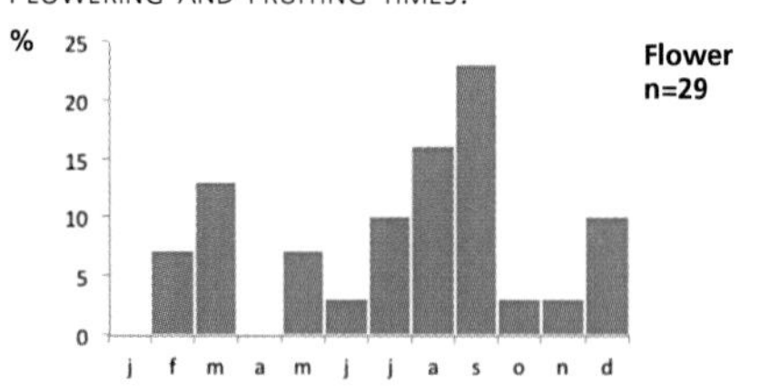

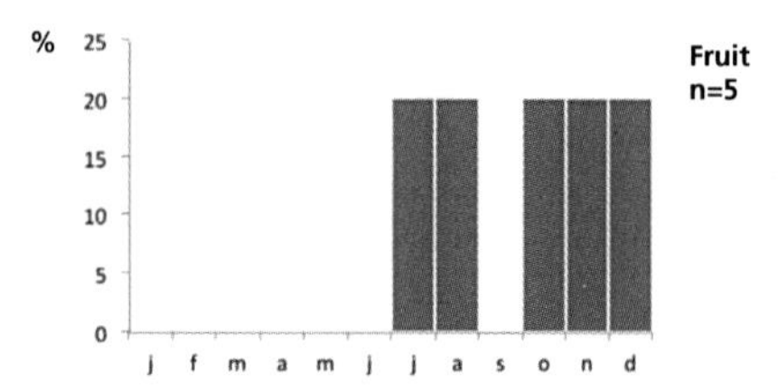

Afrocarpus dawei (Stapf) C. N. Page **Podocarpaceae** LC

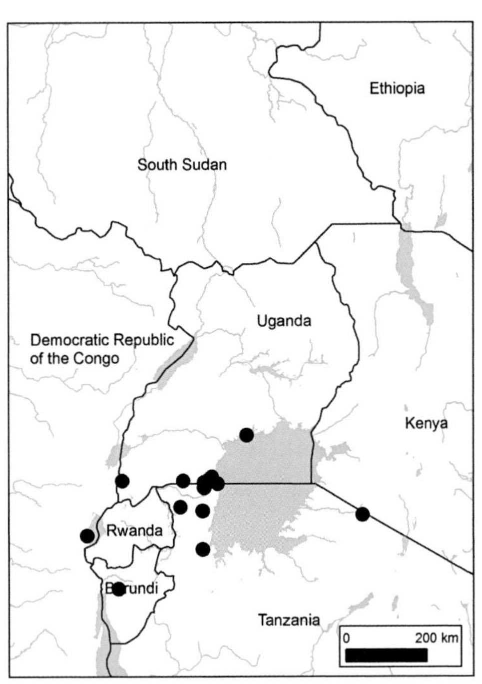

TYPE: Uganda, Masaka, Kagera river, *Dawe* 961 (K holo.)

LITERATURE: Farjon 2001: 252.

SYNONYMS: *Podocarpus usambarensis* Pilg. var. *dawei* (Stapf) Melville; Melville, FTEA 1958: 12. UFT: 75.
Podocarpus dawei Stapf; ITU: 315.

DESCRIPTION: Tree to 35 m high; outer bark of trunk grey-brown to dark brown. Leaves linear to linear-lanceolate, 10–15 mm long in adults, 1.5–2.5 mm wide, to 13 × 8 mm in juveniles. Male cones 4–26 mm long, yellow; female cones solitary. Fruit green to yellowish-green, subglobular, sometimes ellipsoid, 19–35 mm long, 16–30 mm diameter, woody shell 2–8 mm thick.

HABITAT: Swamp forest or riverine forest; 1100–1400(–1650, ?2200) m.

DISTRIBUTION: Kigezi, Masaka. Also in Tanzania, DRC and Burundi.

PROTECTED AREA OCCURRENCE: Kaiso Forest & Malabigambo Forests both in Sango Bay Forest Reserve, Minziro Forest Reserve.

LOCAL NAMES AND USES: Used for firewood, charcoal, timber/furniture, poles, tools, carvings/utensils, ornamental. In 1948 the main timber for a sawmill in Minziro Forest.

RED LIST ASSESSMENT: Locally common to co-dominant in Malabigambo Forest in 1950–1953. Assessed as Least Concern (LC) in 2007 by Bachman, Farjon, Gardner, Thomas, Luscombe & Reynolds with an EOO of 97,974 km^2 and an AOO of 18,215 km^2.

Our mapping has led us to an EOO 134,005 km^2; AOO 42,588 km^2, with 33 collections from 17 locations. Assessed here as Least Concerened (LC), because its EOO and AOO are large, and the number of locations is considerable. There is widespread use of the species for timber and charcoal, especially in Sango Bay/Minziro (probably the largest sub-population of the species in the wild).

FLOWERING AND FRUITING TIMES:

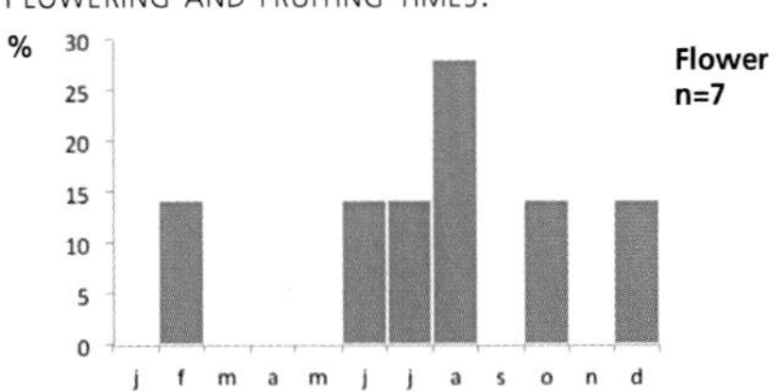

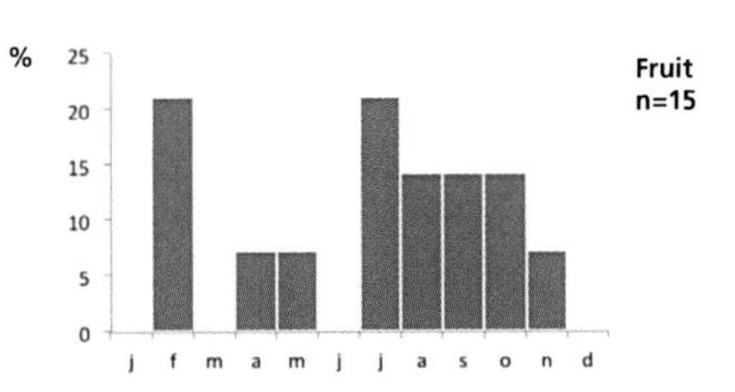

LC **_Drypetes ugandensis_** (Rendle) Hutch. **Putranjivaceae**

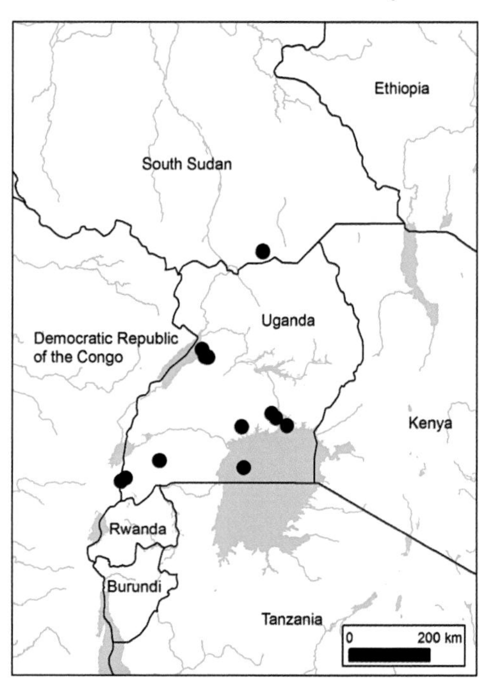

TYPE: Uganda, Victoria Nyanza, Buvuma Island, *Bagshawe* 613 (BM holo., K iso.)

LITERATURE: Radcliffe-Smith, FTEA 1987: 93. ITU: 123. UFT: 138.

SYNONYM: *Drypetes sp.* of ITU: 124.

DESCRIPTION: Shrub or tree to 9 m high; bark smooth to rough. Leaves elliptic to oblong, to 15 × 5 cm, with asymmetric base. Flowers (separate male and female) in groups on bosses on older branches or trunk, 'evil'-smelling. Fruit subglobose, yellow, orange or brown, ± 3 cm across.

HABITAT: Moist forest, mature, open or secondary; 900–1650 m.

DISTRIBUTION: Bunyoro, Masaka, Mengo, Kigezi; also in South Sudan (Imatong Mts only) (according to FTEA also in eastern DRC, but we have failed to find any specimens).

PROTECTED AREA OCCURRENCE: Bwindi Impenetrable National Park, Budongo and Mpanga Forest Reserves, little protected Imatong Mts Forest Reserve.

LOCAL NAMES AND USES: None recorded.

RED LIST ASSESSMENT: Widespread with EOO 104,047 km^2; AOO 34,114 km^2, 30 collections from 15 locations. Assessed here as Least Concern (LC) due to the fairly large EOO and AOO, coupled to the range of habitat types and the altitude range and absence of known specific threats.

FLOWERING AND FRUITING TIMES:

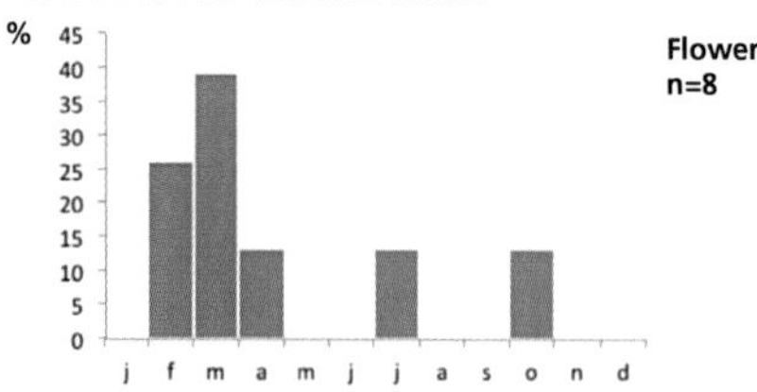

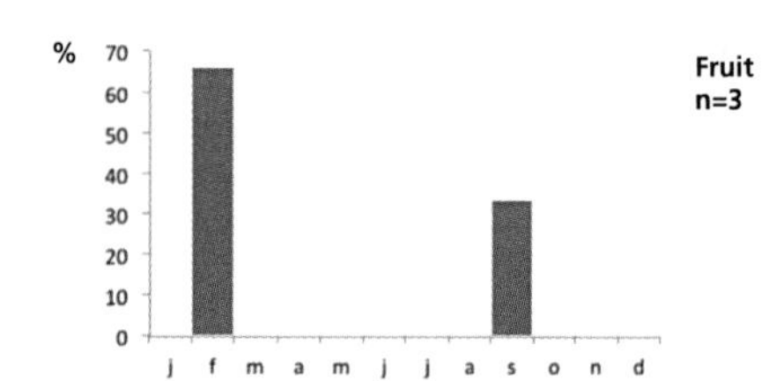

Ixora seretii De Wild. — **Rubiaceae** LC

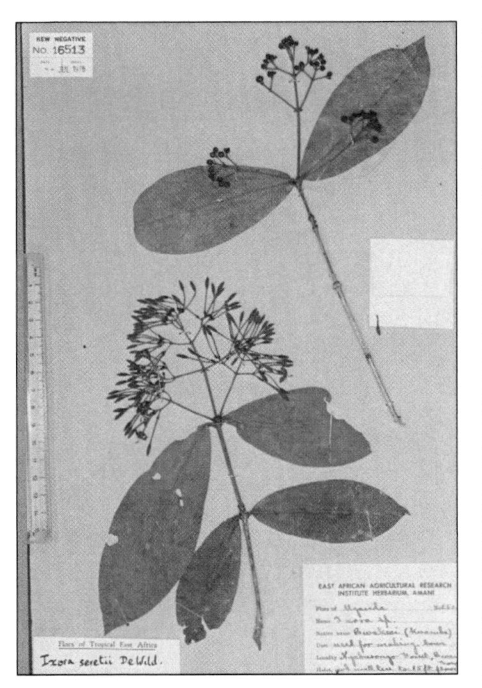

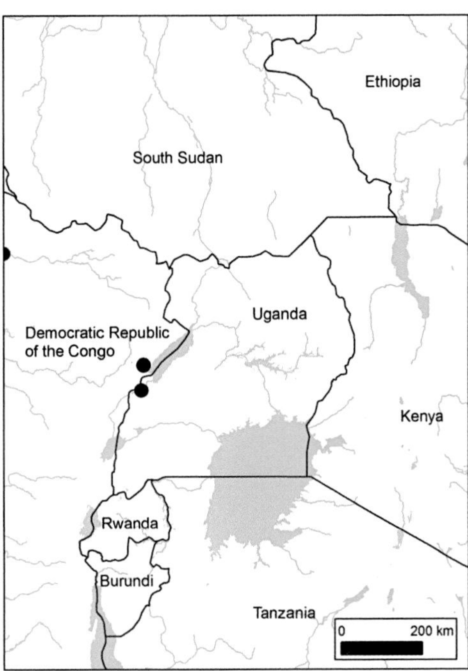

TYPE: DRC, case Koko, left bank of Uele river, *Seret* 381 (BR holo.)

LITERATURE: Bridson & Verdcourt, FTEA 1988: 616.

DESCRIPTION: Shrub or tree 4.5 m high. Young branches glabrous, older branches covered with greenish-buff ± smooth bark. Leaves glabrous, elliptic, sometimes oblong, 8–19.5 cm long, 3.3–6 cm wide, petiole 0.4–1 cm long. Corymbs 6–8.3 cm across, sessile. Corolla white, tube 1.4–1.8 cm long. Mature fruit unknown.

HABITAT: Forest and riverine forest; ± 900 m

DISTRIBUTION: Toro; also in DRC.

PROTECTED AREA OCCURRENCE: Semuliki National Park.

LOCAL NAMES AND USES: Used for making bows.

RED LIST ASSESSMENT: EOO 136,826 km^2; AOO 46,709 km^2; 5 collections from 4 locations; large area of distribution; no known specific threats; used for making bows but this is not believed to be a big threat; assessed here as Least Concern (LC).

FLOWERING AND FRUITING TIMES:

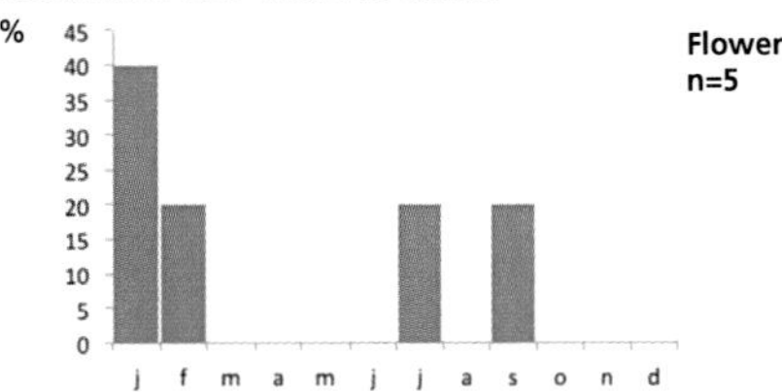

Fruit – no data

LC

Pavetta urundensis Bremek. **Rubiaceae**

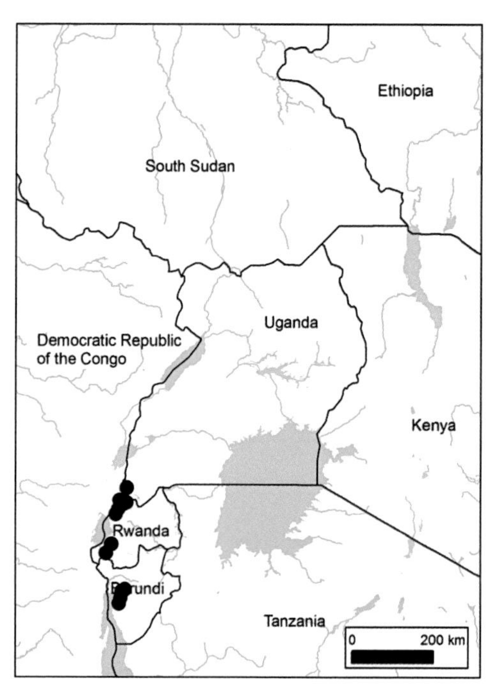

TYPE: Burundi, West Burundi, *Meyer* 1037 (B† holo.)

LITERATURE: Bridson & Verdcourt, FTEA 1988: 632.

DESCRIPTION: Shrub or tree 2–7 m high. Young branches puberulous, soon becoming glabrous; older branches with pale brown shiny bark. Leaf blades oblong-elliptic to broadly elliptic, 8.5–20 cm long, 3–6.5 cm wide, with domatia; petiole 0.5–3 cm long. Corymbs terminal. Corolla tube 5–9 mm long, densely bearded at throat. Fruit black, ± 7 mm across, sparsely hairy.

HABITAT: Evergreen forest, secondary forest, riverine forest, once collected in wooded grassland; 1900–3000 m.

DISTRIBUTION: Kigezi. Also in DRC, Rwanda, Burundi.

PROTECTED AREA OCCURRENCE: Bwindi Impenetrable and Nyungwe National Parks, Mgahinga Gorilla National Park (Sabinio).

LOCAL NAMES AND USES: None recorded.

RED LIST ASSESSMENT: EOO 7,947 km^2; AOO 4,608 km^2; 19 collections from 14 locations; an Albertine Rift endemic; protected in Rwanda in Nyungwe National Park, and in Uganda (Bwindi). Here assessed as Least Concern (LC) due to its EOO and AOO, its range of habitats and altitude range; its occurrence in protected areas; and the lack of a specific major threat.

FLOWERING AND FRUITING TIMES:

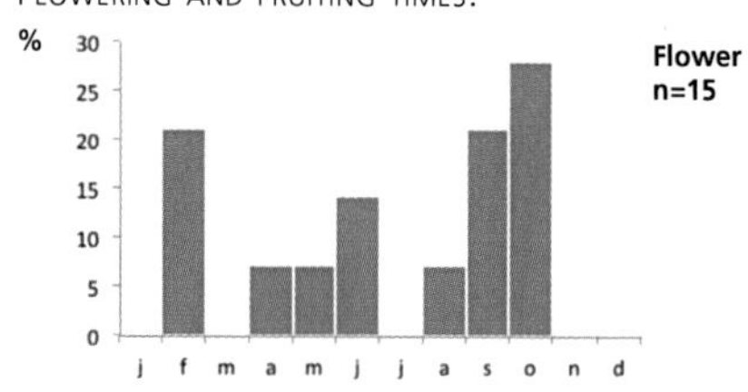

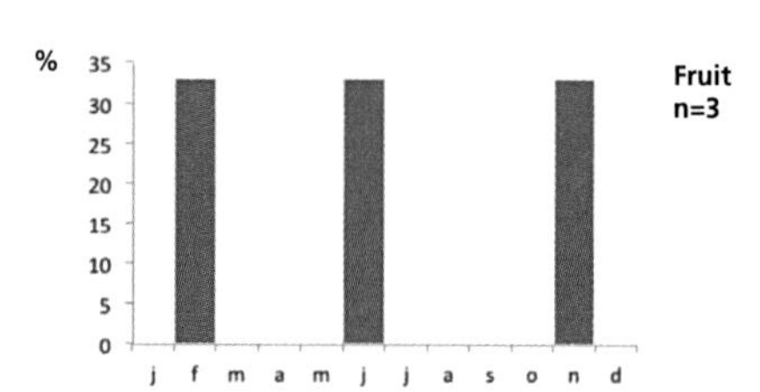

Psychotria bagshawei E. M. A. Petit **Rubiaceae** LC

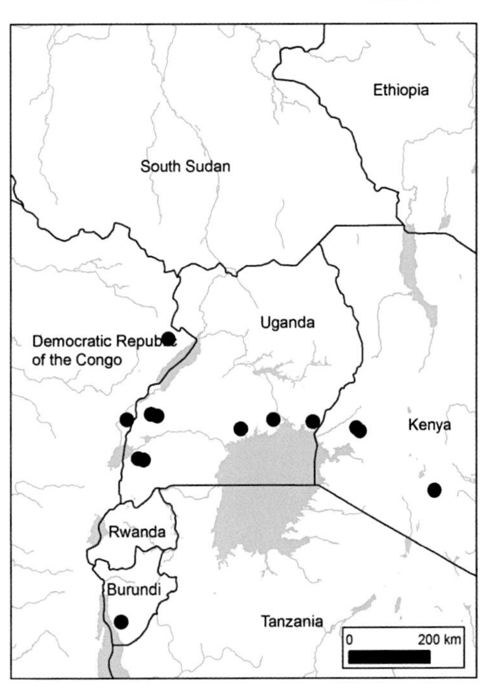

TYPE: Uganda, Toro, Isunga, *Bagshawe* 1092 (B lecto.)

LITERATURE: Verdcourt, FTEA 1976: 43.

DESCRIPTION: Shrub or 'small' tree. Young stems glabrous. Leaf blades elliptic, 4–20 cm long, 2–8 cm wide, with domatia; petiole 0.6–5 cm long. Flowers pentamerous, in panicles. Corolla yellow or white. Drupes red, subglobose, 7 mm in diameter.

HABITAT: Riverine forest and (secondary) evergreen forest; 1200–2300 m.

DISTRIBUTION: Toro, Ankole, Mengo. Also in East DRC and West Kenya.

PROTECTED AREA OCCURRENCE: Mpanga and Kakamega Forest Reserves.

LOCAL NAMES AND USES: None recorded.

RED LIST ASSESSMENT: EOO 284,074 km²; AOO 70,694 km²; 14 collections from 13 locations; mostly in protected areas with no specific threats. Here assessed as Least Concern (LC) due to its EOO and AOO, its common habitats and its altitude range; its occurrence in protected areas; and the lack of a specific major threat.

FLOWERING AND FRUITING TIMES:

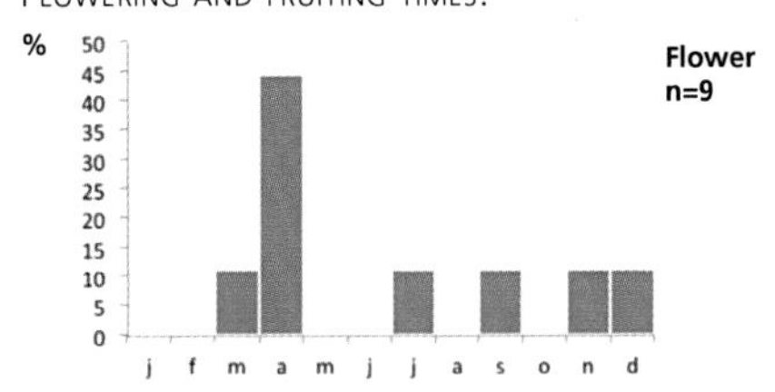

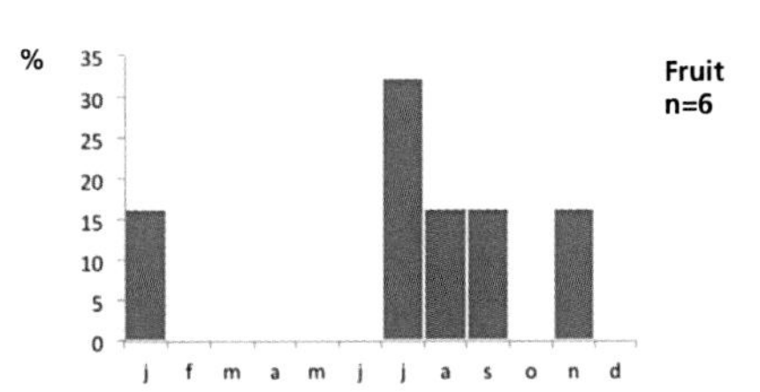

LC

Rytigynia acuminatissima (K. Schum.) Robyns **Rubiaceae**

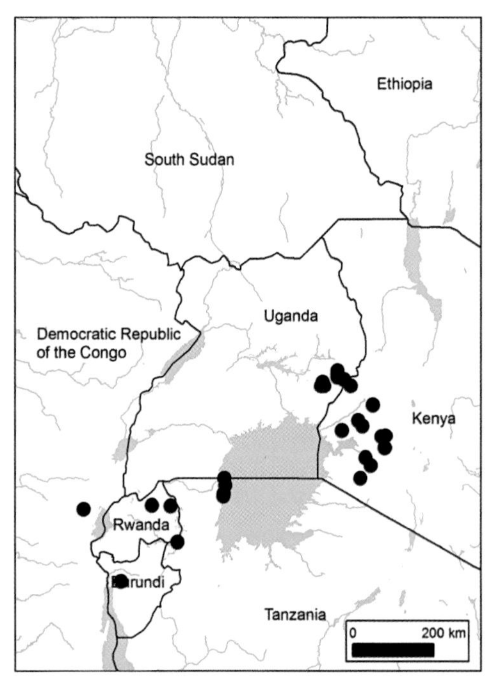

TYPE: Tanzania, Bukoba, *Stuhlmann* 1014 (B† holo.)

LITERATURE: Verdcourt & Bridson, FTEA 1991: 811.

DESCRIPTION: Shrub or tree 2–9 m high. Branches often with dark purplish red bark. Branchlets with lenticels, ridged. Leaf blades oblong-elliptic, elliptic or obovate, 3–14.5(–15) cm long, 0.6–6(–8) cm wide; petiole 3–6 mm long. Inflorescences (1–)2–6(–13)-flowered, often borne at leafless nodes. Corolla whitish or pale green. Fruit subglobose, 0.7–1.3 cm long, 6–10 mm wide.

HABITAT: Forest and forest edges, riverine forest, secondary bushland; 850–2400 m.

DISTRIBUTION: Mbale. Also in DRC, W Kenya, NW Tanzania, Rwanda, Burundi.

PROTECTED AREA OCCURRENCE: Akagera National Park; Kakamega, Tinderet, Minziro and Rubare Forest Reserves.

LOCAL NAMES AND USES: Kamonwet (Kony); Munyabwita (Kihaya), Apindi (Luo); used for cordage.

RED LIST ASSESSMENT: EOO 282,867 km^2; AOO 123,207 km^2; 39 collections from 29 locations; a number of populations in protected areas. In Uganda some sub-populations are not protected and the species occurs in less than five Ugandan locations. While this taxon is under threat in Uganda, its global assessment must be Least Concern (LC) due to its large EOO and AOO and many extra-Ugandan populations occurring in protected areas.

FLOWERING AND FRUITING TIMES:

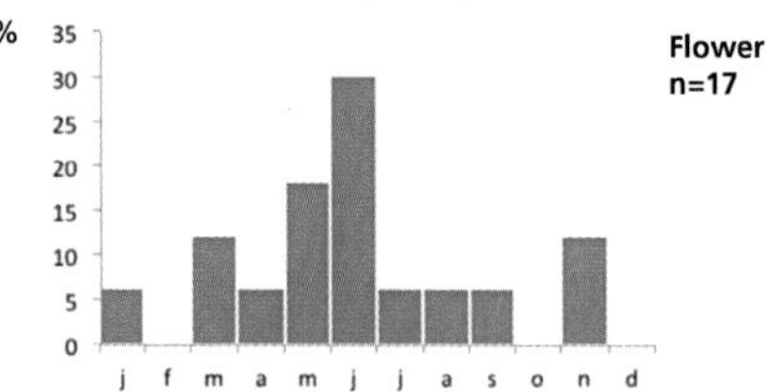

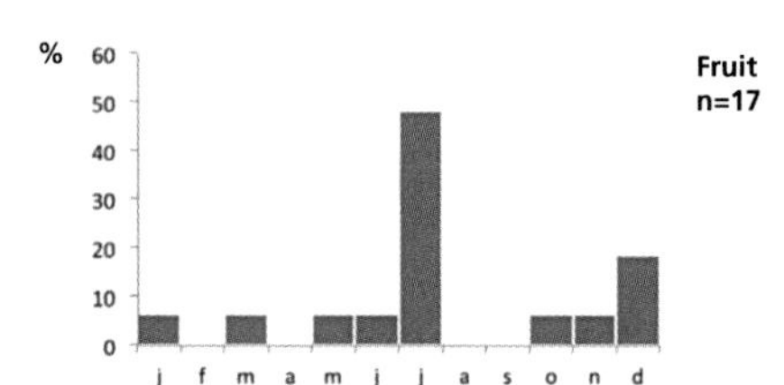

Rytigynia kigeziensis Verdc. **Rubiaceae** LC

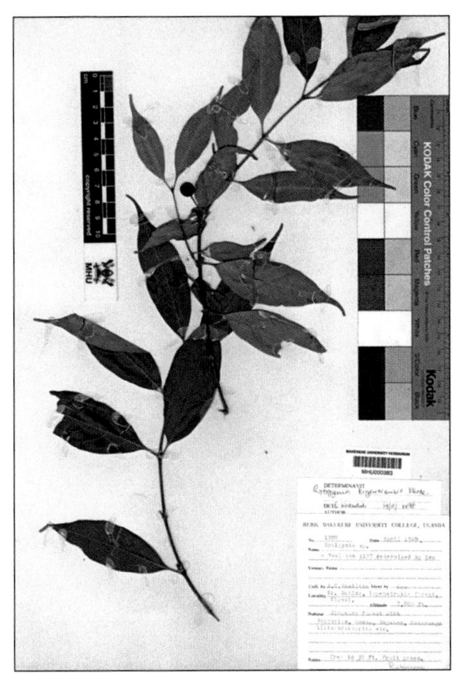

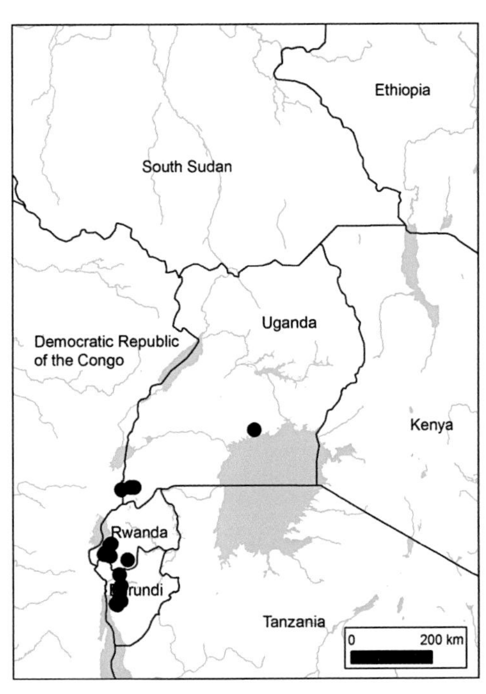

TYPE: Uganda, Kigezi, Luhizha, *Purseglove* 3665 (K holo., EA iso.)

LITERATURE: Verdcourt & Bridson, FTEA 1991: 815, fig. 144.

DESCRIPTION: Shrub or tree 1–8 m high. Bark grey smooth on trunk; stems glabrous, lenticellate, with dark red or greenish flaking bark. Leaf blades narrowly ovate to elliptic-lanceolate, 1.5–9(–12) cm long, 0.7–3.8(–5) cm wide, paler beneath, often shiny above; petiole 1.5–6 mm long. Inflorescence sweet-scented, 2-flowered. Corolla white or tinged green. Fruit 1–1.2 cm long, 6–10 mm wide.

HABITAT: Evergreen forest, riverine forest, bamboo forest, less often in wooded grassland; 1100–2500 m.

DISTRIBUTION: Kigezi, Mengo (Bwindi, Mafuga, Entebbe). Also in DRC, Rwanda and Burundi.

PROTECTED AREA OCCURRENCE: Bwindi, Nyungwe and Kibira National Parks, Mafuga Central Forest Reserve.

LOCAL NAMES AND USES: 'Umushabarara', Uruhamanyungwe (Kinyarwanda). Used as medicine in Uganda.

RED LIST ASSESSMENT: EOO 50,301 km^2; AOO 23,270 km^2; 34 collections from 19 locations. Though rare in Uganda, elsewhere this species is found at a wide range of altitude from 1200 to 2500 m; most sub-populations in Rwanda and Burundi are protected in Nyungwe and Kibira National Parks respectively. Here assessed as Least Concern (LC) due to its EOO and AOO, its range of habitats and altitude range; its occurrence in protected areas; and the lack of a specific major threat; medicinal use in Uganda is not a threat since it is regulated by Bwindi National Park authorities.

FLOWERING AND FRUITING TIMES:

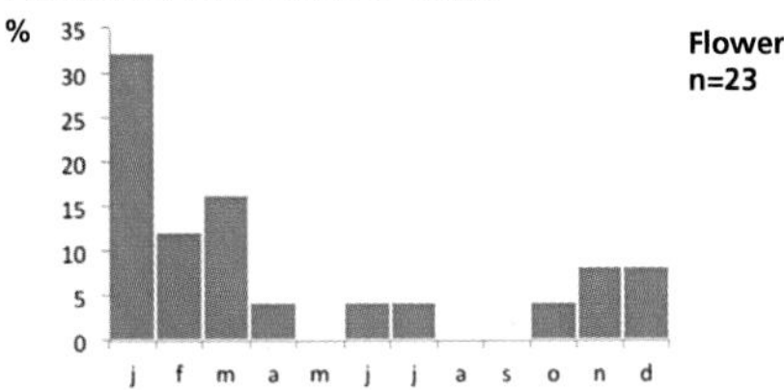

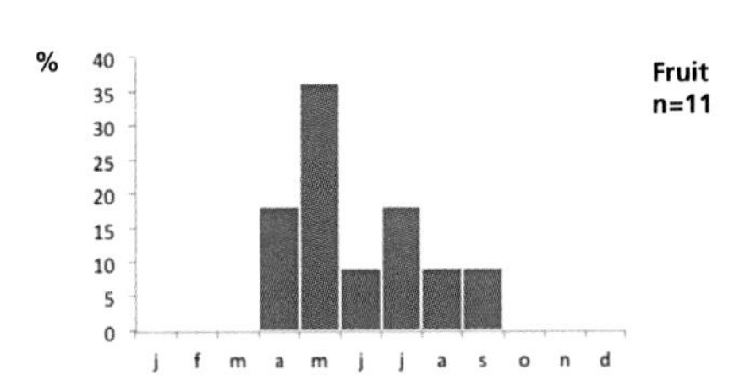

Rytigynia ruwenzoriensis (De Wild.) Robyns **Rubiaceae**

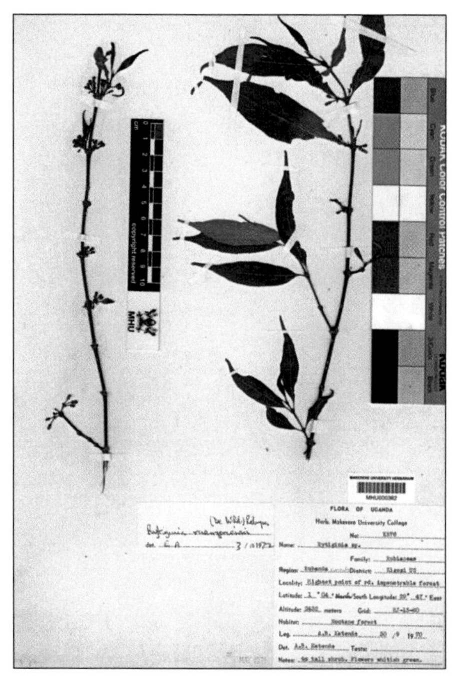

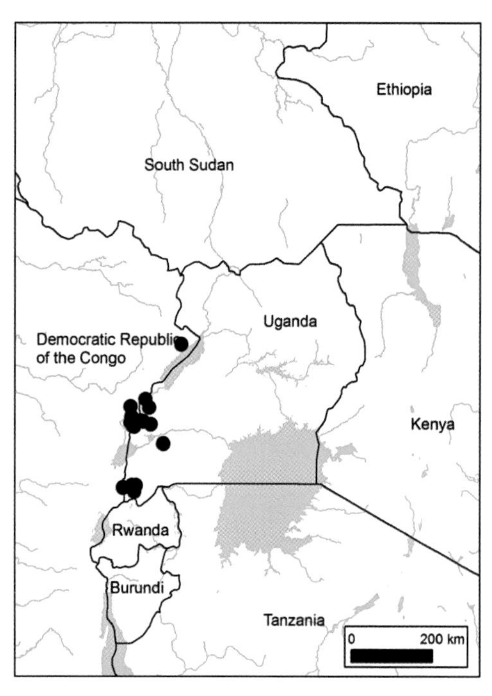

TYPE: DRC, Ruwenzori, Lamia valley, *Bequaert* 4299 (BR lecto.)

LITERATURE: Verdcourt & Bridson, FTEA 1991: 814.

DESCRIPTION: Shrub or tree 2–6(–15) m high. Older stems dark reddish purple with fine longitudinal fissures, lenticellate, glabrous; young shoots olive; slash red or chocolate-coloured. Leaf blades narrowly elliptic to oblong-lanceolate or lanceolate, (2–)5–13 cm long, 1–3.5 cm wide; petiole 4–10 mm long. Cymes (2–)4–7(–12)-flowered. Corolla green, greenish-white or cream, lobes oblong. Ovary 2(rarely 3)-locular. Fruit 1.1–1.2 cm long, 1–1.2 cm wide.

HABITAT: Evergreen forest, forest edges, bamboo; 1400–2450 m.

DISTRIBUTION: Toro, Kigezi. Uncommon in Uganda; known from Bwindi, Bwamba, Rwenzori. Also in DRC.

PROTECTED AREA OCCURRENCE: Bwindi Impenetrable and Rwenzori Mts National Parks.

LOCAL NAMES AND USES: Used for building poles.

RED LIST ASSESSMENT: EOO= 22,412 km², AOO= 15,227 km²; 22 collections from ± 20 locations. In Uganda, the Bwindi and Rwenzori populations are protected although the lower slopes where it grows may not be quite free from illegal extraction. Here assessed as Least Concern (LC) due to its EOO and AOO, its common habitats and altitude range; its occurrence in protected areas; and the lack of a specific major threat.

FLOWERING AND FRUITING TIMES:

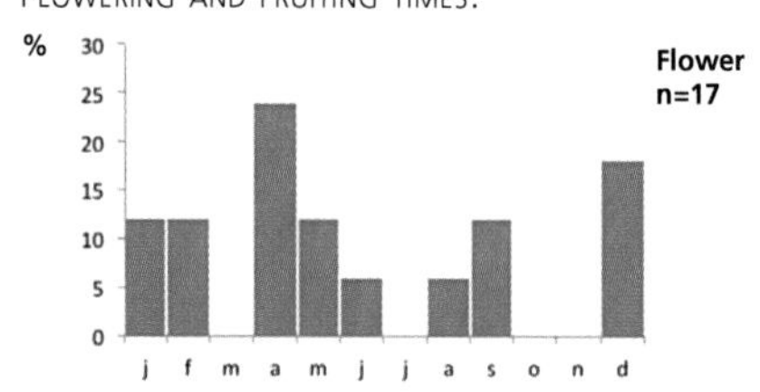

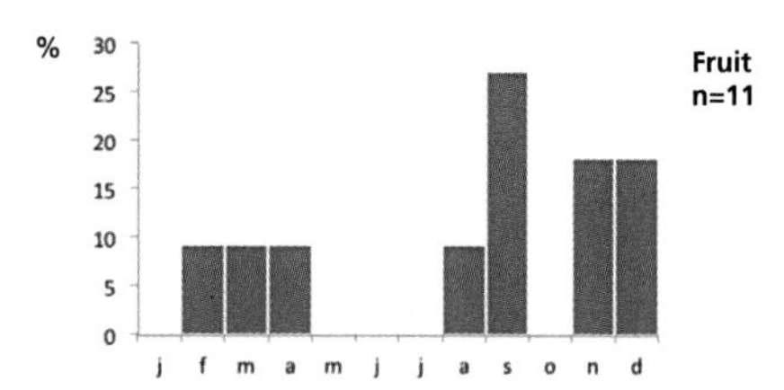

Rytigynia sp. B of FTEA

Rubiaceae

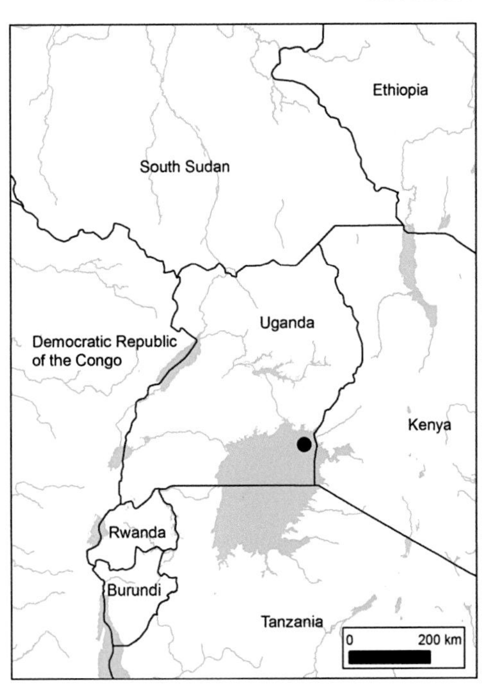

TYPE: none (yet), based on Uganda, Lake Victoria, Lolui Island, *Jackson* 12641.

LITERATURE: Verdcourt & Bridson, FTEA 1991: 809.

DESCRIPTION: 'Small tree'. Bark dark purple-red; branches glabrous, ridged, lenticellate. Leaf blades narrowly oblong-elliptic to elliptic-lanceolate, 2.5–4.5 cm long, 0.5–1.5 cm wide, glabrous, without domatia; petiole ± winged, 2–4 mm long. Inflorescence very short, 2–5-flowered. Corolla greenish. Ovary 3-locular.

HABITAT: Rocky shore; 1130 m.

DISTRIBUTION: Busoga, only known from a single specimen.

PROTECTED AREA OCCURRENCE: No protection to our knowledge.

LOCAL NAMES AND USES: None recorded.

RED LIST ASSESSMENT: Only known from a single specimen, not collected again since 1964; uncertain if still extant. Data Deficient (DD) due to uncertainty over taxonomy.

FLOWERING AND FRUITING TIMES:

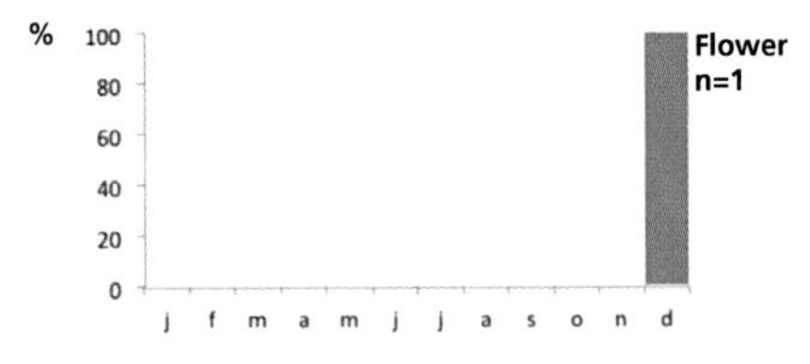

Fruit – no data

LC

Tricalysia bagshawei S. Moore subsp. ***bagshawei*** **Rubiaceae**

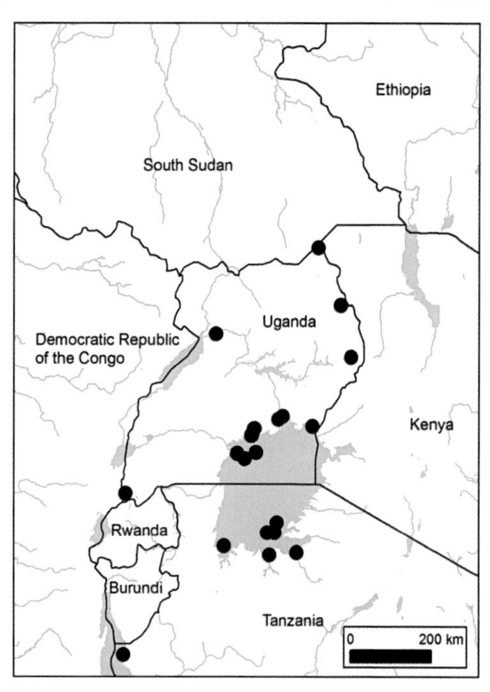

TYPE: Uganda, Entebbe, *Bagshawe* 792 (BM holo.)

LITERATURE: Bridson & Verdcourt, FTEA 1988: 550.

DESCRIPTION: Shrub or tree 1–6 m high. Young twigs puberulous or shortly hirsute. Leaf blades mostly obovate, sometimes elliptic or ovate, (3–)5–9(–11) cm long, (1–)2–4(–5) cm wide; petiole 1–5(–7) mm long, puberulous or shortly hirsute. Flowers 6(–7)-merous, sweetly scented, in (1–)3–5-flowered subsessile inflorescences. Corolla white. Fruit red, ellipsoid, c. 7 mm long, 5 mm wide. Seeds 1 or 2 in each chamber, ellipsoidal.

HABITAT: Forest, riverine or lakeside forest, evergreen thicket, secondary bushland; 650–1200 m.

DISTRIBUTION: Karamoja, Busoga, Masaka, Acholi, Mengo. Endemic to Uganda and Tanzania.

PROTECTED AREA OCCURRENCE: Murchison Falls National Park, Mabira Forest Reserve.

LOCAL NAMES AND USES: 'Kasakimba' (Ssese); roots pounded for sores; roots pounded and used externally for stomach pain (N Tanzania); used against rashes and itches (Ukara Is., Tanzania).

RED LIST ASSESSMENT: Said to have been common in Kyewaga Forest near Entebbe in 1949, and in Mabira Forest in 1922; not rare in Uganda from old collections but no records since 1962. EOO 871,330 km²; AOO 309,300 km²; 39 collections from 23 locations; many of its recorded localities of occurrence in Uganda are badly degraded or converted; the use of its roots for medicine in Tanzania leads us to infer a real threat to the population; but due to its EOO and AOO, the species is assessed here as Least Concern (LC).

FLOWERING AND FRUITING TIMES:

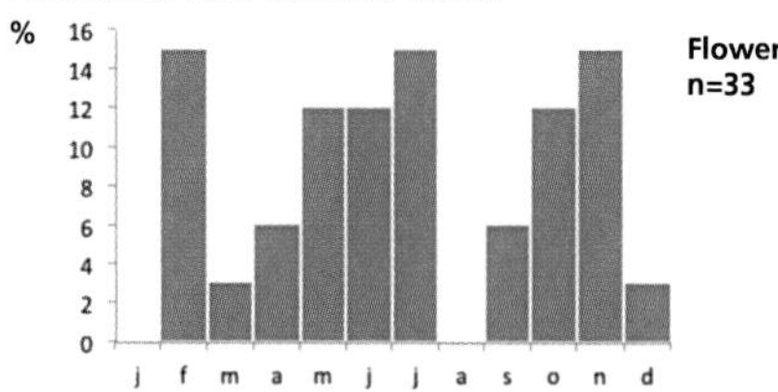

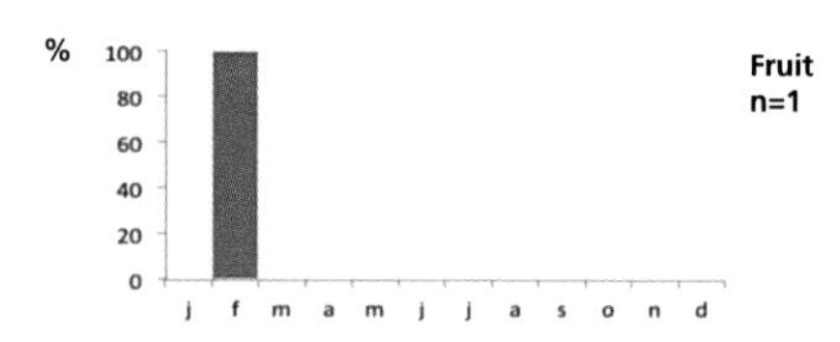

Aeglopsis eggelingii M. Taylor **Rutaceae**

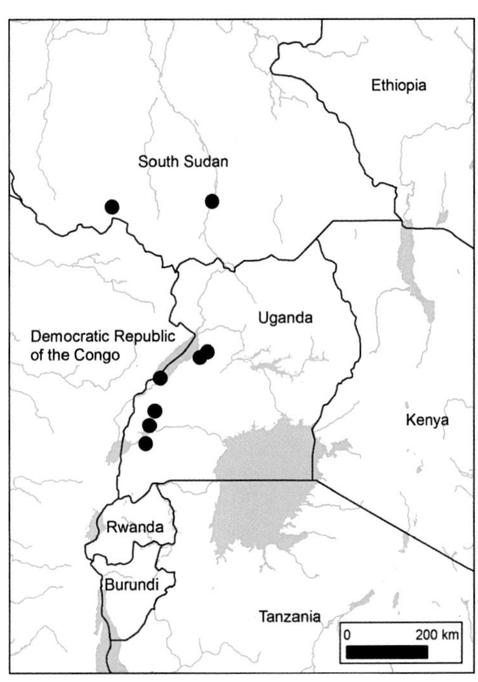

TYPE: Uganda, Bunyoro, Siba forest, *Eggeling* 3006 (K holo., ENT iso.)

LITERATURE: Kokwaro, FTEA 1982: 7, fig. 2. ITU: 359. UFT: 135.

DESCRIPTION: Shrub or tree 2–4(–7) m high, spiny. Bark dark grey and glabous; spines solitary or paired and axillary, 2–3.5 cm long, straight. Leaves 1(–2–3)-foliolate, alternate; petiole 5–18 mm long, grooved above towards lamina; lamina elliptic to ovate-lanceolate, 7–17 cm long, 3–6.5 cm wide, glabrous. Panicles 2–5 cm long. Flowers 3–4-merous, greenish-white. Ovary 5–8-locular. Fruit spherical or pear-shaped, 9–11 cm long, 7–8 cm in diameter, aromatic, yellow-green, woody, each locule many-seeded.

HABITAT: Forest, forest edge, riverine forest; 900–?1500 m.

DISTRIBUTION: Bunyoro, Toro, Ankole, Mengo. Also in DRC and South Sudan.

PROTECTED AREA OCCURRENCE: Kibale National Park, Budongo Forest Reserve.

LOCAL NAMES AND USES: None recorded.

RED LIST ASSESSMENT: Forest obligate. EOO 96,738 km^2; AOO 21,687 km^2. 13 collections from 8 locations. Here assessed as Least Concern (LC) due to extent of occurrence, coupled with a range of forest habitats, a wide range of altitude, and the occurrence in some protected areas.

FLOWERING AND FRUITING TIMES:

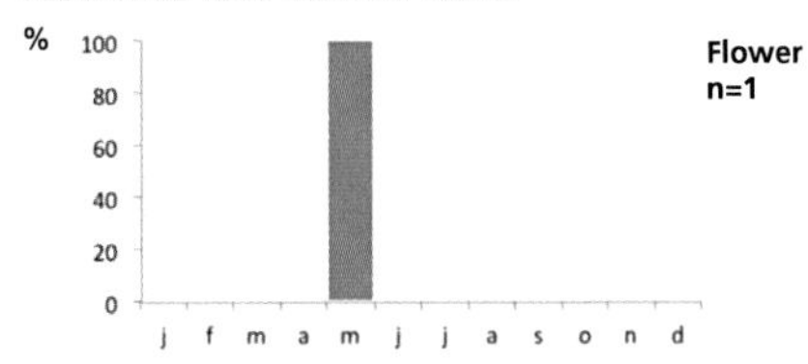

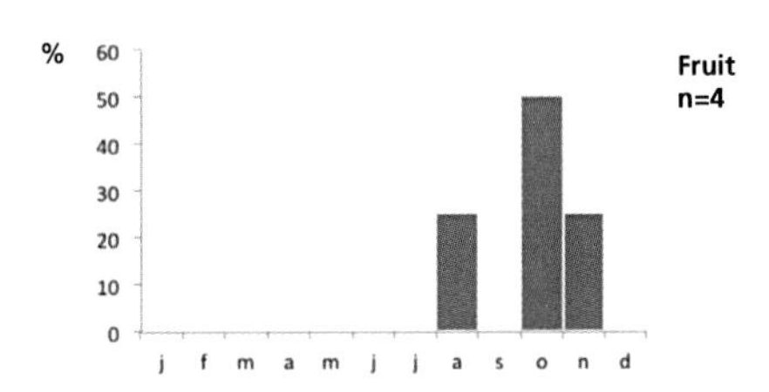

LC ***Balsamocitrus dawei*** Stapf **Rutaceae**

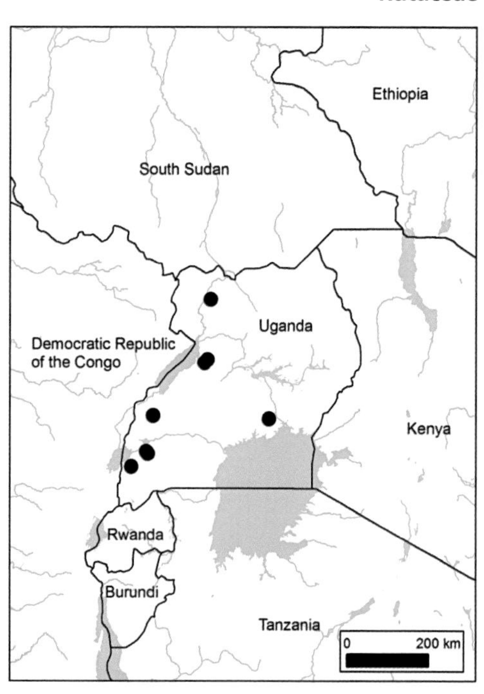

TYPE: Uganda, Bunyoro, Budongo forest, *Dawe* 788 (K holo., ENT iso.)

LITERATURE: Kokwaro, FTEA 1982: 10, fig. 3. ITU: 359, fig. 74. UFT: 194.

DESCRIPTION: Tree 6–21 m high. Bark rough, dark grey or brown; branchlets armed with single or paired axillary spines, occasionally unarmed; spines 1–4.3 cm long, straight, if paired, one is shorter than the other. Leaves trifoliolate, some occasionally unifoliolate by abortion, glabrous; petiole 1.5–3.5 (–4.5) cm long; leaflets elliptic to ovate-lanceolate, terminal leaflet larger than laterals. Flowers greenish-yellow to white. Fruit spherical or ovoid-globose, 11–13 cm long, 9–12 cm in diameter, aromatic, yellow-orange when ripe. Seeds brown and glossy.

HABITAT: Rainforest, riverine forest; 750–1400 m.

DISTRIBUTION: West Nile, Kigezi, Mengo, Bunyoro, Madi, Toro, Ankole. Endemic to Uganda.

PROTECTED AREA OCCURRENCE: Kibale National Park; Budongo, Mabira, Maramagambo, Zoka, Kasyoha-Kitomi Forest Reserves.

LOCAL NAMES AND USES: Fruits eaten by elephant; used as an aphrodisiac.

RED LIST ASSESSMENT: May be locally common. EOO 58,472 km^2; AOO 11,914 km^2; 11 collections from 8 locations, with several from protected Forest Reserves. Here assessed as Least Concern (LC), because of the lack of threat and the occurrence in protected areas.

FLOWERING AND FRUITING TIMES: No data [the single fruiting specimen only has a year, not a month].

Vepris eggelingii (Kokwaro) Mziray **Rutaceae** VU

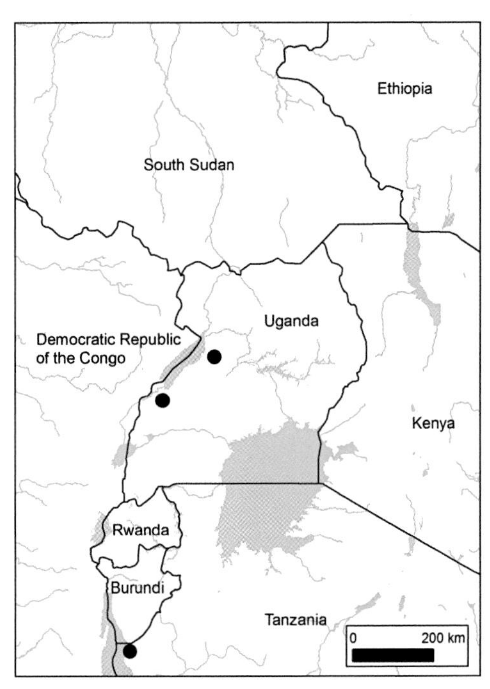

TYPE: Uganda, Toro, Itwara forest, *Eggeling* 5269 (EA holo., ENT iso.)

LITERATURE: Mziray in *Symb. Bot. Upsal.* 30: 1–95 (1992).

SYNONYMS: *Teclea eggelingii* Kokwaro; Kokwaro, FTEA 1982: 29.
Diphasia angolensis in the sense of ITU: 362.
Teclea sp. nov. 1 of ITU: 370.

DESCRIPTION: Shrub or tree 3–9 m high. Branchlets, inflorescence and petioles golden tomentose, older branches grey and with lenticels. Leaves trifoliolate, occasionally 1–2-foliolate by abortion; petiole 1–7 cm long, golden or rusty tomentose; leaflets subsessile, oblanceolate, 4–13.5 cm long, 1.5–6 cm wide. Flowers yellow-white, subsessile. Fruit ellipsoid, 10–13 mm long, 7–8 mm in diameter, red, glandular-bullate, 1-seeded.

HABITAT: Evergreen forest, riverine forest; 1200–1500 m.

DISTRIBUTION: Toro, Mengo. Also in W Tanzania.

PROTECTED AREA OCCURRENCE: Mabira, Itwara Forest Reserves.

LOCAL NAMES AND USES: 'Alimee' (Tongwe).

RED LIST ASSESSMENT: EOO 45,853 km²; AOO 26,956 km²; known from only 4 collections from 4 locations; presumably also in DRC, but no specimens seen from there. Three of the collections are from before 1963, one was collected from Mabira between 1993 and 1995 by the Forest Department. Here assessed as Vulnerable (D2) based on the four locations and the threat implied on the label of one of the Tanzanian collections ('patch of relict forest'), but possibly Endangered [EN B2a,b(iii)].

FLOWERING AND FRUITING TIMES:

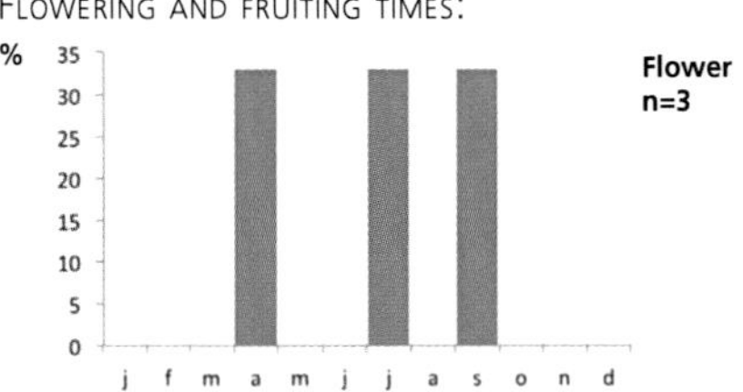

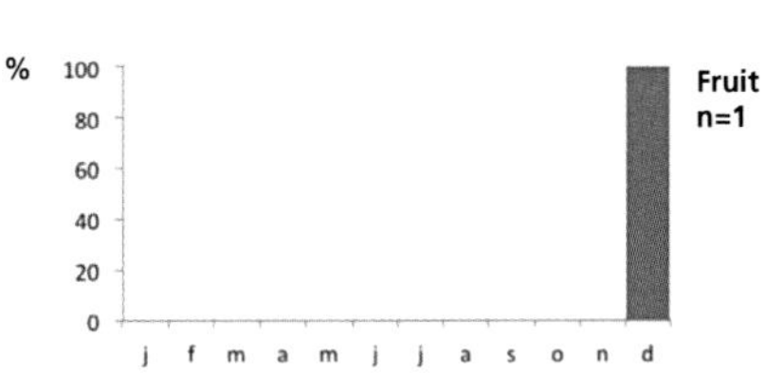

LC **_Zanthoxylum mildbraedii_** (Engl.) P. G. Waterman **Rutaceae**

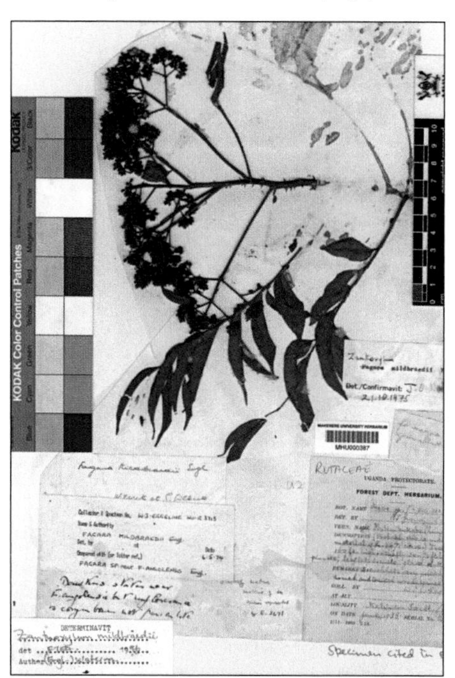

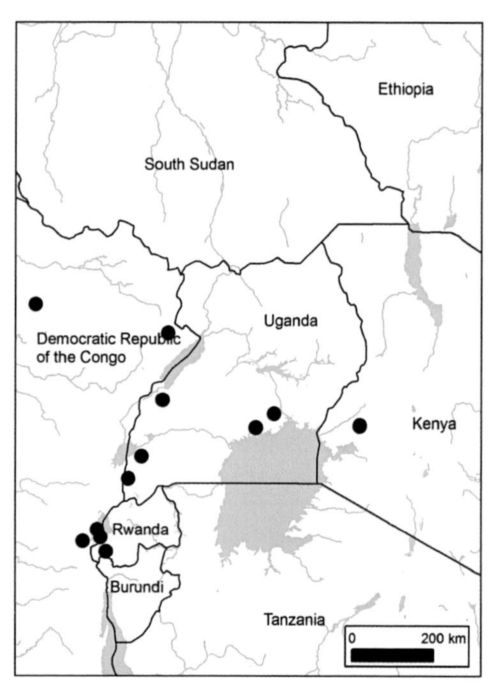

TYPE: DRC, Lake Kivu, Idjwi island, *Mildbraed* 1196 (B† holo., BM sketch of holo.)

LITERATURE: Kokwaro, FTEA 1982: 40.

SYNONYMS: *Fagara mildbraedii* Engl.; UFT: 206.
Fagara sp. near angolensis of ITU: 365.

DESCRIPTION: Tree 12–30 m high. Trunk usually straight and unbranched for several metres, armed with woody spines; bark generally smooth, yellowish-brown and aromatic; branchlets with small black conical prickles 1.5–3 mm long. Leaves 15–37 cm long; rachis unarmed, rusty puberulent; leaflets 5–11 pairs, subopposite towards base, progressively more opposite above. Fruit of paired rhomboid follicles 8–10 mm long, 6–7 mm in diameter. Seeds shiny black, 4–5 mm in diameter.

HABITAT: Rainforest; 1200–2100 m.

DISTRIBUTION: Toro, Ankole, Mengo, Kigezi. Also in DRC, Rwanda, W Kenya.

PROTECTED AREA OCCURRENCE: Bwindi Impenetrable National Park; Kalinzu, Itwara and Kakamega Forest Reserves.

LOCAL NAMES AND USES: 'Mulemankobe' (LunyAnkole), 'Mbithethe', 'Sembi' (Kilende/Kilendu); 'Umwasa' (Kinyarwanda), 'Kanyabumbu' (Kitembo).

RED LIST ASSESSMENT: Forest obligate; EOO 249,397 km²; AOO 65,931 km²; 19 collections from 13 locations; widespread and protected in forest reserves. Assessed here as Least Concern (LC) due to its EOO and AOO, its altitude range; its occurrence in protected areas; and the lack of a specific major threat.

FLOWERING AND FRUITING TIMES:

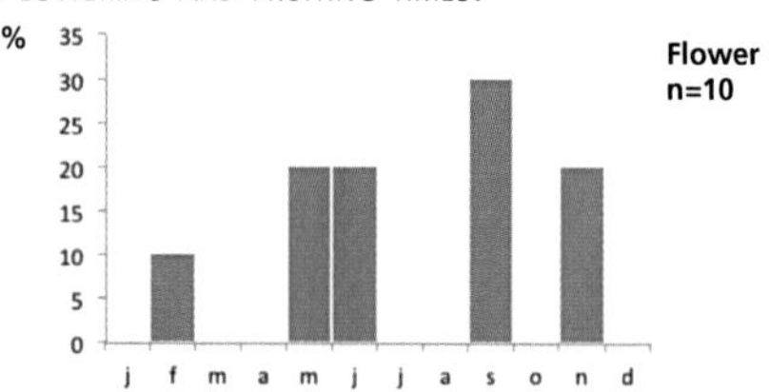

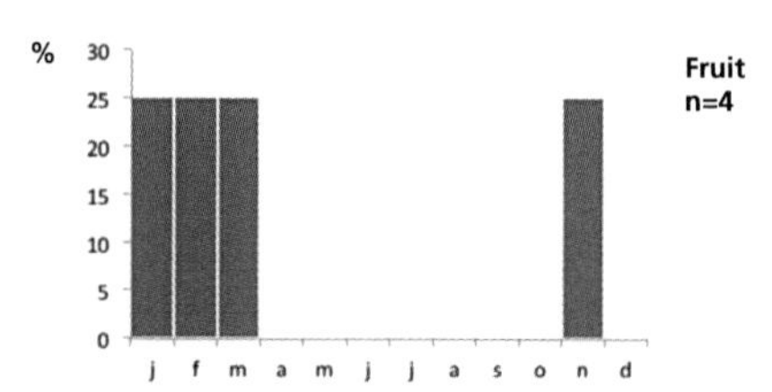

Chrysophyllum muerense Engl. — **Sapotaceae** LC

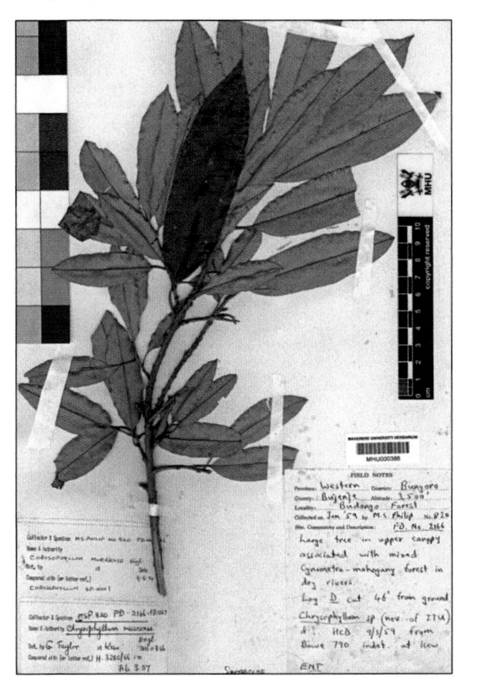

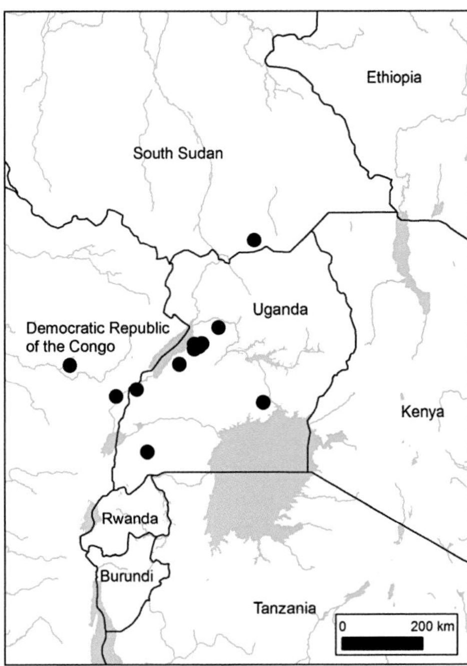

TYPE: DRC, Beni, near Muera, *Mildbraed* 2243 (B† holo.)

LITERATURE: Hemsley, FTEA 1968: 15. UFT: 84.

SYNONYMS: *Chrysophyllum sp. nov.* of ITU: 395.
Gambeya muerensis (Engl.) Liben in *Bull. Jard. Bot. Natl. Belg.* 59: 480 (1989).

DESCRIPTION: Tree to 35 m high. Young branches, buds and petioles with greyish-brown to tawny-brown indumentum. Leaf lamina narrowly elliptic to oblong-elliptic, 6–16(–20) cm long, 1.7–3.5(–5) cm wide, lower side with silvery-grey or fawn silky indumentum. Flowers axillary. Fruits on short woody stalks to 6 mm long, yellow at maturity, subglobose to obovoid.

HABITAT: Rainforest, moist semi-deciduous forest dominated by *Cynometra*; 900–1400 m.

DISTRIBUTION: Bunyoro, Toro, Mengo. Also in East DRC and South Sudan.

PROTECTED AREA OCCURRENCE: Semuliki, Murchison Falls (Rabongo Forest) & Kibale National Parks; Budongo, Ituri & Mabira Forest Reserves.

LOCAL NAMES AND USES: Munyamata' (Lunyoro).

RED LIST ASSESSMENT: Several times noted as 'not common'; reported by Eilu *et al.* (2004) as rare with restricted range and low density in the Albertine Rift area; within Uganda protected in Forest Reserves and National Parks; widespread with EOO=134,280 km², AOO= 29,442 km²; 30 collections from 13 locations. Here assessed as Least Concern (LC) due to its EOO and AOO; its occurrence in protected areas; and the lack of a specific major threat.

FLOWERING AND FRUITING TIMES:

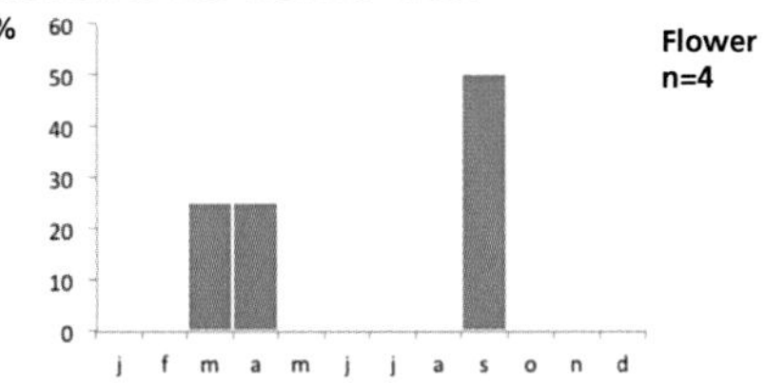

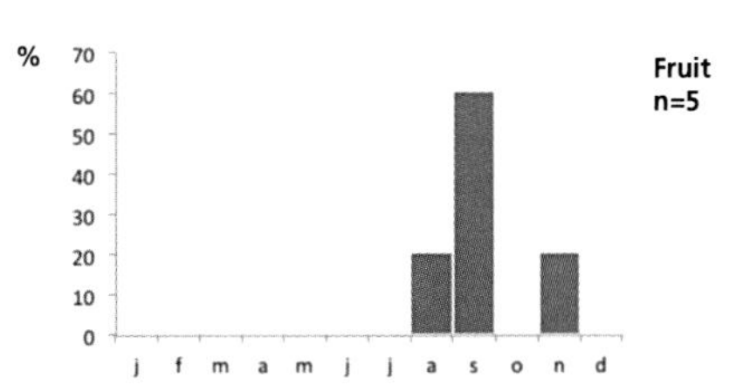

VU ***Dicranolepis incisa*** A. Robyns **Thymelaeaceae**

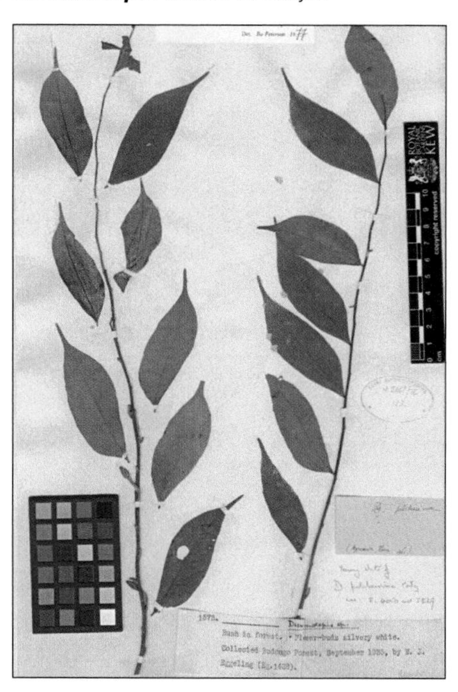

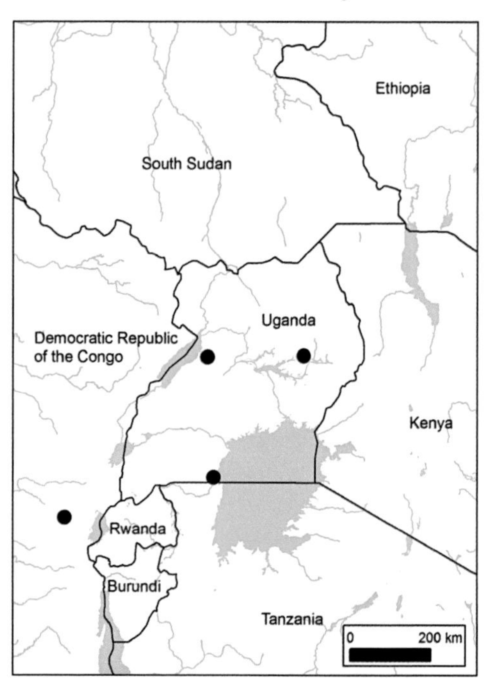

TYPE: DRC, Walikale, Mifuti, *Léonard* 1856 (BR holo.)

LITERATURE: Peterson, FTEA 1978: 2.

DESCRIPTION: Shrub or tree to 3 m high. Young branches densely pubescent, later glabrous. Leaf lamina oblong, unequal-sided, 40–80(–93) mm long, 20–35 mm wide, glabrous above, sparsely appressed-pubescent below. Flowers silvery-white, usually solitary; petals white.

HABITAT: Rainforest; 1200–1300 m.

DISTRIBUTION: Bunyoro, Masaka, Teso. Also in DRC.

PROTECTED AREA OCCURRENCE: Budongo Forest Reserve, Sango Bay Forest Reserve (Tero Forest).

LOCAL NAMES AND USES: None recorded.

RED LIST ASSESSMENT: EOO 88,600 km²; AOO 20,086 km²; 6 collections from 4 locations, with the Teso population likely to have disappeared or at best in very degraded habitat; we assess this here as Vulnerable D2, due to the number of locations and the degradation of at least the Ugandan habitat.

FLOWERING AND FRUITING TIMES:

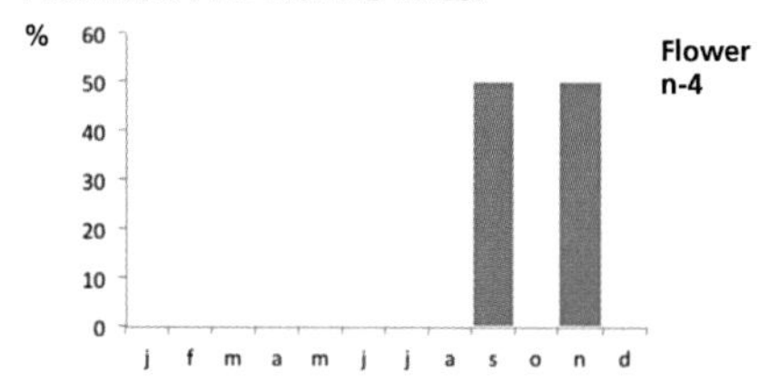

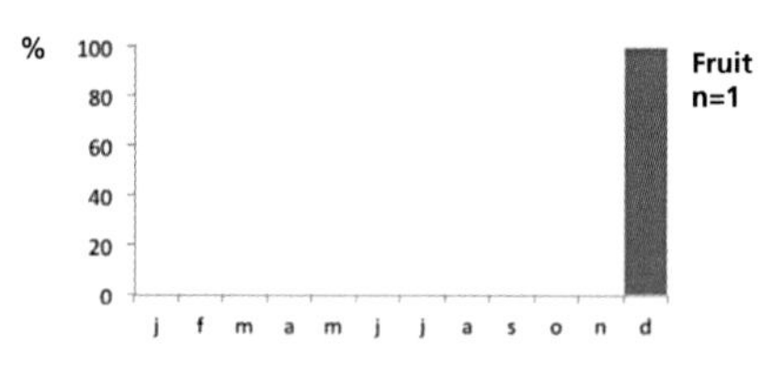

Rinorea beniensis Engl. — Violaceae — LC

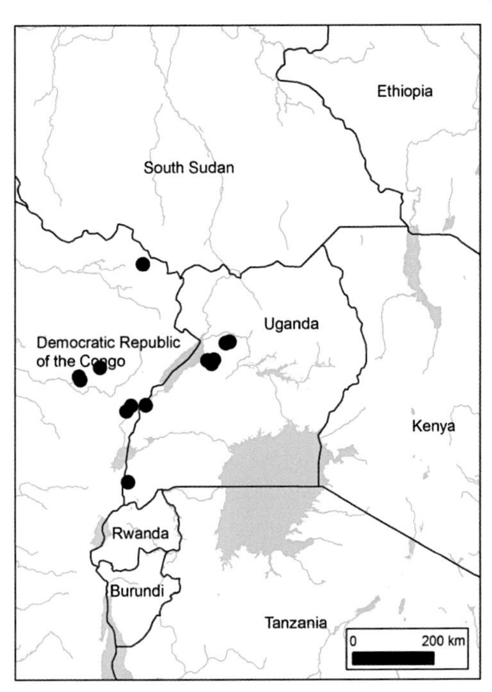

TYPE: DRC, Beni, Mwera, *Mildbraed* 2406 & 2768 and Mokoko, *Mildbraed* 2935 (all B† syn.)

LITERATURE: Grey-Wilson, FTEA 1986: 17.

SYNONYM: *Rinorea ardisiiflora* in the sense of ITU: 446, fig. 94. UFT: 130.

DESCRIPTION: Shrub or tree to 12 m high. Branches sparsely pubescent at first. Leaf lamina elliptic to elliptic-lanceolate, 3.6–11.4 cm long, 1.6–4.2 cm wide, glabrous or with a few hairs on midrib and lateral veins beneath at first, margin crenate-serrate to near-entire. Inflorescence few-flowered raceme, or solitary or fascicled. Flowers white, fragrant. Capsule shallowly 3-lobed, 9–12 mm long, glabrous, greenish-yellow when ripe.

HABITAT: Rainforest, secondary forest, forest edge, often associated with *Cynometra*; (350–)750–1400 m.

DISTRIBUTION: Bunyoro, Toro, Mengo, Madi, almost exclusively known from Budongo; also in DRC.

PROTECTED AREA OCCURRENCE: Semuliki National Park; Budongo Forest Reserve.

LOCAL NAMES AND USES: 'Buji' (Ituri forest); wood useless.

RED LIST ASSESSMENT: Said to be dominant at Budongo, fide Waibira (1962); fairly common in Budongo (1932, 1935, 1995); very common in Semliki Forest (1951); Eilu *et al.* (2004) described it as rare in the forests of western Uganda.

Widespread with EOO 812,670 km^2, AOO 377,050 km^2; no known specific threats; 63 collections from 22 locations; assessed here as Least Concern (LC) due to its EOO and AOO, its altitude range; its occurrence in protected areas; and the lack of a specific major threat.

FLOWERING AND FRUITING TIMES:

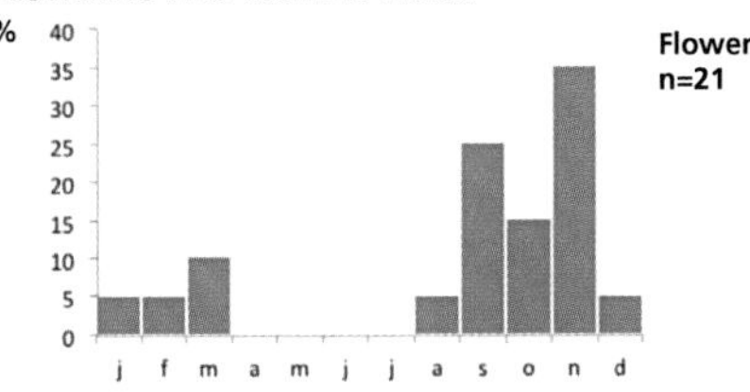

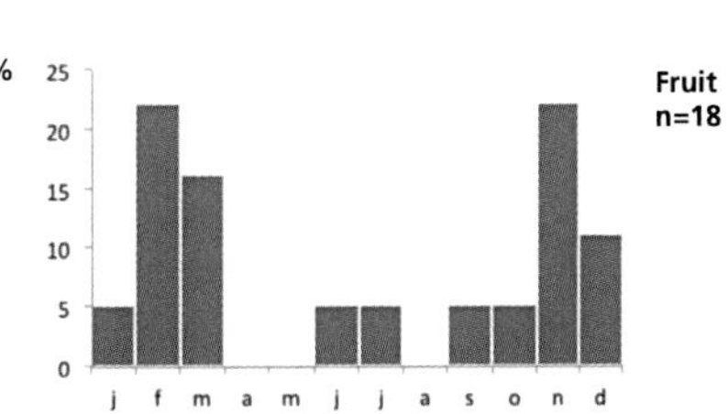

LC

Rinorea tschingandaensis Taton **Violaceae**

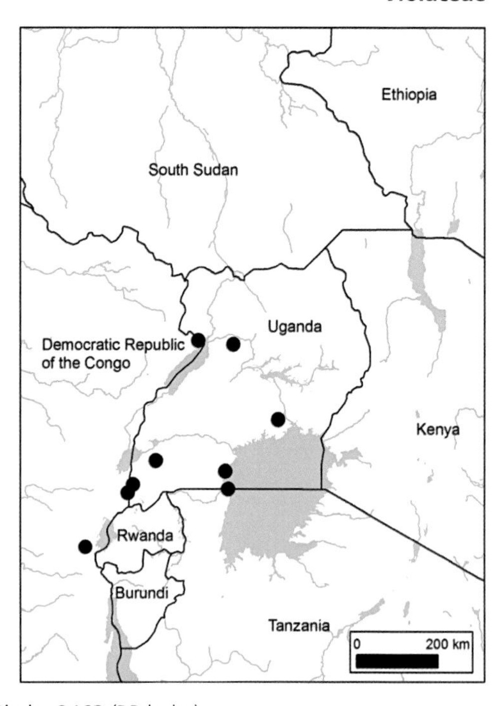

TYPE: DRC, Kavumu-Walikale, Tshinganda river, *Pierlot* 2463 (BR holo.)

LITERATURE: Grey-Wilson, FTEA 1986: 15, fig. 4.

SYNONYM: *Rinorea dentata* of ITU: 446, partly.

DESCRIPTION: Shrub or tree to 15 m high. Young branches glabrous or sparsely pubescent. Leaf lamina elliptic to elliptic-oblanceolate, 4.8–14.5 cm long, 1.6–4.8 cm wide, glabrous, margin shallowly serrate-crenate to subentire; petiole 6–14 mm long, glabrous. Flowers yellow or brownish yellow, nodding at maturity. Capsule 3-lobed, 14–18 mm long, finely rugose.

HABITAT: Evergreen forest; 900–1900 m.

DISTRIBUTION: Kigezi, Ankole, Bunyoro, Mengo, Masaka. Also in East DRC.

PROTECTED AREA OCCURRENCE: Bwindi Impenetrable, Murchison Falls (Rabongo Forest) National Parks; Kasyoha-Kitomi, Minziro, Mabira, Kirala Forest Reserves.

LOCAL NAMES AND USES: 'Chokpadyasi' (Kilendu).

RED LIST ASSESSMENT: EOO 104,551 km^2; AOO 33,404 km^2; 13 collections from 9 locations. No known specific threats; assessed here as Least Concern (LC) due to its EOO and AOO, its altitude range; its still common habitat; and the lack of a specific major threat.

FLOWERING AND FRUITING TIMES:

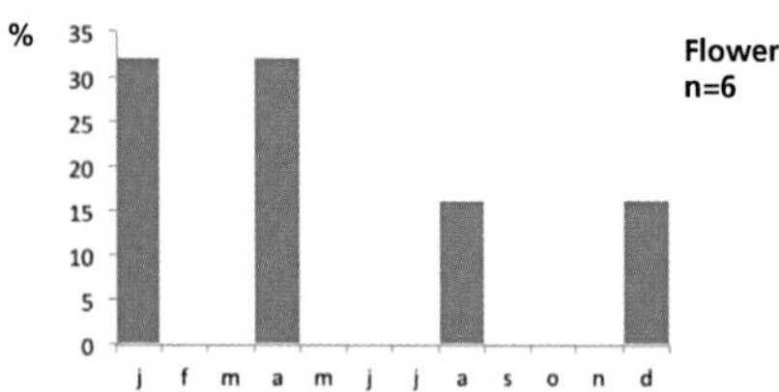

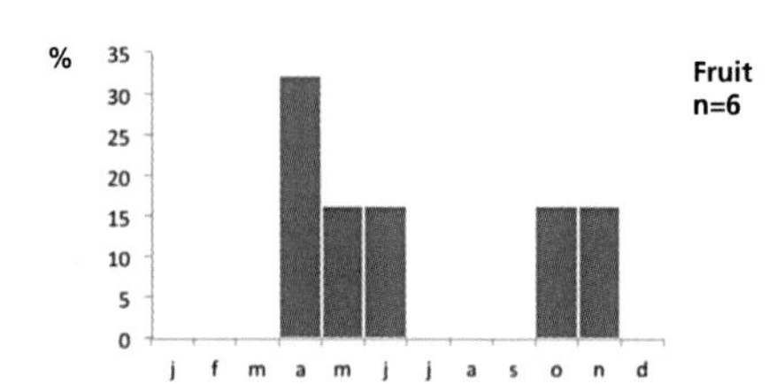

Encephalartos septentrionalis Schweinf. **Zamiaceae** LC

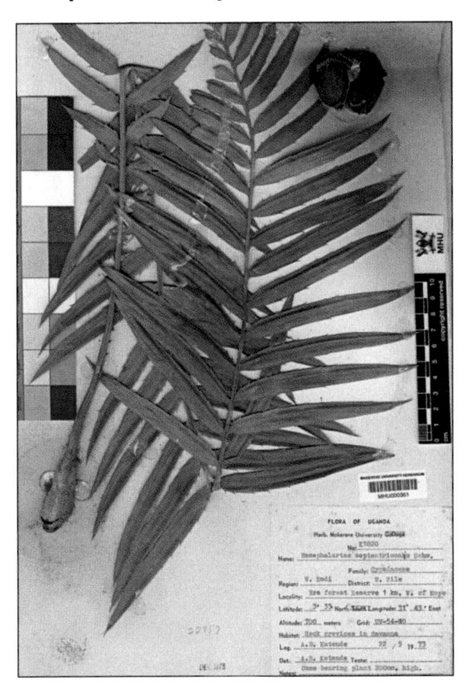

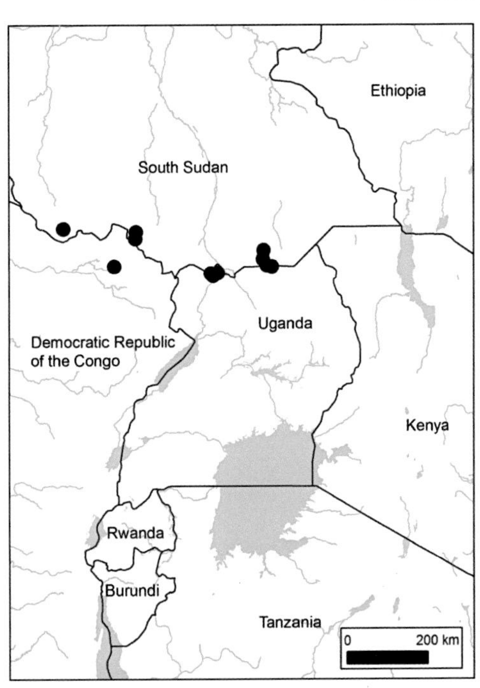

TYPE: South Sudan, Gumanga, *Schweinfurth* 2952 (K syn.) & near Nganye, *Schweinfurth* 3992 (K syn.)

LITERATURE: Melville, FTEA 1958: 6. ITU: 104.

SYNONYM: *Encephalartos barteri* in the sense of Melville, FTEA 1958: 5.

DESCRIPTION: Tree 1–3 m high, with stem to 70 cm in diameter, less inclined to clump than other Ugandan species. Leaves many per crown, to 2 m long, with many leathery leaflets, each with a toothed margin. Fruit a yellow-green to brown ovoid cone. Plants are either male or female.

HABITAT: Rocky places in *Loudetia/Hyparrhenia/Combretum* grassland and open bushland, also among bamboo; 600–2400 m.

DISTRIBUTION: Acholi, West Nile, Madi; also in South Sudan, CAR, DRC.

PROTECTED AREA OCCURRENCE: Era, Imatong Mts, Agoro Forest Reserves.

LOCAL NAMES AND USES: 'Bofetebe', 'Muriepai', 'Noupia' (Zande).

RED LIST ASSESSMENT: Assessed as Data Deficient (DD) in 2003 by J. Donaldson, and as Near threatened (NT) by Bösenberg in 2009. CITES Appendix I listed.

Limited range in Uganda — Imatong Mts., Moyo, & Agoro only; Imatongs not surveyed for several years owing to civil war hence status uncertain; subject to regular grass fires (Goode 2001: 275) but flourishing; ± 200 plants in Uganda (Goode, l.c.) with good seedling regeneration. EOO 1,320,910 km^2; AOO 239,169 km^2, 24 collections from 16 locations. We assess it here as Least Concern (LC), because there seems to be no threat, and where the species occurs it is said to be locally common, as well as resistant to fires. Heibloem (1999) states the South Sudan populations differ in several characters and represent a different, undescribed species. If this is true, this assessment would have to be recast.

FLOWERING AND FRUITING TIMES:

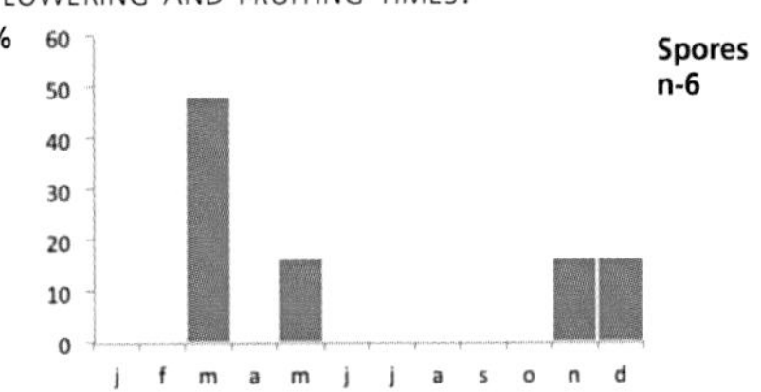

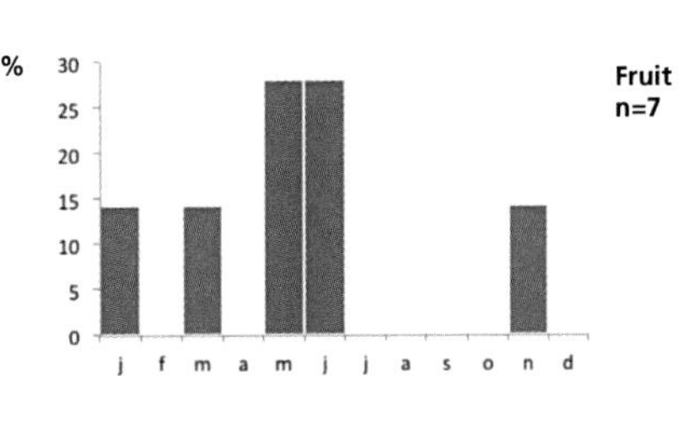

Encephalartos whitelockii P. J. H. Hurter **Zamiaceae**

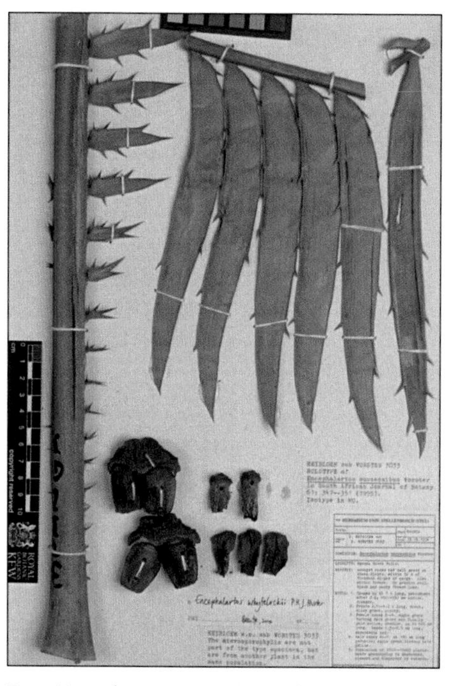

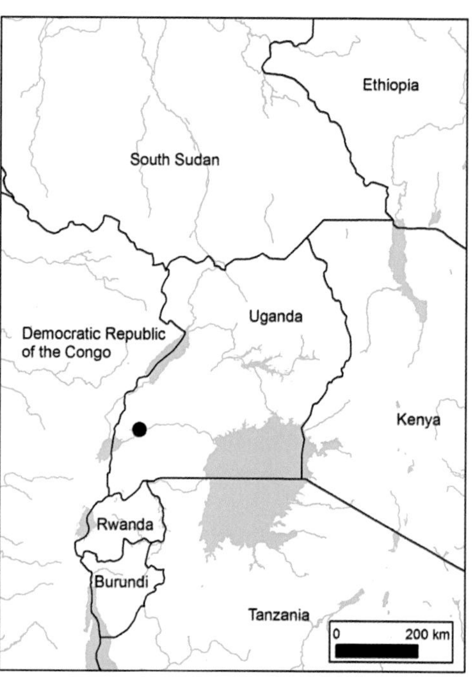

TYPE: Uganda, Mpanga River Falls, 27 Oct. 1994, *Hurter* 94U/3a (PRE holo.)

LITERATURE: Goode 2001: 247.

SYNONYMS: *Encephalartos hildebrandtii* of Melville, FTEA 1958: 6, partly.
Encephalartos laurentianus in the sense of ITU: 104.
Encephalartos successibus Vorster in *S. African J. Bot.* 61: 347 (1995).

DESCRIPTION: Tree with trunks to 5 m high and 1 m in diameter, and a 'shuttlecock' of leaves to 4 m long, each with many leathery leaflets with toothed margins. Fruit a green ovoid cone.

HABITAT: Rocky steep slopes of narrow gorge, open grassland on outskirts of riverine forest; ± 1000 m.

DISTRIBUTION: Toro. Ugandan Endemic.

PROTECTED AREA OCCURRENCE: Unprotected; only a negligibly small fraction (far less than 2%) of the population occurs inside the Queen Elizabeth National Park.

LOCAL NAMES AND USES: 'Omuhure' (Lutoro language); harvesting by communities for food, sale to collectors, and building materials.

RED LIST ASSESSMENT: As there were many seedlings when Goode visited (Goode, *Cycads of Africa* 2001: (247)) and he perceived 'no threat', Goode assessed the species as Least Concern. Assessed as Vulnerable, VU B1ab(iii,v)+2ab(iii,v); D2, by Golding & Hurter in 2003, and as Critically Endangered (CR, B1ab(ii,iii,v)+2ab(ii,iii,v)) by Kalema in 2009. CITES Appendix I listed. EOO is ± 6 km², AOO is 1 km²; with about 8000–10000 individuals in the single known site, which we assess as a single location. Recently Dr Kalema has assessed this taxon as Critically Endangered (CR), because the single location is threatened by the ongoing construction of a hydro-electric scheme, with a network of roads and camps associated with the project; especially as some roads are along the contour line where the cycads are concentrated, damaging a series of cycads. This scheme and its infrastructure has had impact on both habitat quality and actual level of exploitation. Communities burn the habitat and graze their livestock in the area, degrading the habitat. We uphold this assessment.

FLOWERING AND FRUITING TIMES:

Spores – no data

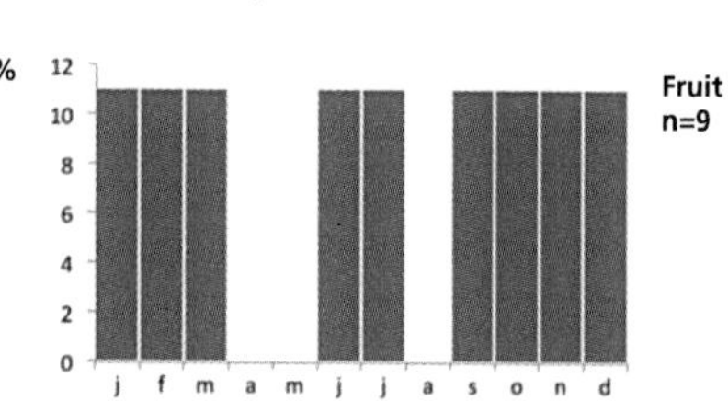

ACKNOWLEDGEMENTS

We are grateful to the Mohamed bin Zayed Species Conservation Fund for supporting the publication of this work. Special thanks to Steve Bachman (Royal Botanic Gardens, Kew) who worked hard to prepare the distribution maps, EOO and AOO data of the tree species. We are indebted to Dr Geoffrey Mwachala (National Museums of Kenya, Nairobi) and Quentin Luke (East African Herbarium, Nairobi, and formerly of Fairchild Tropical Botanic Garden, Miami) who made both critical and helpful comments on the work, which helped us greatly to improve it. Collins Bulafu (Makerere University Herbarium, Dept. of Botany), Frances Crawford and Tim Harris (Herbarium, Royal Botanic Gardens, Kew) helped in scanning the herbarium specimens. Prof. Remigius Bukenya-Ziraba (Dept. of Botany, Makerere University) strongly encouraged the first author to take up the challenge. Axel Poulsen provided the image of *Ficus katendei* and its discoverer. A number of people have provided useful information about the distribution, population size, threats and conservation efforts of trees in Uganda and beyond. These include Dr Gerald Eilu and Dr Fred Babweteera (both from Forestry, Makerere University), Dr David Hafashimana (National Forestry Resources Research Institute NAFORRI), Dr Henry Ndangalasi (Dept. of Botany, University of Dar es Salaam), Quentin Luke (National Museums of Kenya, Nairobi) and Roy Gereau (Missouri Botanical Garden, USA).

REFERENCES

Angiosperm Phylogeny Group (2009). An update of the Angiosperm Phylogeny Group classification for the orders and families of flowering plants: A.P.G. III. *Bot. J. Linn. Soc.* 161(2): 105–121.

Bamps, P., Robson, N. & Verdcourt, B. (1978). Guttiferae. In: Polhill R. M. (ed.), *Flora of Tropical East Africa*. Crown Agents, London.

Beentje, H. J. (1989). Bombacaceae. In: Polhill R. M. (ed.), *Flora of Tropical East Africa*. Balkema, Rotterdam.

Beentje, H. J. (1993). Pandanaceae. In: Polhill R. M. (ed.), *Flora of Tropical East Africa*. Balkema, Rotterdam.

Beentje, H. J. (2002). Compositae part 2. In: Beentje, H. J. (ed.), *Flora of Tropical East Africa*. Balkema, Rotterdam.

Beentje, H. J. (2006). Ericaceae. In: Beentje, H. J. & Ghazanfar, S. A. (eds), *Flora of Tropical East Africa*. Royal Botanic Gardens, Kew.

Beentje, H. J. (2010). *The Kew Glossary*. Royal Botanic Gardens, Kew.

Beentje, H. J., Jeffrey, C. & Hind, D. J. N. (2005). Compositae part 3. In: Beentje H. J. & Ghazanfar, S. A. (eds), *Flora of Tropical East Africa*. Royal Botanic Gardens, Kew.

Berg, C. C. & Hijman, M. E. E. (1989). Moraceae. In: Polhill R. M. (ed.), *Flora of Tropical East Africa*. Balkema, Rotterdam.

Bidgood, S., Verdcourt, B. & Vollesen, K. (2006). Bignoniaceae. In: Beentje, H. J. & Ghazanfar, S. A. (eds), *Flora of Tropical East Africa*. Royal Botanic Gardens, Kew.

Boffa, J. M., Yaméogo, G., Nikiéma, P. & Knudson, D. M., (1996). Shea nut (*Vitellaria paradoxa*) production and collection in agroforestry parklands of Burkina Faso. In: Leakey, R. R. B., Temu, A. B., Melnyk, M. & Vantomme P., (eds) *Domestication and commercialization of non-timber forest products in agroforestry systems. Non-wood Forest Products* 9: 110–122. FAO, Rome.

Brenan, J. P. M. (1959). Leguminosae, Mimosoideae. In: Hubbard, C. E. & Milne-Redhead, C. E. (eds), *Flora of Tropical East Africa*. Crown Agents, London.

Brenan, J. P. M. (1967). Leguminosae, Caesalpinioideae. In: Milne-Redhead, C. E. & Polhill, R. M. (eds), *Flora of Tropical East Africa*. Crown Agents, London.

Breteler, F. J. (ed.) (1989). The Connaraceae. In: *Agric. Univ. Wageningen Pap.* 89, 6.

Breteler, F. J. (1988). Dichapetalaceae. In: Polhill, R. M. (ed.), *Flora of Tropical East Africa*. Balkema, Rotterdam.

Breteler, F. J. (2003). The African genus Sorindeia (Anacardiaceae): a taxonomic revision. *Adansonia* ser. 3, 25: 93–113.
Breteler, F. J. (2004). The genus *Trichoscypha* (Anacardiaceae) in Lower Guinea and Congolia: a synoptic revision. *Adansonia* ser. 3, 26: 97–127 (2004).
Bridson, D. & Verdcourt, B. (1988). Rubiaceae part 2. In: Polhill R. M. (ed.), *Flora of Tropical East Africa*. Balkema, Rotterdam.
Bruce, E. A. & Lewis J. (1960). Loganiaceae. In: Hubbard, C. E. & Milne-Redhead, C. E. (eds), *Flora of Tropical East Africa*. Crown Agents, London.
Brummitt, R. K & Marner, S. K. (1993). Proteaceae. In: Polhill R. M. (ed.), *Flora of Tropical East Africa*. Balkema, Rotterdam.
Carter, S. & Radcliffe-Smith, A. (1988). Euphorbiaceae part 2. In: Polhill, R. M. (ed.), *Flora of Tropical East Africa*. Balkema, Rotterdam.
Carter, S. (1994). Aloaceae. In: Polhill, R. M. (ed.), *Flora of Tropical East Africa*. Balkema, Rotterdam.
Cheek, M. (1998). *The Plants of Mount Cameroon, a Conservation Checklist*. Royal Botanic Gardens, Kew.
Clayton, W. D. (1970). Gramineae. In: Milne-Redhead, C. E. & Polhill, R. M. (eds), *Flora of Tropical East Africa*. Crown Agents, London.
Cufodontis, G. (1966). Pittosporaceae. In: Milne-Redhead, C. E. & Polhill, R. M. (eds), *Flora of Tropical East Africa*. Crown Agents, London.
Darbyshire, I., Vollesen, K. & Ensermu K. (2010). Acanthaceae part 2. In: Beentje, H. J. (ed.), *Flora of Tropical East Africa*. Royal Botanic Gardens, Kew.
Davies, F. G. & Verdcourt, B. (1998). Sapindaceae. In: Polhill, R. M. (ed.), *Flora of Tropical East Africa*. Balkema, Rotterdam.
de Wilde, W. J. J. O. (1975). Passifloraceae. In: Polhill, R. M. (ed.), *Flora of Tropical East Africa*. Crown Agents, London.
Djossa, B. A., Fahr, J., Wiegand, T., Ayihouénou, B. E., Kalko, E. K. V. & Sinsin, B. A. (2008). Land use impact on *Vitellaria paradoxa* C. F. Gaertn. Stand structure and distribution patterns: A comparison of Biosphere Reserve of Pendjari in Atacora district in Benin. *Agroforestry Systems* 72(3): 205–220.
Dransfield, J. (1986). Palmae. In: Polhill R. M. (ed.), *Flora of Tropical East Africa*. Balkema, Rotterdam.
Edmonds, J. (2012). Solanaceae. In: Beentje, H. J. (ed.) *Flora of Tropical East Africa*. Royal Botanic Gardens, Kew.
Edwards, P. J. (2005). Cyatheaceae. In: Beentje, H. J. & Ghazanfar, S. A. (eds), *Flora of Tropical East Africa*. Royal Botanic Gardens, Kew.
Eggeling, W. J. & Dale, I. R. (1952). *The indigenous trees of the Uganda Protectorate*, 2nd edition. Government Printer, Entebbe.
Eilu, G., Hafashimana, D. L. N. & Kasenene, J. M. (2004). Density and diversity of tree species in forests of the Albertine Rift, western Uganda. *Diversity and Distributions* 10: 303–312.
Elffers, J., Graham, R. A. & DeWolf, G. P. (1964). Capparidaceae. In: Hubbard, C. E. & Milne-Redhead, C. E. (eds), *Flora of Tropical East Africa*. Crown Agents, London.
Farjon, A. (2001). *World checklist and bibliography of conifers*. Royal Botanic Gardens, Kew.
Fondoun, J. M. & Onana, J. (2001) Ethnobotany and Importance of Three Local Species in Northern Cameroon. In: Pasternak, D. & Schlissel, Arnold (eds), *Combating Desertification with Plants*. Springer Verlag.
Forest Department (1995). *National Biomass Review Mission Report*. Kampala, Uganda.
Forest Department (2003). *National Biomass Study, Technical Report of 1996–2002*. Ministry of Water Lands and Environment, Kampala.
Friis, I. (1989). Urticaceae. In: Polhill, R. M. (ed.), *Flora of Tropical East Africa*. Balkema, Rotterdam.
Friis, I. (2008). The Tropical East African Urticaceae as a sample plant biodiversity study: what does it show about taxa, habitats, existing scientific collections and protected areas? *African J. Ecol.* 46 (Suppl.1), 122–124.
Gillett, J. B. (1991). Burseraceae. In: Polhill, R. M. (ed.), *Flora of Tropical East Africa*. Balkema, Rotterdam.
Gillett, J. B., Polhill, R. M. & Verdcourt, B. (1971). Leguminosae, Papilionoideae parts 1 & 2. In: Milne-Redhead, C. E. & Polhill, R. M. (eds), *Flora of Tropical East Africa*. Crown Agents, London.

Golding, J. S. & Hurter, P. J. H. (2003). A Red List account of Africa's cycads and implications of considering life-history and threats. *Biodiversity & Conservation* 12: 507–528.
Goode, D. (2001). *Cycads of Africa*. Cycads of Africa publishers, Gallo Manor.
Graham, R. A. (1960). Rosaceae. In: Hubbard, C. E. & Milne-Redhead, C. E. (eds), *Flora of Tropical East Africa*. Crown Agents, London.
Green, P. S. (2002). A revision of *Olea* L. *Kew Bull.* 57: 93–99.
Groombridge, B. & Jenkins, M. D. (2002). *World Atlas of biodiversity*. Prepared by the UNEP World Conservation Monitoring Centre. University of California Press, Berkeley, USA
Grey-Wilson, C. (1986). Violaceae. In: Polhill, R. M. (ed.), *Flora of Tropical East Africa*. Balkema, Rotterdam.
Halliday, P. (1984). Myrsinaceae. In: Polhill, R. M. (ed.), *Flora of Tropical East Africa*. Balkema, Rotterdam.
Hamilton, A. (1981). *A field guide to Uganda forest trees*. Privately published, Kampala.
Harris, J. G. & Harris, M. W. (1994). *Plant Identification terminology: An illustrated glossary*. Spring Lake Publishing, Utah.
Heibloem, P. (1999). *The cycads of central Africa*. PACSOA, Brisbane.
Hemsley, J. H. (1956). Connaraceae. In: Turrill, W. B. & Milne-Redhead, C. E. (eds), *Flora of Tropical East Africa*. Crown Agents, London.
Hemsley, J. H. (1968). Sapotaceae. In: Milne-Redhead, C. E. & Polhill, R. M. (eds), *Flora of Tropical East Africa*. Crown Agents, London.
IUCN. (2001). *IUCN Red List categories and criteria*. IUCN, Gland & Cambridge.
IUCN (2011). IUCN Red List of Threatened Species, version 2011.2. Downloaded 6 Feb. 2012.
Jeffrey, C. & Beentje, H. J. (2000). Compositae part 1. In: Beentje, H. J. (ed.), *Flora of Tropical East Africa*. Balkema, Rotterdam.
Johnston, M. C. (1972). Rhamnaceae. In: Milne-Redhead, C. E. & Polhill, R. M. (eds), *Flora of Tropical East Africa*. Balkema, Rotterdam.
Kalema, J. (2006). The significance of Important Bird Areas for conservation of plants in Uganda. In: Ghazanfar, S. A. & Beentje, H. J. (eds), *Taxonomy and Ecology of African Plants, their Conservation and Sustainable Use*, pp. 457–472. Royal Botanic Gardens, Kew
Kalema, J. (2008). The use of herbarium plant databases in identifying areas of biodiversity concentration: the case of family Acanthaceae in Uganda. *African J. Ecol.* 46 (Suppl.1), 1–2.
Kalema, J. & Bukenya-Ziraba, R. (2005). Patterns of plant diversity in Uganda. *Biol. Skr.* 55: 331–341.
Kalema, J., Namaganda, M. & Mulumba, J. W. (Unpublished). *Vegetation and Climate change in Eastern Africa (VECEA): The case of Uganda.* A report submitted to VECEA Project, April 2009. Funded by The Rockefeller Foundation.
Kalema, J., Namaganda, M., Ssegawa, P., Kabuye, C., Maganyi, O. & Muncuguzi, P. (2010). Status of higher plants in Uganda. In: Kaddu, J. B. & Busuulwa, H. (eds) *Baseline report on state of biodiversity in the Nile Uganda* 2010. A production of the Wetlands and Biodiversity Conservation Component of the Nile Transboundary Environmental Action Project. Nile Basin Initiative Secretariat.
Kalema, J. & Mucunguzi, P. (Unpublished). On-site assessment of the conservation status of Uganda's endemic cycad, *Encephalartos equatorialis* P. J. H. Hurter & Glen (Zamiaceae) in Mayuge District, Uganda. A report submitted to *Nature* Uganda, Kampala, January 2009.
Kalema, J. & Ssegawa, P. (2007). The flora of highly degraded and vulnerable wetland ecosystems of Nyamuriro and Doho, Uganda, *African J. Ecol.* 45 (Suppl. 1): 28–33.
Katende, A. B., Birnie, A. & Tengnas, B. (1995). Useful trees and shrubs for Uganda: Identification, propagation and management for agricultural and pastrol communities. *Technical Handbook* No 10. Regional Soil Conservation Unit, RSCU/SIDA, Kenya.
Keay, R. W. J. (1954). *Flora of West Tropical Africa*, 2nd ed. 1. Crown Agents, London.
Kokwaro, J. O. (1982). Rutaceae. In: Polhill, R. M. (ed.), *Flora of Tropical East Africa*. Balkema, Rotterdam.
Kokwaro J. O. 1986. Anacardiaceae. In: Polhill R. M. (ed.), *Flora of Tropical East Africa*. Balkema, Rotterdam.
Langdale-Brown, I., Osmaston, H. A. & Wilson, J. G. (1964). *The vegetation of Uganda and its bearing on land uses*. Uganda Government Printer, Entebbe.

Lebrun, J.-P. & Stork, A. L. (1991). *Enumeration des plantes a fleurs d'Afrique tropicale*, vol. 1. Conserv. & Jard. Botanique, Genève.
Lebrun, J.-P. & Stork, A. L. (1992). *Enumeration des plantes a fleurs d'Afrique tropicale*, vol. 2. Conserv. & Jard. Botanique, Genève.
Lebrun, J.-P. & Stork, A. L. (1995). *Enumeration des plantes a fleurs d'Afrique tropicale*, vol. 3. Conserv. & Jard. Botanique, Genève.
Lebrun, J.-P. & Stork. A. L. (1997). *Enumeration des plantes a fleurs d'Afrique tropicale*, vol. 4. Conserv. & Jard. Botanique, Genève.
Lebrun, J.-P. & Stork. A. L. (2010). *Enumeration des plantes a fleurs d'Afrique tropicale*, vol. 5. Conserv. & Jard. Botanique, Genève.
Leeuwenberg,. A. J. M. (1961). The Loganiaceae of Africa I. *Anthocleista*. In: *Acta Bot. Neerl.* 10: 1–53.
Lewis, J. (1956). Rhizophoraceae. In: Turrill, W. B. & Milne-Redhead, C. E. (eds), *Flora of Tropical East Africa*. Crown Agents, London.
Lock, J. M. (1989). *Legumes of Africa: a checklist*. Royal Botanic Gardens, Kew.
Lucas, G. L. (1968). Icacinaceae. In: Milne-Redhead, C. E. & Polhill, R. M. (eds), *Flora of Tropical East Africa*. Crown Agents, London.
Lucas, G. L. (1968). Olacaceae. In: Milne-Redhead, C. E. & Polhill, R. M. (eds), *Flora of Tropical East Africa*. Crown Agents, London.
Lye, K. Å., Bukenya-Ziraba, R., Tabuti, J. R. S. & Waako, P. J. (2008). *Makerere Herbarium Handbook no. 2: Plant-Medicinal dictionary for East Africa*. Department of Botany, Makerere University, Kampala.
Melville, R. (1958). Gymnospermae. In: Turrill, W. B. & Milne-Redhead, C. E. (eds), *Flora of Tropical East Africa*. Crown Agents, London.
Mbuya, L. P., Msanga, H. P., Ruffo, C. K., Birnie, A. & Tengnaas, B. (1994). Useful trees and shrubs for Tanzania: Identification, propagation and management for agricultural and pastroral communities. *Technical Handbook* No. 6. Sida Regional Soil Conservation Unit, RSCU, Swedish International Development Authority.
Milne-Redhead, E. (1953). Hypericaceae. In: Turrill, W. B. & Milne-Redhead, C. E. (eds), *Flora of Tropical East Africa*. Crown Agents, London.
Mittermeier, R. A., Myers, N.,Thomsen, J. B., da Fonseca, G. A. B. & Olivieri, S. (1998). Biodiversity hotspots and major tropical wilderness areas: approaches to setting conservation priorities. *Conservation Biol.* 12: 516–520.
Moore, J. L., Manne, L., Brooks, T., Burgess, N. D., Davies, R., Rahbek, C., Williams, P. & Balmford, A. (2002). The distribution of cultural and biological diversity in Africa. *Proc. Roy. Soc. London, Ser. B, Biol. Sci.* 269:1645–1653.
NEMA (2007). *State of Environment Report for Uganda*. NEMA, Kampala.
Odebiyi, J. A., Bada, S. O, Omoloye, A. A., Awodoyin, R. O., Oni, P. I. (2004) Vertebrate and insect pests and hemi-parasitic plants of *Parkia biglobosa* and *Vitellaria paradoxa* in Nigeria. *Agroforestry Systems* 60: 51–59.
Okullo, J. B. L., Obua, J. & Okello, G. (2004). Use of indigenous knowledge in predicting fruit production of shea butter tree in agroforestry parklands of north-eastern Uganda. *Uganda J. Agric. Sci.* 9(1): 360–366.
Omino, E. A. (2002). Apocynaceae. In: Beentje, H. J. (ed.), *Flora of Tropical East Africa*. Balkema, Rotterdam.
Paiva, J. (2007). Polygalaceae. In: Beentje, H. J. & Ghazanfar, S. A. (eds), *Flora of Tropical East Africa*. Royal Botanic Gardens, Kew.
Pennington, T. D. (1991). *The genera of Sapotaceae*. Royal Botanic Gardens, Kew & New York Botanical Garden.
Peters, C. M. (1994). *Sustainable harvest of non-timber plant resources in tropical moist forest: an ecological primer*. USAID Biodiversity Support Programme, Washington DC.
Peterson, B. (1978). Thymeleaceae. In: Polhill, R. M. (ed.), *Flora of Tropical East Africa*. Crown Agents, London.
Pitman, N. A. & Jorgensen, P. M. (2002). Estimating the size of the World's threatened flora. *Science* 298: 989.

Polhill, R. M. & Verdcourt, B. (2000). Myricaceae. In: Beentje, H. J. (ed.), *Flora of Tropical East Africa*. Balkema, Rotterdam.
Polhill, R. M. (1966). Ulmaceae. In: Hubbard, C. E. & Milne-Redhead, C. E. (eds), *Flora of Tropical East Africa*. Crown Agents, London.
Polhill, R. M. (2005). Santalaceae. In: Beentje, H. J. & Ghazanfar, S. A. (eds), *Flora of Tropical East Africa*. Royal Botanic Gardens, Kew.
Radcliffe-Smith, A. (1987). Euphorbiaceae part 1. In: Polhill R.M. (ed.), *Flora of Tropical East Africa*. Balkema, Rotterdam.
Robson, N. K. B., Hallé, N., Mathew, B. & Blakelock, R. (1994). Celastraceae. In: Polhill, R. M. (ed.), *Flora of Tropical East Africa*. Balkema, Rotterdam.
Sacande, M., Sanou, L. & Beentje, H. J. (in press). *Guide de terrain des arbres du Burkina Faso*. Royal Botanic Gardens, Kew.
Sands, M. J. S. (2003). Balanitaceae. In: Beentje H. J. (ed.), *Flora of Tropical East Africa*. Balkema, Rotterdam.
Seyani, J. H. (1991). *Dombeya* in Africa. *Opera Bot. Belg.* 2. National Botanic Garden of Belgium, Meise, Belgium.
Syngellakis, K. & Arudo, E. (2006). *Energy for water, health, education*. Energy sector Policy review paper.
Sleumer, H. (1975). Flacourtiaceae. In: Polhill R. M. (ed.), *Flora of Tropical East Africa*. Crown Agents, London.
Smith, D. L. (1966). Linaceae. In: Hubbard, C. E. & Milne-Redhead, C. E. (eds), *Flora of Tropical East Africa*. Crown Agents, London.
Ssegawa, P. & Kasenene, J. (2007). Medicinal plant diversity and uses in the Sango Bay area, Southern Uganda. *J. Ethnopharmacol.* 113: 521–540.
Stannard, B. (2000). Simaroubaceae. In: Beentje, H. J. (ed.), *Flora of Tropical East Africa*. Balkema, Rotterdam.
Styles, B. T. & White, F. (1991). Meliaceae. In: Polhill, R. M. (ed.), *Flora of Tropical East Africa*. Balkema, Rotterdam.
Tabuti, J. R. S., Lye, L. A. & Dhillion, S. S. (2003). Traditional herbal drugs of Bulamogi, Uganda: plants, use and administration. *J. Ethnopharmacol.* 88: 19–44.
Tennant, J. R. (1968). Araliaceae. In: Milne-Redhead, C. E. & Polhill, R. M. (eds), *Flora of Tropical East Africa*. Crown Agents, London.
Timberlake, J. R., Golding, J. S. & Smith, P. (2006). A preliminary analysis of endemic and threatened plants of the Flora Zambesiaca area. In: Ghazanfar, S. A. & Beentje, H. J. (eds), *Taxonomy and ecology of African Plants: their conservation and sustainable use*. Pp. 749–760. Royal Botanic Gardens, Kew.
Townsend, C. C. (1989). Umbelliferae. In: Polhill, R. M. (ed.), *Flora of Tropical East Africa*. Balkema, Rotterdam.
Turrill, W. B. (1952). Oleaceae. In: Turrill, W. B. & Milne-Redhead, C. E. (eds), *Flora of Tropical East Africa*. Crown Agents, London.
Uganda Government (2002). *The Energy Policy for Uganda*. Ministry of Energy and Mineral Development.
Verdcourt, B. (1956). Canellaceae. In: Turrill, W. B. & Milne-Redhead, C. E. (eds), *Flora of Tropical East Africa*. Crown Agents, London.
Verdcourt, B. (1958). Alangiaceae. In: Turrill, W. B. & Milne-Redhead, C. E. (eds), *Flora of Tropical East Africa*. Crown Agents, London.
Verdcourt, B. (1958). Melianthaceae. In: Hubbard, C. E. & Milne-Redhead, C. E. (eds), *Flora of Tropical East Africa*. Crown Agents, London.
Verdcourt, B. (1962). Theaceae. In: Hubbard, C. E. & Milne-Redhead, C. E. (eds), *Flora of Tropical East Africa*. Crown Agents, London.
Verdcourt, B. (1968). Aquifoliaceae. In: Milne-Redhead, C. E. & Polhill, R. M. (eds), *Flora of Tropical East Africa*. Crown Agents, London.
Verdcourt, B. (1968). Monimiaceae. In: Milne-Redhead, C. E. & Polhill, R. M. (eds), *Flora of Tropical East Africa*. Crown Agents, London.

Verdcourt, B. (1968). Salvadoraceae. In: Milne-Redhead, C. E. & Polhill, R. M. (eds), *Flora of Tropical East Africa*. Crown Agents, London.
Verdcourt, B. (1968). Scytopetalaceae. In: Milne-Redhead, C. E. & Polhill, R. M. (eds), *Flora of Tropical East Africa*. Crown Agents, London.
Verdcourt, B. (1971). Annonaceae. In: Milne-Redhead, C. E. & Polhill, R. M. (eds), *Flora of Tropical East Africa*. Crown Agents, London.
Verdcourt, B. (1971). Hamamelidaceae. In: Milne-Redhead, C. E. & Polhill, R. M. (eds), *Flora of Tropical East Africa*. Crown Agents, London.
Verdcourt, B. (1975). Oliniaceae. In: Polhill, R. M. (ed.), *Flora of Tropical East Africa*. Crown Agents, London.
Verdcourt, B. (1976). Rubiaceae part 1. In: Polhill, R. M. (ed.), *Flora of Tropical East Africa*. Crown Agents, London.
Verdcourt, B. (1984). Erythroxylaceae. In: Polhill, R. M. (ed.), *Flora of Tropical East Africa*. Balkema, Rotterdam.
Verdcourt, B. (1984). Ixonanthaceae. In: Polhill, R. M. (ed.), *Flora of Tropical East Africa*. Balkema, Rotterdam.
Verdcourt, B. (1986). Moringaceae. In: Polhill, R. M. (ed.), *Flora of Tropical East Africa*. Balkema, Rotterdam.
Verdcourt, B. (1991). Boraginaceae. In: Polhill, R. M. (ed.), *Flora of Tropical East Africa*. Balkema, Rotterdam.
Verdcourt, B. (1992). Verbenaceae. In: Polhill, R. M. (ed.), *Flora of Tropical East Africa*. Balkema, Rotterdam.
Verdcourt, B. (1994). Lythraceae. In: Polhill, R. M. (ed.), *Flora of Tropical East Africa*. Balkema, Rotterdam.
Verdcourt, B. (1996). Lauraceae. In: Polhill, R. M. (ed.), *Flora of Tropical East Africa*. Balkema, Rotterdam.
Verdcourt, B. (1997). Myristicaceae. In: Polhill, R. M. (ed.), *Flora of Tropical East Africa*. Balkema, Rotterdam.
Verdcourt, B. (1998). A new species of *Ficus* (Moraceae) from Uganda. *Kew Bull.* 53: 233–236.
Verdcourt, B. (2001). Myrtaceae. In: Beentje H. J. (ed.), *Flora of Tropical East Africa*. Balkema, Rotterdam.
Verdcourt, B. (2005). Ochnaceae. In: Beentje, H. J. & Ghazanfar, S. A. (eds), *Flora of Tropical East Africa*. Royal Botanic Gardens, Kew.
Verdcourt, B. & Bridson, D. (1988). Rubiaceae part 3. In: Polhill, R. M. (ed.), *Flora of Tropical East Africa*. Balkema, Rotterdam.
Vollesen, K. (2008). Acanthaceae part 1. In: Beentje, H. J. & Ghazanfar, S. A. (eds), *Flora of Tropical East Africa*. Royal Botanic Gardens, Kew.
WCMC (1992). *Global biodiversity: status of the Earth's living resources*. Chapman & Hall, London.
White, F. (1983). *The vegetation of Africa*. UNESCO, Paris.
Whitehouse, C. M., Cheek, M., Andrews, W. & Verdcourt, B. (2001). Tiliaceae. In: Beentje H. J. (ed.), *Flora of Tropical East Africa*. Balkema, Rotterdam.
Wickens, G. E. (1973). Cornaceae. In: Polhill R. M. (ed.), *Flora of Tropical East Africa*. Crown Agents, London.
Wickens, G. E. (1975). Melastomataceae. In: Polhill, R. M. (ed.), *Flora of Tropical East Africa*. Crown Agents, London.
Wilmot-Dear, C. M. (1985). Salicaceae. In: Polhill, R. M. (ed.), *Flora of Tropical East Africa*. Balkema, Rotterdam.

WEBSITES CONSULTED

www.iucnredlist.org/: IUCN official Red List

www.ipni.org/ipni: International Plant Names Index

www.ville-ge.ch/musinfo/bd/cjb/africa/recherche.php: African Plant Database

THE RED LIST SYSTEM

The **IUCN Red List of threatened species™** is an international and globally accepted standard for evaluating the conservation status of plant and animal species. It uses a scientifically rigorous approach to determine risks of extinction based on a collaboration of the IUCN Survival Commission (SSC) with a network of scientists and partner organisations working in almost every country in the world.
Assessments are done on a global scale, that is, the complete distribution of the species under consideration is taken into account. Criteria are quantitative in nature, based on sizes of geographic range combined with patterns of habitat occupancy, combined with an evaluation of the threat(s) to either habitat or species (or both). Assessments are documented and state the criteria that are being met to meet the requirements; these criteria include, *inter alia*, population size, inter-population distance or fragmentation, number of mature individuals, decline in numbers, decline in habitat. Important definitions are the Extent Of Occurrence (EOO), the area containing all known species individuals and measured by a 'minimum convex polygon' (for precise definitions and working method, see IUCN (2001)); and Area Of Occupancy (AOO), the area actually occupied by a species. EOO and AOO differ in that EOO encompasses the whole distribution area, including any unsuitable areas within the distribution area *not* occupied by the species; while AOO is narrowed down to the actual occupation area. Further specialised criteria such as location (a distinct area in which a single threat event can affect all individuals of the species present) and precise definitions of the following threat categories can be found in IUCN (2001) as well.

The categories of threat are, in declining order of seriousness:

Extinct (EX) — when there is no reasonable doubt that the last individual of a species has died.

Critically endangered (CR) — facing an extremely high risk of extinction in the wild.

Endangered (EN) — facing a very high risk of extinction in the wild.

Vulnerable (VU) — facing a high risk of extinction in the wild.

Near threatened (NT) — close to qualifying for one of the above categories, or likely to qualify in the near future.

Least concern (LC) — not qualifying for one of the above categories. Such species may be decining, but are not considered at immediate risk of extinction; or they may be widespread and abundant.

Data deficient (DD) — information is inadequate to make an assessment based on distribution and population status.

The category of threat provides an assessment of the extinction risk under current circumstances; it does not, by itself, provide priorities for conservation action. Setting such priorities needs to be done in-country, and will include other factors such as costs, logistics, legal requirements, chance of success, and biological characteristics of the species concerned.

An example of the criteria required for a red list assessment is given below for the Endangered (EN) category, to give an indication how rigorous the process is:

A taxon is Endangered when the best available evidence indicates that it meets any of the following criteria (A to E), and it is therefore considered to be facing a very high risk of extinction in the wild:

A. Reduction in population size based on any of the following:

1. An observed, estimated, inferred or suspected population size reduction of ≥70% over the last 10 years or three generations, whichever is the longer, where the causes of the reduction are clearly reversible AND understood AND ceased, based on (and specifying) any of the following:
   (a) direct observation
   (b) an index of abundance appropriate to the taxon
   (c) a decline in area of occupancy, extent of occurrence and/or quality of habitat
   (d) actual or potential levels of exploitation
   (e) the effects of introduced taxa, hybridisation, pathogens, pollutants, competitors or parasites.

2. An observed, estimated, inferred or suspected population size reduction of ≥50% over the last 10 years or three generations, whichever is the longer, where the reduction or its causes may not have ceased OR may not be understood OR may not be reversible, based on (and specifying) any of (a) to (e) under A1.

3. A population size reduction of ≥50%, projected or suspected to be met within the next 10 years or three generations, whichever is the longer (up to a maximum of 100 years), based on (and specifying) any of (b) to (e) under A1.

4. An observed, estimated, inferred, projected or suspected population size reduction of ≥50% over any 10 year or three generation period, whichever is longer (up to a maximum of 100 years in the future), where the time period must include both the past and the future, and where the reduction or its causes may not have ceased OR may not be understood OR may not be reversible, based on (and specifying) any of (a) to (e) under A1.

B. Geographic range in the form of either B1 (extent of occurrence) OR B2 (area of occupancy) OR both:

1. Extent of occurrence estimated to be less than 5000 km^2, and estimates indicating at least two of a–c:

- a. Severely fragmented or known to exist at no more than five locations.
- b. Continuing decline, observed, inferred or projected, in any of the following:
  - (i) extent of occurrence
  - (ii) area of occupancy
  - (iii) area, extent and/or quality of habitat
  - (iv) number of locations or subpopulations
  - (v) number of mature individuals.
- c. Extreme fluctuations in any of the following:
  - (i) extent of occurrence
  - (ii) area of occupancy
  - (iii) number of locations or subpopulations
  - (iv) number of mature individuals.

2. Area of occupancy estimated to be less than 500 km^2, and estimates indicating at least two of a–c:

- a. Severely fragmented or known to exist at no more than five locations.
- b. Continuing decline, observed, inferred or projected, in any of the following:
  - (i) extent of occurrence
  - (ii) area of occupancy
  - (iii) area, extent and/or quality of habitat
  - (iv) number of locations or subpopulations
  - (v) number of mature individuals.
- c. Extreme fluctuations in any of the following:
  - (i) extent of occurrence
  - (ii) area of occupancy
  - (iii) number of locations or subpopulations
  - (iv) number of mature individuals.

C. Population size estimated to number fewer than 2500 mature individuals and either:

1. An estimated continuing decline of at least 20% within five years or two generations, whichever is longer, (up to a maximum of 100 years in the future) OR

2. A continuing decline, observed, projected, or inferred, in numbers of mature individuals AND at least one of the following (a–b):

- a. Population structure in the form of one of the following:
  - (i) no subpopulation estimated to contain more than 250 mature individuals, OR
  - (ii) at least 95% of mature individuals in one subpopulation.
- b. Extreme fluctuations in number of mature individuals.

D. Population size estimated to number fewer than 250 mature individuals.

E. Quantitative analysis showing the probability of extinction in the wild is at least 20% within 20 years or five generations, whichever is the longer (up to a maximum of 100 years).

INDEX

(Main entry page listed first)